4th International Conference of the Transdisciplinary Research on Environmental Problems in Southeast Asia (TREPSEA 2021)

IOP Conference Series: Earth and Environmental Science
Volume 1027

Online
16 – 18 September 2021

ISBN: 978-1-7138-5466-1
ISSN: 1755-1307

Printed from e-media with permission by:

Curran Associates, Inc.
57 Morehouse Lane
Red Hook, NY 12571

Some format issues inherent in the e-media version may also appear in this print version.

Additional copies of this publication are available from:

Curran Associates, Inc.
57 Morehouse Lane
Red Hook, NY 12571 USA
Phone: 845-758-0400
Fax: 845-758-2633
Email: curran@proceedings.com
Web: www.proceedings.com

TABLE OF CONTENTS

IOP Conf. Series: Earth and Environmental Science 1027 (2022) 011001 doi:10.1088/1755-1315/1027/1/011001

About TREPSEA 2021

In the last half century, Asia has experienced the fastest economic growth and rapid increase of human prosperity, compared to other continents. However, ecological risks due to significant land use changes, global environmental change and climate change-related disasters have also occurred rapidly in Asia. The global deterioration of soils and landscape poses a significant threat to poverty reduction, sustainable food production, and climate change adaptation and mitigation. A recent study by the International Food Policy Research Institute (IFPRI) has estimated that the annual cost of land degradation due to "land use and cover change" (LUCC) and the use of land-degrading management practices on static crop and grazing land is about $300 billion. This does not include the costs related to deterioration of ecosystem services, which may increase the estimates significantly. However, global hunger also increases, reaching 815 million people in 2017, where about 2 billion people suffer from micronutrient-deficiencies.

In Indonesia, agriculture grows at 3.87 percent per year, which is not very low, although it does not contribute much to poverty reduction and job creation in the country. The poverty level in September 2019 was 24.79 (9.22 percent of total population), where the majority (14.93 million or 12.60 percent) of them are living in rural area, and the rest 9.86 million (6.56 percent) are living in urban area. Income inequality in Indonesia is considered high, where the Gini coefficient generally has increased in the past decade or so, from 0.32 in 1998 to 0.38 in 2007, increasing to 0.41 in 2012 and started to decrease in recent years to 0.382 in September of 2019 (BPS, 15 January 2020). The hunger level in Indonesia has also decreased significantly from 44.1 million in 2004-2006 to 22.0 million in 2016-2018. Interestingly, the hunger level in urban area is 8.2 percent of total population, whereas that in rural area is 7.57 percent. The level of stunting among children has also decreased from 32.9 percent in 2013 to 29.9 percent in 2018, whereas the level of anemia among pregnant women has increased from 37.1 percent in 2013 to 48.9 percent in 2018.

Higher ecological risks and more severe climate change-related disasters have more serious consequences on food and nutritional security in Indonesia and other parts of Asia. Hunger and malnutrition have multiple causes, and in facts many of these challenges transcend national borders. Thus, sustainable, integrated and trans-disciplinary innovative solutions to such problems require a more holistic approach in formulating the development strategies to diversify agricultural and food production, increase productivity, hence generating income and employment. Tran-disciplinary approach in mitigating and adapting the climate change related disaster are also required for more sustainable development strategies, coupled with better strategies for disaster risk reduction and management, workable and effective programs to sustainable development and environmental conservation, and better access to food and improved nutritional security.

Every other year, the International Conference of Transdisciplinary Research on Environmental Problems in Southeast Asia (TREPSEA) is conducted in several cities in Indonesia. The first TREPSEA was conducted in Makassar in 2014, the second was in Bandung in 2016, and the third was in Gorontalo in 2018. The fourth TREPSEA Conference will be conducted as an online conference on September 16-18, 2021. The purpose of International Conference of the Transdisciplinary Research on Environmental Problems in Southeast Asia (TREPSEA) is to conduct integrative research of interactions between natural environment and human-social systems in Southeast Asia to solve the environmental problems in Southeast Asia. Its scope thus includes topics of sustainable development, agricultural and resource economics, economics, social science, environmental science, engineering, medicine, sociology, education, etc. Transdisciplinary Research (TDR) is defined as research efforts conducted by investigators from different disciplines and non-academic participants working jointly to create new conceptual, theoretical, methodological, and transnational innovations. Related stakeholders include academics, governments, development organizations, business and industries, sponsoring institutions, civil society organizations, the media, etc.

The theme of 4th International Conference of TREPSEA will address important themes of "Managing Ecological Risks and Natural Disasters in Southeast Asia: Challenges for Food Security, Public Health, and Economic Welfare". The conference will serve as an arena for stimulating academic exercise and policy dialogues on trans-disciplinary dimensions of environmental problems in Southeast Asia, primarily on the themes of managing ecological risk, disaster mitigation, environmental conservation, food and nutritional security, social and economic welfare, and other related issues in sustainable development in general. The 4th International Conference of TREPSEA will be held as an online conference on September 16-18 of 2021.

Objectives

The objectives of the 4th International Conference of TREPSEA will address important themes of "Managing Ecological Risks and Natural Disasters in Southeast Asia: Challenges for Food Security, Public Health, and Economic Welfare" are as follows:

1. to stimulate academic exercise and policy dialogues on academic exercise and policy dialogues on trans-disciplinary dimensions of environmental problems in Southeast Asia;

2. to encourage trans-disciplinary research dimensions on of environmental problems, primarily on the themes of managing ecological risk, disaster mitigation, environmental conservation, food and nutritional security, and

3. to update some research progress and new knowledge on managing ecological risk, disaster mitigation, environmental conservation, food and nutritional security, and other related issues in sustainable development in general.

Topics of TREPSEA 2021

"***Managing Ecological Risks and Natural Disasters in Southeast Asia: Challenges for Food Security, Public Health, and Economic Welfare***" as the main theme of the conference and sub-themes are as follows:

1. Environmental Issues and Public Health
2. Ecological Risks and Food Security
3. Disaster Risks Reduction (DRR) and Disaster Management (DM)
4. Sustainable Agriculture and Equitable Development
5. Resources Management and Global Value Chain
6. Legal Issues and Business Ethics in Resource Management
7. Heavy Metal Problems and Renewable Energy
8. Urban Management and Community Development
9. Sustainable Development Goals (SDGs)

TREPSEA 2021 Organizers

TREPSEA 2021 International Conference is organized by

- University of Lampung (UNILA) as the host organization,

and co-organized by

- Research Institute for Humanity and Nature (RIHN),
- Bandung Institute of Technology (ITB),
- Gorontalo State University (UNG) and
- SRIREP Project (RIHN).

Supporting Organization / Institution

The conference is supported by

- IAGI (Ikatan Ahli Geologi Indonesia) and
- Future Earth.

General Committee

Professor Dr. Karomani, M.Si.
Chairperson of TREPSEA 2021
University of Lampung (UNILA)

Professor Masayuki Sakakibara
Vice-Chairperson of TREPSEA 2021
Research Institute for Humanity and Nature (RIHN)
&
Ehime University (EU)

Professor Dr. Dwia Aries Tina Pulubuhu, M.A.
Hasanuddin University (UNHAS)

Professor Dr. Emmy Suparka
Bandung Institute of Technology (ITB)

Professor Dr. Eduart Wolok, ST, MT
Rector of State University of Gorontalo (UNG)

Professor Dr. Arif Satria, SP, M.Si
Bogor Agricultural University (IPB)

Professor Dr. H. Gufran Darma Dirawan, M. EMD.
State University of Makassar (UNM)

Scientific Committee

Professor Dr. Ir. Bustanul Arifin, M.Sc.
University of Lampung (UNILA)

Professor Dr. Hasriadi Mat Akin, M.P.
University of Lampung (UNILA)

Dr. Lusmeilia Afriani, D.E.A.
University of Lampung (UNILA)

Professor Dr. Ir. Irwan Sukri Banuwa, M.Si.
University of Lampung (UNILA)

Professor Ir. Benjamin Sapiie, Ph.D.
Bandung Institute of Technology (ITB)

Professor dr. Budu, Ph.D, Sp.M(K), M.Med.
Hasanuddin University (UNHAS)

Dr. Hayati Sari Hasibuan, ST., MT.
University of Indonesia (UI)

Dr. Heriansyah Putra
Bogor Agricultural University (IPB)

Dr. Irwan Meilano ST, M.Sc.
Dean of Faculty of Earth Sciences and Technology,
Bandung Institute of Technology (ITB)

Dr. Mirzam Abdurrachman
Bandung Institute of Technology (ITB)

Dr. Idham Andri Kurnuawan
Bandung Institute of Technology (ITB)

Professor Katsuya Tanaka
Shiga University

Professor Mai Trong Nhuan
Vietnam National University, Hanoi (VNUH)

Dr. rer. nat. Mohamad Jahja, S.Si., M.Si.
State University of Gorontalo (UNG)

Prof. Dr. dr. Muhammad Nadjib Bustan, MPH
State University of Makassar (UNM)

Dr. Zaenal Abidin, M.Agr
Bogor Agricultural University (IPB)

Local Committee

Professor Dr. Karomani, M.Si.
University of Lampung (UNILA)

Dr. dr. Asep Sukohar, M.Kes.
University of Lampung (UNILA)

Professor Dr. Ir. Bustanul Arifin, M.Sc.
University of Lampung (UNILA)

Dr. Ir. R. Hanung Ismono, M.P.
University of Lampung (UNILA)

Professor Dr. Ir. Irwan Sukri Banuwa, M.Si.
University of Lampung (UNILA)

Dr. Lusmeilia Afriani, D.E.A.
University of Lampung (UNILA)

Dr. Warsono
University of Lampung (UNILA)

Dr. Junaidi, S.Si., M.Sc.
University of Lampung (UNILA)

Dr. Endang Linirin Widiastuti
University of Lampung (UNILA)

Dr. Muhammad Ibnu, S.P., M.Sc.
University of Lampung (UNILA)

Dr. Ir. F.E. Prasmatiwi, M.P.
University of Lampung (UNILA)

Dr. Melya Riniarti
University of Lampung (UNILA)

Professor Ryohei Kada
Shijyonawate Gakuen University,
Osaka, Japan

Professor Dr. Heryandi, S.H., M.S.
University of Lampung (UNILA)

Professor Dr. Suharso, M.Si.
University of Lampung (UNILA)

Professor Dr. Yulianto, M.Si.
University of Lampung (UNILA)

Dr. Yulia Rahma Fitriana, S.Hut., M.Sc.
University of Lampung (UNILA)

Professor Dr. Ir. Hasriadi Mat Akin, M.P.
University of Lampung (UNILA)

Professor Rudy, S.H., LL.M., LL.D
University of Lampung (UNILA)

Dr. Lukmanul Hakim, S.T., M.Sc., Eng
University of Lampung (UNILA)

Dr. Ir. Slamet Budi Yuwono, M.S.
University of Lampung (UNILA)

Dr. Novi Rosanti
University of Lampung (UNILA)

Dr. Ir. Dyah Aring Hepiana Lestari, M.Si
University of Lampung (UNILA)

Dr. Ir. Agus Hudoyo, M.Sc.
University of Lampung (UNILA)

Dr. Hendra Prasetia
University of Lampung (UNILA)

Conference Secretariat

Ms. Tika Leoni Putri
University of Lampung (UNILA)

Ms. Dian Rahmalia
University of Lampung (UNILA)

Mr. Ibrohim Saputra
University of Lampung (UNILA)

Mr. Abu Hasan As-Sadili
University of Lampung (UNILA)

Mr. Myo Han Htun
Research Institute for Humanity and Nature
(RIHN)

Ms. Ririn Aristiyani
University of Lampung (UNILA)

Ms. Maria Sari
University of Lampung (UNILA)

Mr. Muher Sukmayanto
University of Lampung (UNILA)

Ms. Mari Takehara
Research Institute for Humanity and Nature
(RIHN)

IOP Conf. Series: Earth and Environmental Science 1027 (2022) 011002 doi:10.1088/1755-1315/1027/1/011002

Peer Review Statement

All papers published in this volume have been reviewed through processes administered by the Editors. Reviews were conducted by expert referees to the professional and scientific standards expected of a proceedings journal published by IOP Publishing.

- **Type of peer review:** Single Anonymous
- **Conference submission management system:** Morressier
- **Number of submissions received:** 74
- **Number of submissions sent for review:** 74
- **Number of submissions accepted:** 48
- **Acceptance Rate (Submissions Accepted / Submissions Received × 100):** 64.9
- **Average number of reviews per paper:** 2
- **Total number of reviewers involved:** 49
- **Contact person for queries:**
  Name: MYO HAN HTUN
  Email: myo@chikyu.ac.jp
  Affiliation: Research Institute for Humanity and Nature (RIHN) - SRIREP Project

Central Java regional geochemistry: influence of environmental geology and mineral occurrences

P Sendjaja [1] A Kusniadi [1] and V E Setiawan [1]

[1] Centre for Geological Survey, Geological Agency, Ministry of Energy and Mineral Resources, Jln. Diponegoro 57 Bandung, Indonesia.

Corresponding Author: pppurs@gmail.com

Abstract. Stream sediments geochemical surveys were undertaken in Central Java, Indonesia as a part of the Geological Agency's geochemical mapping project with support from several universities. A total of 1583 stream sediment samples were collected from Central Java and analyzed using X-Ray Fluoresces for 29 elements. Geochemical maps were processed by the GIS software. Based on the overlay between geological and geochemical maps, there are several elemental anomalies, for example, the ultramafic formation from Kebumen showing distinctive Chromium (Cr), Cobalt (Co), and Magnesium (Mg) anomalies. Meanwhile, the potassic igneous rocks from Muria Mountain have Cerium (Ce), Lanthanum (La), and Zirconium (Zr) anomalies. Elemental anomalies related to mineral resources and mining areas such as Arsenic (As), Copper (Cu), and Lead (Pb) are found in Trenggalek, Pacitan, and Kebumen; while Barium (Ba), Iron (Fe), and Manganese (Mn) are scattered along the southern coast of Central Java. Anthropogenic elements (Chlorine (Cl), Ce, and As) spread along the north coast of Central Java probably because of pollution by industrial, farming, and urban areas. Heavy metals pollutions that were identified previously in the Bengawan Solo River flow can be seen on this geochemical map which can convincingly delineate elements anomalies due to geological and environmental conditions.

1. Introduction

A series of regional geochemical atlases were created by the Centre for Geological Survey for the Java Island as part of the technical report of the Java-Kalimantan geochemical survey project from 2011 to 2014. The West Java Geochemical Atlas was published in 2015. [1], while other atlases are still in progress.

Stream sediment sampling in the Central Java area was carried out in 2012 and 2013 involving students and researchers from Gadjah Mada University (UGM), Pembangunan Nasional University (UPN), and IST Akprind. Samples obtained in Central Java reach a total of 1583 samples, one sample represents the density of 25 km^2. The sampling guidelines followed the Geological Map Sheet with a scale of 1:100.00 and the topographic map with a scale of 1:50,000. Fourteen geochemical maps of the elements are shown in this paper. These geochemical maps constitute an unpublished report to complete the 20 geological maps that have been published locally in Central Java.

The Central Java's geochemical maps are bounded by Longitude 108°00'E and 112°00'E; Latitude 6°00' S and 8°00'S. The Quaternary and Tertiary Sediments, Tertiary Volcanic and Plutonic rocks are geological features on Central Java. Gold and base metal deposits are widespread in this region following the low and high epithermal mineralization zone (Figure 1). These geological conditions are

IOP Conf. Series: Earth and Environmental Science 1027 (2022) 012001 doi:10.1088/1755-1315/1027/1/012001

caused by subduction, volcanism, and deposition that have occurred since 90 Ma until recently. The deposition of sedimentary rocks since the Tertiary was much influenced by volcanic activity and sea-level changes which were characterized by thick sequences of volcanic and epiclastic rocks throughout the Late Oligocene to Early Miocene, followed by carbonate formation in the Early Miocene to Middle Miocene. There are no Pliocene or Quaternary deposits in the Southern Mountains zone due to uplift and erosion, different from the northern part of Java, the deposition of Pliocene and Quaternary sediments widespread in the offshore area [2]. This geochemical survey by using sediment streams method is expected to assist in the exploration of mineral resources and determine environmental influences after being overlaid with the geological region of Central Java and current conditions.

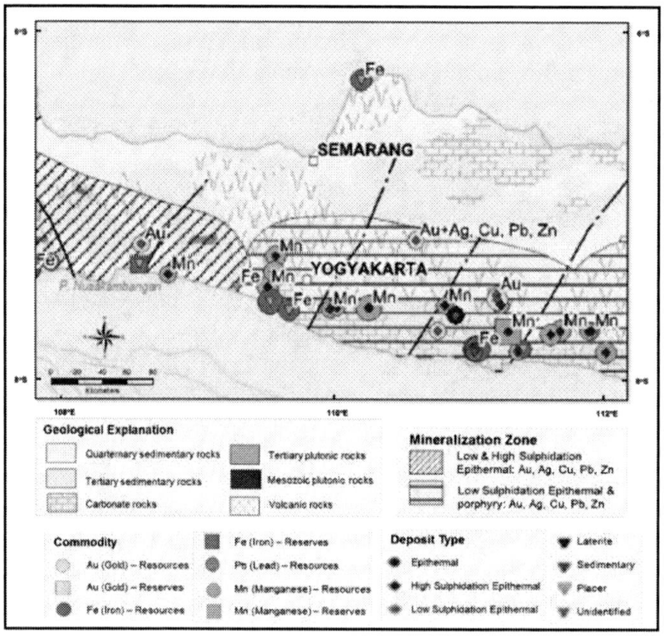

Figure 1. Central Java Metallogenic map (Modified from [3]).

2. Data and Method

Geochemical mapping survey activities included: taking samples of stream sediment in the field, laboratory analysis, and configuring geochemical map by GIS software (ArcGIS and Oasis Montaj). Stream sediment sampling activity followed the standard operating procedure document made by the Centre for Geological Survey and referred to the stream sediment sampling techniques which were commonly used in mineral exploration and the global standard of geochemical mapping [4].

The samples taken consisted of active river sediment that had been sieved through a #40 sieve, an #80 sieve and a grab sample. A total of 1583 sediment stream samples were collected, with sample density, one sample represents a 5x5 km grid cell (25km^2) in Central Java region. These samples were dried and subsequently sieved to the <200 μm fraction. The dried and sieved material was placed in the plastic sample, labelled, and prepared into pellet samples. The pellet is the result of a pressed mixture of powder sample, polyvinyl alcohol, and borate acid. The ARL 9900 Thermo Scientific X-ray fluorescence was used for geochemical analysis. The instruments belonging to the Geological Laboratory of the Centre for Geological Survey could determine 29 elements.

The results of the analysis were then made by a geochemical map of the elements using Kriging method on Esri, ArcMap 10.0 software, concentration increases as indicated by the contour from blue to red. Statistical analysis and frequency diagram used Geosoft Oasis Montaj software [5]. Due to the large number of samples, only 14 geochemical maps are shown for further discussion.

3. Result and Discussion

Figures 2 to 15 show the map of the distribution of elements in alphabetical order. The geology of Central Java which consists of several zones such as sedimentary, volcanic, plutonic and metamorphic could create a distinctive pattern of elemental anomalies.

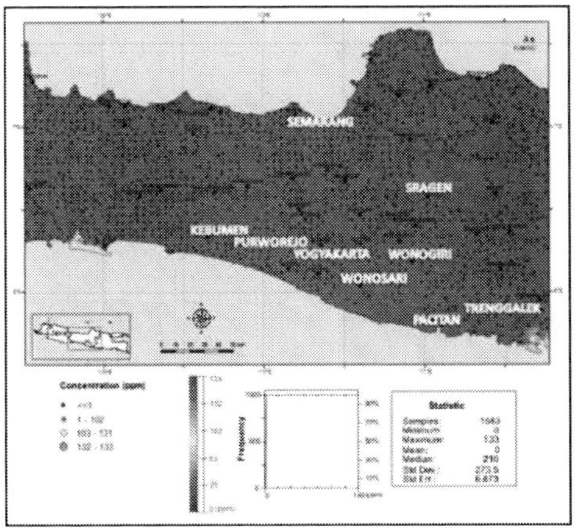

Figure 2. Arsenic (As) geochemical map.

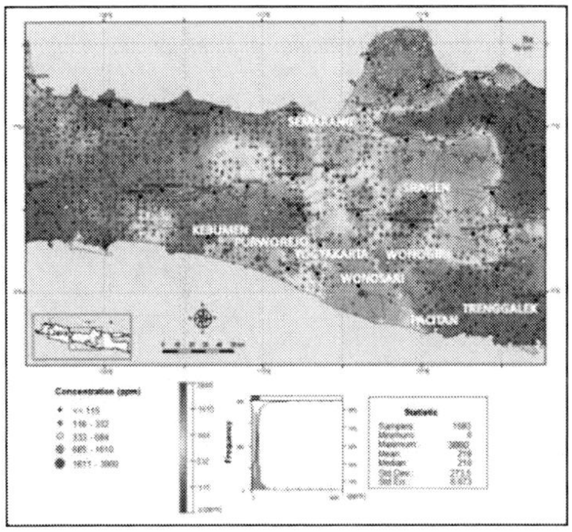

Figure 3. Barium (Ba) geochemical map.

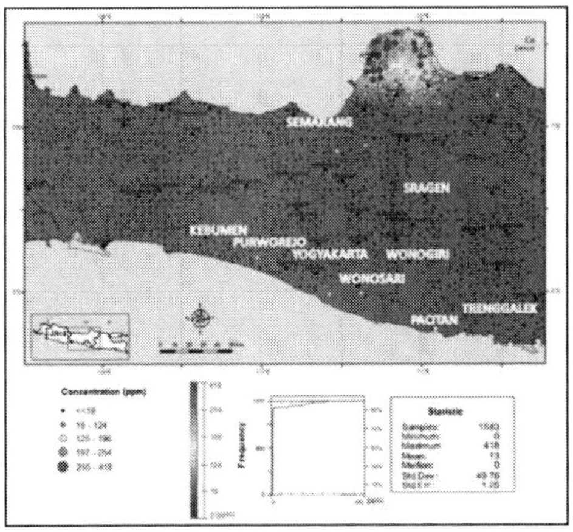

Figure 4. Cerium (Ce) geochemical map.

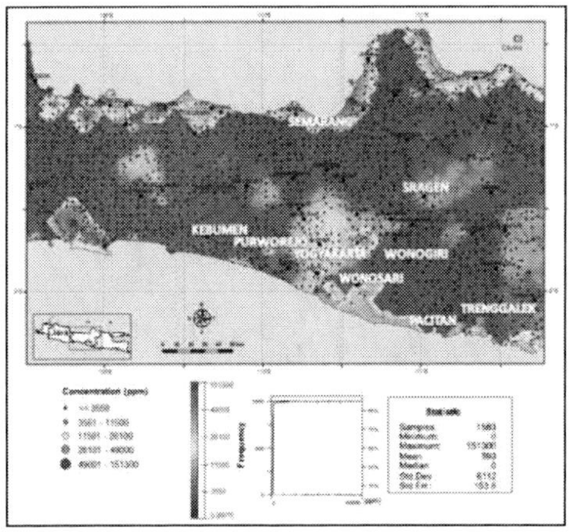

Figure 5. Chlorine (Cl) geochemical map.

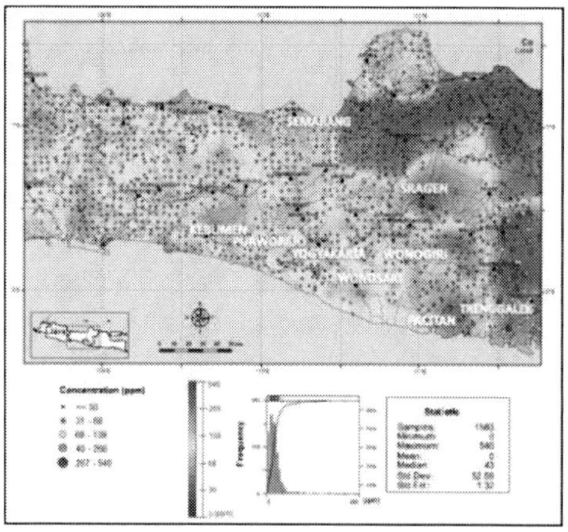

Figure 6. Cobalt (Co) geochemical map.

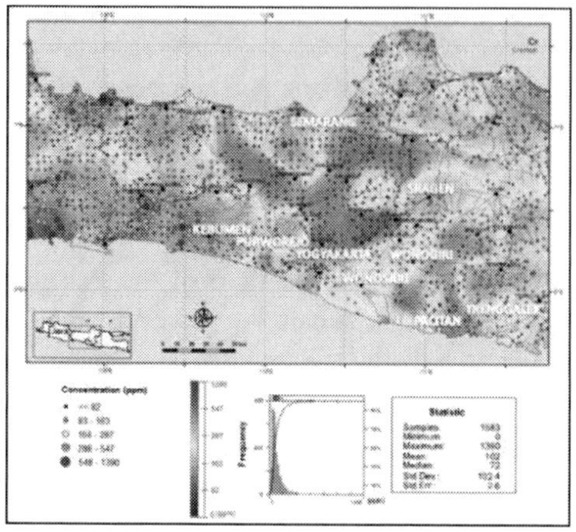

Figure 7. Chromium (Cr) geochemical map.

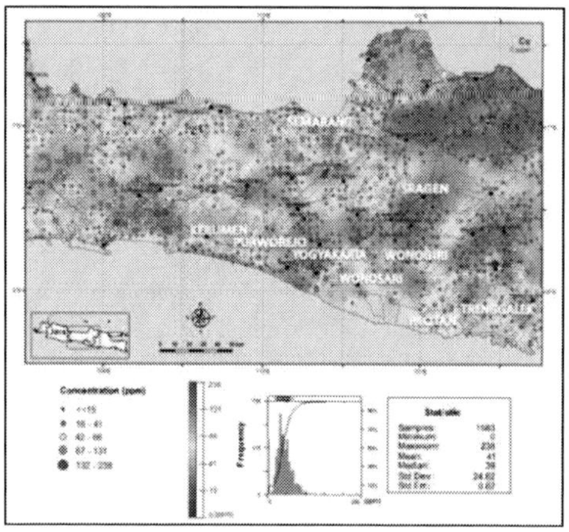

Figure 8. Cupper (Cu) geochemical map.

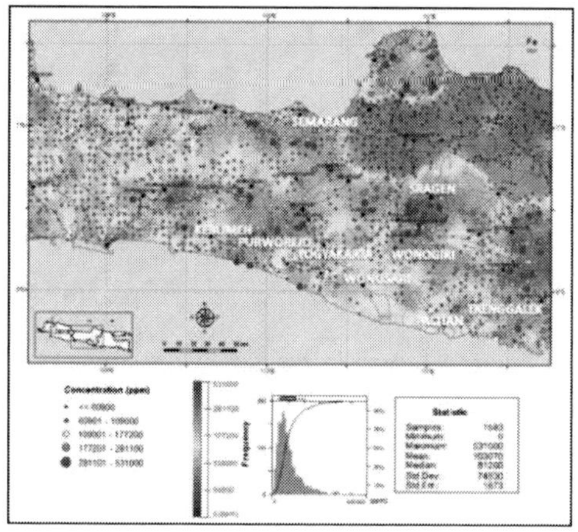

Figure 9. Iron (Fe) geochemical map.

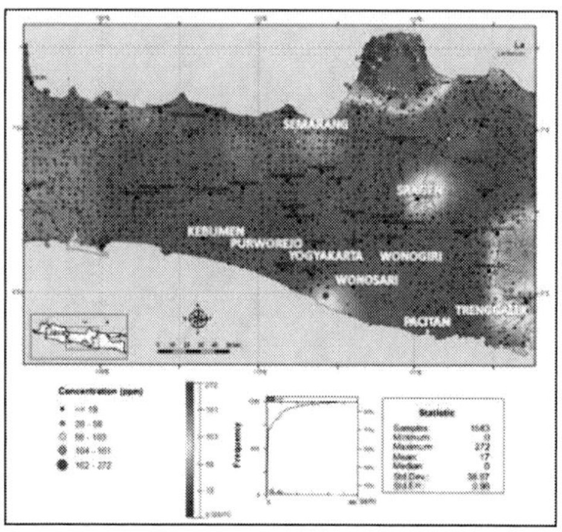

Figure 10. Lanthanum (La) geochemical map.

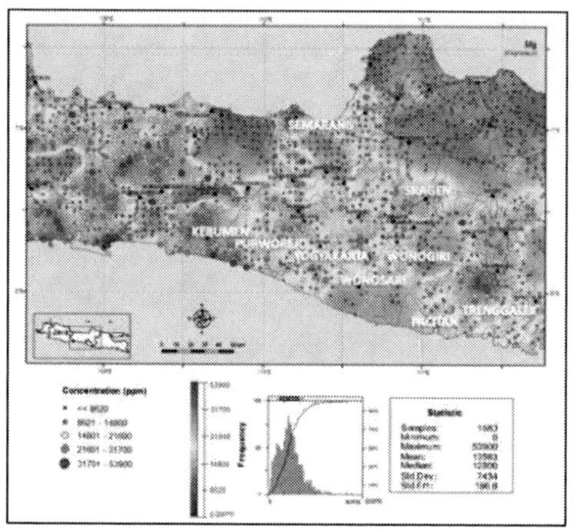

Figure 11. Magnesium (Mg) geochemical map.

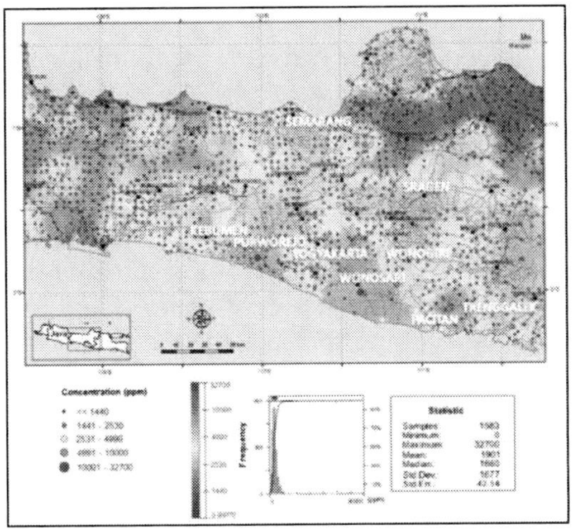

Figure 12. Manganese (Mn) geochemical map.

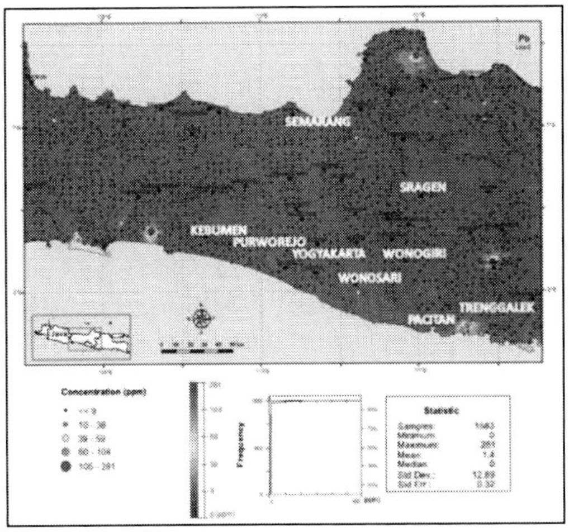

Figure 13. Lead (Pb) geochemical map.

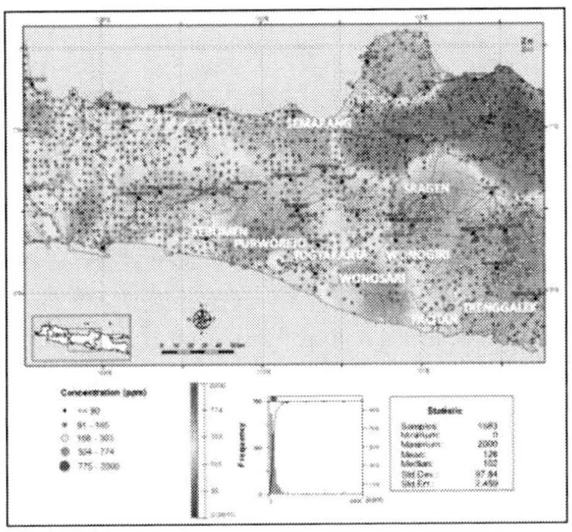

Figure 14. Zinc (Zn) geochemical map.

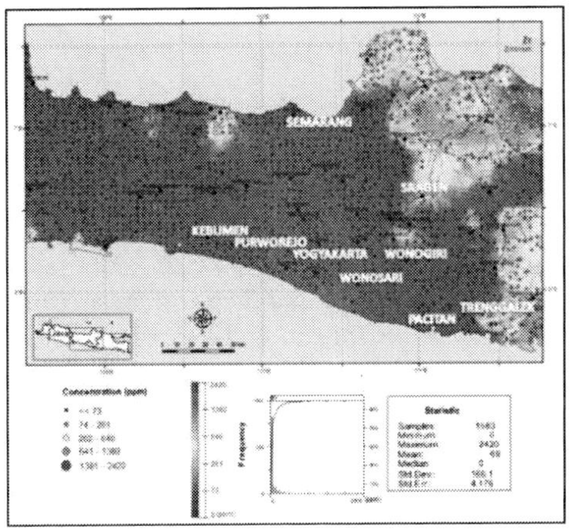

Figure 15. Zirconium (Zr) geochemical map.

As clearly shown by the ultramafic rocks from Kebumen. Enrichment with Cobalt (Co), Chromium (Cr), and Magnesium (Mg) (Figure 6, Figure 7, and Figure 11, respectively) was found in stream sediments collected from Karangsambung ophiolite, which is underlain mainly by the Luk Ulo Melange Complex [6]. Meanwhile, the potassic igneous rocks from the Muria volcano have contrast anomalies of Cerium (Ce), Lanthanum (La) and Zirconium (Zr) (Figure 4, Figure 10, and Figure 15, respectively). Muria volcano is different from other active volcanoes in Central Java, located north of the present volcanic arc in Central Java [7, 8].

Some commodities of mineralization are found in this area; which belongs to several types of mineralization: low & high sulphidation epithermal in the west part of Central Java, and low sulphidation epithermal & porphyry in the east (Figure 1). Elemental anomalies related to mineral resources and mining areas such as Arsenic (As), Copper (Cu) and Lead (Pb) (Figure 2, Figure 8, and Figure 13, respectively) are found in Trenggalek, Pacitan, and Kebumen. The anomalies in Trenggalek may be correlated with gold-mineralized quartz veins, in volcanic rocks of oligo-miocene age close to andesite plug in to the northern of the prospect area [9]. While the elements Ba, Fe, and Mn (Figure 3, Figure 9, and Figure 12, respectively) are scattered along the southern coast of Central Java. The southern coast is part of the Southern Mountains Range formed from the Old Andesite formation which has many mineral discoveries of iron, manganese, and gold deposits.

Initially, geochemical maps were only for mineral exploration, but in its development, it can also be used for environmental assessments [10]. In this case in Central Java, it can be seen that the segment of Wonogiri-Sragen, some industries (textile, plastic, metals, and chemical) disposed their wastewater to Bengawan Solo River. These industries cause an increase in heavy minerals pollution in water and sediments [11]. This phenomenon can be seen on the stream sediment geochemical anomaly map for the elements Cr, Cu, and Zinc (Zn) in the Bengawan Solo River flow between Wonogiri and Sragen city. Anthropogenic elements such as Chlorine (Cl), Ce, and As were spread along the north coast of Central Java probably because of pollution by industrial, farming, and urban locations. The area of cities such as Semarang, Yogyakarta, Purworejo, Wonosari, and Kebumen have increased concentrations of Ce, Cl, and Pb (Figure 4, Figure 5 and Figure 13, respectively).

Commonly, the elemental concentration pattern of the stream sediments reflects faithfully the distribution of geology and mineral deposits. However, the exception is limestone, which insignificantly influences the elemental concentrations of stream sediments [12]. The geochemical map of Fe, Mn, and Zn may be an example of this situation. These elements have a lower concentration than the surrounding area, exception for the element of Zirconium (Zr). These patterns are caused by the presence of carbonate or limestone from the Rembang zone in the south of the alkaline Muria volcano.

4. Conclusions

Regional geochemical surveys that have been carried out on the island of Java provide a typical pattern of elemental anomalies as a result of geological and environmental conditions. In this central part of Java, the geological products of alkaline igneous rocks are clearly shown by Ce and La anomalies, the Karangsambung Ophiolite may consider in Cr and Co pattern. While the consequences of human activities are shown from the Cl, Cu, and Pb patterns. The pattern is random and scattered close to the city, farming, and industrial areas. This stream sediment geochemical map can convincingly delineate elements due to geological and environmental conditions. The elemental anomalies shown are crucial for environmental pollution assessment or as preliminary exploration for mineral resources.

5. References

[1] Sendjaja P, Kusniadi A, Partoyo E, Baharuddin, Hikmat A, Suyono, Andriani S, Rosana M, Safri I and Hendratno A 2015 *Atlas Geokimia Jawa Bagian Barat* Pusat Survei Geologi Badan Geologi Kementerian Energi dan Sumber Daya Mineral ISBN 978-602-6923-06-6 39p

[2] Nugraha S A M, Hall R 2013 Cenozoic Stratigraphy of the East Java Forearc *Indonesian Journal of Sedimentary Geologi* **26** 05/2013 ISBN: 0853-9413

[3] Geological Agency 2013 *Metallogenic Map of Indonesia* Geological Agency Bandung

[4] Darnley A G, Björklund A, Bølviken B, Gustavsson N, Koval P V, Plant J A, Steenfelt A, Tauchid M, Xuejing X, Garrett R G and Hall G E M 1995 A Global Geochemical Database for Environmental and Resource Management: Final report of IGCP Project 259 *Earth Sciences* **19** UNESCO Publishing Paris pp 122

[5] Geosoft 2012 Analysing Geochemical Data Course Manual *http://www.geosoft.com.*

[6] Permana H, Munasri, Mukti M M, Nurhidayati A U and Aribowo S 2018 The origin of oceanic crust and metabasic rocks protolith, the Luk Ulo Mélange Complex, Indonesia *IOP Conf. Series: Earth and Environmental Science 118 (2018) 012004*

[7] Marin J, Winarno T, and Mindasari D 2019 Mineralogy and Geochemistry Variation of Igneous Rocks from Ungaran and Muria Volcano and Its Processes related to Subduction Zone Magmatism of Sunda Arc *The International Conference on Geoscience IOP Conf. Series:*

Earth and Environmental Science 279 (2019) 012017 IOP Publishing

[8] Katili J A and Sudrajat A 1984 *Volcanological Survey of Indonesia* Bandung pp 102

[9] Takahashi R, Shingo Y, Imai A, Watanabe K, Harijoko A, Warmada I W, Idrus A, Setijadji D L, Phoumephone P, Scherstén A and Page L 2014 *Epithermal Gold Mineralization in the Trenggalek District East Java, Indonesia* Resource Geology **64** 2 pp 149–166

[10] Ohta A and Imai N 2011 Comprehensive Survey of Multi-Elements in Coastal Sea and Stream Sediments in the Island Arc Region of Japan: Mass Transfer from Terrestrial to Marine Environments Advanced *Topics in Mass Transfer (Mohamed El-Amin, ed.)* pp 373–398, *InTech.*

[11] Shabrina H A and Roosmini D 2016 Heavy Metals Distribution in Water and Sediment at Bengawan Solo River on Wonogiri-Sragen Segment *Proceedings of the 12th International Symposium on Southeast Asian Water Environment (SEAWE2016) Hanoi Vietnam November 28-30 2016*

[12] Ohta A and Minami M 2013 Less impact of limestone bedrock on elemental concentrations in stream sediments: Case study of Akiyoshi area *Bulletin of the Geological Survey of Japan* **64** pp 121-138

IOP Conf. Series: Earth and Environmental Science 1027 (2022) 012002 doi:10.1088/1755-1315/1027/1/012002

Effects of mentoring housewives assistance on the ability to sort waste in sub-district of Panakukkang, Makassar

Nurleli[1], Sri Syatriani[1] and Sri Fauziah[1]

[1] Sekolah Tinggi Ilmu Kesehatan Makassar, Jl. Maccini Raya No. 197, Makassar 90231, South Sulawesi, Indonesia

Corresponding Author: nurleli.stikma98@gmail.com

Abstract. Waste sorting is separating one type of waste from another. In 2016, the amount of waste in Makassar reached 4,183.41 m3/day, while the amount handled was 3,962.63 m3/day. This study was conducted in 53 families of Sinrijala Village, Panakukkang District, Makassar with 48 housewives in order to evaluate the effects of mentoring housewives on the ability to sort waste using quasi-experimental research with a pretest-posttest non-randomized control group design. The results before mentoring showed that, in the experimental group, there was one respondent (4.20%) who sorted waste, while 23 respondents (95.80%) did not sort the waste. In the control group, there were two housewives (8.30%) who sorted the waste in contrast to 22 respondents (91.70%) who had never sorted the waste before. After being guided, 10 respondents of the experiment group (41.70%) sorted the waste. In the control group, two respondents (8.30%) sorted the waste. Statistical test obtained Z = -3000 and p-value of 0.003. Therefore, it is concluded that there is an effect of mentoring on waste sorting. For further studies, it is recommended to study the practice of recycling waste into goods of economic value.

1. Introduction

The World Health Organization says that waste is something that is not used, not liked, or something that is thrown away that comes from human activities and does not happen by itself [1]. Garbage is a direct consequence of life; therefore, it can be said that garbage has arisen from human life. The emergence of waste along with human activities, starting from the business of taking natural resources as raw materials, continuing to materials that are ready for energy, semi-finished materials for goods and service activities in consuming these goods to achieve the welfare of human life [2]. Garbage is one of the biggest problems in urban areas in Indonesia [3]. Urban areas are one of the areas that are vulnerable to environmental problems. The high population and dense community settlements make urban areas require good urban planning management. One of the environmental problems in urban areas is waste. A large number of residents causes the high generation of waste generated every day [3].

According to Law Number 18 of 2008 concerning waste management, it is explained that waste is a national problem therefore its processing needs to be carried out in a comprehensive and integrated manner from upstream to downstream in order to provide economic benefits, health for the community, and safe environment, and which can also change people's behavior (Ministry of Environment, 2008) [4].

According to Jambeck (2016), Indonesia is ranked second in the world for the marine plastic waste producer, reaching 187.2 million tons after China, which reached 262.9 million tons. The Philippines ranks the third, which produces marine plastic waste reaching 83.4 million tonnes, followed by Vietnam, which reaches 55.9 million tonnes, and Sri Lanka, which reaches 14.6 million tonnes per year [5].

Indonesia is the fourth most populous country in the world with 255,993,674 people [6]. The large population with various activities has an impact on increasing waste piles. Along with population growth, changes in consumption patterns and people's lifestyles also increase the number of waste piles, types, and diversity of characteristics of waste. Until 2015, the number of increased waste piles in Indonesia has reached 175,000 tonnes/day, or the equivalent of 64 million tonnes/year [7].

According to the data from the Regional Environmental Agency of South Sulawesi Province, the amount of waste generation is 4,423,013 kg/day with a population of 8,342,047 people [8]. According to the data from Department of Parks and Sanitation of Makassar City, 2016, Makassar is the tenth-largest city in Indonesia according to its population, which is 1,449,401 people (Indonesian Population Census Data, 2016) [9]. In 2016, the amount of waste produced in the city of Makassar reached 4,183.41 m3 / day, while the amount of waste handled was 3,962.63 m3 / day. In the fruit season, the higher volume of waste can be doubled. The largest amount of waste was contributed by high-population areas such as Rappocini District of 460.69 m3, Panakkukang District of 403.13 m3, and Manggala District of 312.52 m3. From these data, it can be seen that the volume of waste entering the Makassar City landfills final disposal site (TPA) is still quite large, far from the target of reducing waste which is one of the total waste production per day [10].

Households are the largest waste producer at 84%, in addition to institutional waste producers (hospitals, offices, and other agencies). Most of the household waste is organic waste, such as kitchen waste, flour residue, vegetable scraps, fruit peels, and leaves. One of the efforts to solve the waste problem is to carry out waste management which is usually carried out with the 3R principle (reduce, reuse and recycle). The 3R principle is the main principle of managing waste from its source, through various steps that can reduce the amount of waste disposed of in TPA [11]. The main step is sorting it from the source. According to Aswadi (2011) [12], the key to success and waste management lies in sorting it out. Without segregation, waste processing becomes difficult, expensive, and has a high risk of polluting the environment and endangering health. Sorting is separating one type of waste from another. Minimum separation can manage two types of waste such as organic and inorganic waste [13].

Waste segregation is more effective at the level of the source of waste generation (e.g., household) because the components of waste at the source level, especially for inorganic types of waste, still possess pure properties or have not been mixed and contaminated with other waste. The composition of the waste tends to be homogeneous and it also makes it easier to collect certain types of waste needed for recycling, with the goods in good condition. At the final stage in the waste management system, this waste sorting activity can also help reduce the volume of waste that must be transported to the TPA [14]. The waste management system through sorting is also contained in government regulation number 81 of 2012 concerning the management of sorting household waste. Article 17 (1) states that the sorting is done by everyone at the source of the waste [15]. The results of research conducted by Bambang (2011) stated that Community-Based Household Waste Management in Sambiroto Semarang, has been successfully implemented with the 3R principle (Reduce, Reuse, Recycle) through the waste sorting process. The applied model is able to reduce the volume of waste disposed of by up to 70 percent [16].

According to the information obtained from the initial data collection in Sinrijala Village, Panakkukang District, there were 5 Hamlet and 15 neighbourhood with a population of 4,221 people at the end of January 2018 with a male population of 1,999 and a female population of 2,222. Based on the results of the initial survey in Hamlet 002 Neighbourhood 001, there were 53 households with 48 housewives, who claimed they had never carried out waste sorting. Due to the fact that the largest source of waste generation was on a household scale, it was necessary to research the influence assisting housewives on the ability to sort waste in the Panakkukang District, Makassar City.

2. Materials and Methods

This research is quasi-experimental research with a non-randomized control group pretest-posttest design. The research was conducted from June 29, 2018, to July 28, 2018. The research location was in Neighbourhood 001 Hamlet 002 Sinrijala Village, Panakkukang District, Makassar City. The population size of the study was 48 people consisting of RT 001 and RW 002, Sinrijala Village. The sample in this study was all population of housewives, which were divided into an experimental group of 24 people, and a control group of 24 people. The sampling technique used was saturated sampling[17]. Primary data was obtained by conducting field observations by comparing data between the control group and the experimental group [17].

2.1 Research procedure:

a. On the first and second day, the delivery of the aims and objectives of mentoring to housewives and providing an agreement sheet as a form of agreement between the assistant and the housewife before assistance was carried out. The researcher himself conducted scheduling of visits, managing the companion of this study, and taking measurements by giving a pre-test questionnaire whether housewives do the sorting before mentoring.
b. From the third to the seventh day, the researcher provided materials for sorting waste to housewives.
c. In the second week (8th day – 14th day), the researcher conducted assistance in the practice of sorting waste twice a week.
d. In the third week (15th day – 21st day, housewives practiced independently how to sort waste.
e. In the fourth week, the evaluation of the results of housewife assistance on sorting waste was performed.

3. Results and Discussion

3.1 Characteristics of Respondents

Table 1 shows that the majority of the experimental group were in the 41-50 years old group; ten persons (41.70%), while the fewest were in the 20-30 years old group, with three persons (12.50%). In the control group, most respondents were aged 31-40 years and ≥ 51 years, with eight persons each (33.30%), and the least respondents aged 20-30 years old and 41-50 years, with four persons each (16.70%). The highest level of education is high school graduation in the experimental group as many as 11 persons (45.80%), in the control group as ten persons (41.70%), and the least number in the experimental group is not going to school as many as two persons (8.30%), the control has at least one person (4.20%) graduated from higher education. The maximum length of stay in the studied area is ≥ 21 years in the experimental group as many as 17 persons (70.80%) and the control group as many as 19 persons (79.28%), and the minimum length of stay is 11-20 years in both groups with one person (4.20%). The majority of the working type was housewives with 23 persons (95.80%) in the experimental group , and with 20 persons (83.30%) the control group, whereas, the fewest was the civil servant with one person each in the experimental group and in the control group (4.20%). The highest number of family members was 4-6 persons; 12 persons (50.00%) in the experimental group and 13 persons in the control group (54.20%) and the smallest number in the group with ≥ 7 persons; with two persons each in the experiment group and the control group (8.30%).

Table 1. Distribution of Respondents Characteristics in Neighbourhood 001 Hamlet 002 Sinrijala Village, Panakukkang District, Makassar City.

Variables	Experiment group		Control group	
	n	%	n	%
Age (Year)				
20-30	3	12.50	4	16.70
31-40	6	25.00	8	33.30
41-50	10	41.70	4	16.70

≥51	5	20.80	8	33.30
Education level				
Graduated from college	0	0.00	1	4.20
Graduated from elementary school	4	16.70	10	41.70
Graduated from high school	11	45.80	10	41.70
Graduated from junior high school	4	16.70	3	12.50
No school	2	8.30	0	0.00
Not completed in primary school	3	12.50	0	0.00
Length of stay				
1-10 years	6	25.00	4	16.70
11-20 years	1	4.20	1	4.20
≥ 21 years	17	70.80	19	79.20
Profession				
Housewife	23	95.80	20	83.30
Civil servant	1	4.20	1	4.20
Employee	0	0.00	3	12.50
Number of Family Members				
1-3 person	10	41.70	9	37.50
4-6 person	12	50.00	13	54.20
≥ 7 persons	2	8.30	2	8.30

Source: Primary Data

3.2 Distribution of housewives' waste sorting practice before and after assistance
Table 2 shows the practice of sorting housewives' waste before and after mentoring in the experimental group who did not sort waste before mentoring as many as 23 respondents (95.80%), and after mentoring who did not sort waste were 14 respondents (58.30%).

Table 2. Distribution of housewives' waste sorting practice before and after assistance in Neighbourhood 001 Hamlet 002 Sinrijala Village, Panakukkang District, Makassar City.

Sample Group	Pretest				Post-test				Amount	
	Sorting		Not sorting		Sorting		Not sorting			
	n	%	n	%	n	%	n	%	n	%
Experiment	1	2.10	23	47.90	10	20.80	14	29.20	**48**	**100.0**
Control	2	4.20	22	45.80	2	4.20	22	45.80	**48**	**100.0**
Amount	**3**	**3.20**	**45**	**46.80**	**12**	**12.50**	**36**	**37.50**	**96**	**100.0**

Source: Primary Data

Regarding the practice of sorting housewives' waste before / pre-test and after / post-test, there were 22 respondents (91.70%) who did not sort the waste before / pre-test and after / post-test, it remained or did not change due to the absence of assistance or treatment.

3.2.1 The Practice of Sorting Housewives' Waste Before Assistance. The mother in the household is the person who has significant involvement in managing the house. Mother as a role model as well as a teacher in different stages of a child's life is expected to be able to teach and provide necessary explanation or education such as the importance of hygiene, cleaning of household and environment such as sorting waste among matters. Through the role of housewives, it is hoped that all elements of the household can be involved to be able to participate in sorting waste.

Waste sorting is waste grouping at its source based on the type of waste, namely organic and inorganic. Without segregation, waste processing becomes difficult, expensive, and has a high risk of polluting the environment and endangering health.

The results of the research in Neighbourhood 001 Hamlet 002 Sinrijala Village Panakukkang District, Makassar City before the mentoring showed that in the experiment group, one respondent (4.20%) sorted the waste, and 23 respondents (95.80%) did not sort the waste. In the control group, there were 2 respondents (8.30%) who sorted the waste, and 22 respondents (91.70%) did not sort the waste. One of the causes is the lack of knowledge and information of respondents about sorting waste, this is following the initial data before assistance. Most of the respondents have not sorted household waste because they have never heard counseling. This is in line with the research of Sri Pangesti Dewi (2015) [18] in Cetan Village, Ceper District, Klaten Regency, specifically, the extension of the demonstration method affects the actions of mothers in household waste management with p-value = 0.017 [18]. Factors that influence the practice of sorting housewives' waste before assistance are age, education level, length of stay, occupation, and the number of family members. In addition, a stimulus is given in the form of facilities to be willing to be a waste sorter.

The results of the interview showed that regarding the "does the mother sort organic and inorganic waste?" question; respondents who did not sort the waste in the experimental group were 23 respondents (95.80%) with reasons; "it is troublesome" as many as 12 respondents (50.00%), "do not have two types of trash cans" with 9 respondents (37.50%), "remote disposal locations and are not used to" one respondent (4.20%). In the control group, as many as 22 respondents did not sort the waste (91.7%) with reasons; "it is troublesome" as many as 19 respondents (79.20%), "do not have two types of trash cans" for one respondent (4.20%), "are not useful for personal use and do not have the budget to provide separated trash cans" for one respondent (4.20%).

The number of respondents who sorted waste on the grounds of helping environmental preservation was one respondent (4.20%) in the experimental group, and two respondents (8.30%) in the control group. They understand that the waste problem can cause environmental damage that leads to a decrease in health quality, but they hope that all components of society can do it simultaneously so that a significant clean environment is achieved.

Regarding the availability of facilities in the research location, all respondents answered that they did not get sorted trash facilities in the experimental group as many as 24 respondents (100.00%) and the control group as many as 24 (100.00%).

3.2.2 Housewife Waste Sorting Practices After Assistance. Assistance is an activity that is carried out and can mean coaching, teaching, the direction in groups that have the connotation of mastering and controlling. The results of the research in RT 001 RW 002 Sinrijala Village, Panakukkang District, Makassar City after mentoring were that those who sorted the waste from the experimental group were 10 respondents (41.70%), the control group were two respondents (8.30%) and those who did not sort the waste were 14 respondents (58.30%) in the experimental group and 22 respondents (91.70%) in the control group.

According to the results of the interview, the answers to the question "does the mother sort organic and inorganic waste?" showed that the respondents who did not sort the waste in the experimental group were 14 respondents (58.30%), with reasons; "troublesome". In the control group as many as 22 respondents (91.70%) with reasons; "troublesome" as many as 17 respondents (70.80%), "do not have two types of places and are not useful for personal trash" two respondents (8.30%), "do not have the budget to provide separated trash cans" as one respondent (4.20%).

The number of respondents who sorted waste in the experimental group were 10 respondents (41.70%) with reasons; "help to conserve the environment" as many as one respondent (4.20%) and "benefits for yourself" nine respondents (37.50%). In the control group as many as two respondents (8.30%) with reasons; "help to conserve the environment" and "benefits for yourself" as one respondent each (4.20%). In the experimental group, 24 (100.0%) answered yes to the category "to get a trash can facility". In the control group, 24 (100.0%) answered that they did not get trash cans.

This is in line with the research of Dewa Ayu Posmaningsih (2016) in East Denpasar, which together with knowledge, attitudes, availability of facilities, local institutions, and economic benefits, have a contribution value (65.10%) to community participation in waste management [19].

Community participation in the waste management of the Hanjuang KSM program in The Margaluyu environment, Cicurug Village, is running according to the plan from the previous stage planning of waste management activities and implementation stages of management activities waste by empowering the community in the KSM program waste management Hanjuang. The results of community participation can be seen from the awareness of citizens to carry out waste sorting business, and in the manufacture of recycled products recycle from trash [20].

Armadi's research, (2021) said that community participation in Denpasar City is quite effective in waste management activities through composting and 3R (reuse, reduce and recycle) [21]. This is in line with Wardi's research in the Gianyar, Bandung, and Denpasar areas which suggests strategies for implementing environmental (waste) management regulations effectively through reward (for those who merit) and Punishment (for those who violate) [22].

3.3 The Effect of Housewife Assistance on Waste Sorting

Table 3 shows that based on the results of the Wilcoxon signed-rank test calculation, the calculated Z value is -3,000, while the Z table value is obtained from the Z table with α 5% (0.05) Z table value 0.0011 with p-value (Asymp. Sig 2 tailed) amounting to 0.003 which is less than 0.05 so that the hypothesis decision has an effect of housewife assistance on the ability to sort.

Table 3. The Effect of Housewife Assistance in the Experiment Group on Waste Sorting in Neighbourhood 001 Hamlet 002 Sinrijala Village, Panakukkang District, Makassar City.

Test Statistics	
	Post test Experiment - Pre test Experiment
Z	-3.000^{b}
Asymp. Sig. (2-tailed)	.003

Mann Whitney U test is a nonparametric test used to determine the difference in the median of two independent or independent groups if the dependent variable data scale is ordinal or interval/ratio but not normally distributed [17]. This test is an alternative test of the independent T-test.

Based on the results of the Mann-Whitney calculation between the post-experiment and control, there is a p-value of 0.008 less than 0.05, therefore there is a significant difference between post-experiment and control. This means that the experimental group that was given treatment had a greater value than the untreated control. In this case, the assistance carried out in the experimental group can change the behavior of respondents from those who are not given treatment/control (Table 4).

Table 4. The Effect of Assistance on Waste Sorting in Neighbourhood 001 Hamlet 002 Sinrijala Village, Panakukkang District, Makassar City.

Test Statistics	
	Result

Mann-Whitney U	192.000
Wilcoxon W	492.000
Z	-2.639
Asymp. Sig. (2-tailed) .008	

The practical assistance program for sorting waste by extension methods and the practice of sorting organic and inorganic waste through individual or group approaches can influence the behavior of household wives sorting. Counseling and the practice of sorting organic and inorganic waste were carried out routinely and continuously for nine visits to each target, both individually and in groups. According to Muljono (2007) [23], counseling will change people's awareness and behavior (knowledge, attitudes, and skills) in a better direction and can achieve a more prosperous life. The results of the research by Sri Pangesti, et al (2014) show that, in general, extension with the demonstration method affects the increase in knowledge and actions in managing waste ($p = 0.037$) [24].

The theory of Ebbinghaus and Boreas in Prasetyaningsih (2005) states that the power to remember of humans is decreasing over time, which in the end humans will experience forgetfulness [25]. The frequency of home visits and counseling carried out by researchers after mentoring decreased in the second week to twice visits, which caused the response of housewives to the materials that had been given during mentoring also gradually decreased. This phenomenon shows that the extension process and the practice of sorting organic and inorganic waste after mentoring, especially for targeted housewives, must be carried out continuously by the local government.

The results of research in Neighbourhood 001 Hamlet 002 Sinrijala Village, Panakukkang District, Makassar City, showed that, in the experimental group, there was an effect of housewives' assistance on waste sorting because of the treatment given such as counseling, practicing sorting waste, and providing waste sorting facilities, before assisting housewives who sort 1 respondent (4.2%) and 9 respondents (37.5%) after mentoring.

In this study, although there was an effect of housewife assistance on waste sorting, there were still many respondents who did not sort the waste. In details, 36 respondents (75.00%) did not sort the waste for reasons; "troublesome" as many as 31 respondents (64.58%), "not personally use and do not have two types of trash cans" by four respondents (4.20%), "do not have the budget to provide" with one respondent (2.10%) segregated trash cans, and those who sorted the waste were 12 respondents (25.00%) with reasons; "benefits for yourself" 10 respondents (21.00%), "helping to preserve the environment" two respondents (4.16%). Overall, the answers were more dominant, the respondents answered "troublesome", this has a significant effect on being an important factor for respondents in deciding not to sort or sort the waste. This follows the opinion of Vincent (1997) which affects the perception that humans tend to reject offers that are not following what they expect [26].

4. Conclusions

Based on the results of research on the effect of housewives' shelter on the ability to sort waste in the Panakukkang District of Makassar city, it can be concluded as follows: before mentoring, it showed that there was only one respondent (4.20%) who sorted the waste in the experimental group whereas 23 respondents (95.80%) did not sort the waste, however, after conducting assistance sorting waste in the experimental group, as many as 10 respondents (41.70%) sorted the waste whereas 14 respondents (58.30%) still did not sort the waste. Therefore, there is an effect of housewife assistance on waste sorting. For further studies, it is expected to research the practice of recycling waste into goods of economic value.

Acknowledgements

This project is supported by the Makassar College of Health Sciences (Sekolah Tinggi Ilmu Kesehatan Makassar).

References

[1] Dobiki and Joflius 2018 Analysis of availability of waste infrastructure in Kumo Island and kakara island in north Halmahera Regency. *J. Spasial.* **5** (2) ISSN : 2442-3262
[2] Selintung, Mary and Rombe I R R D R 2015. Study on integrated waste management at the kelurahan level in Makassar City (case study: Paropo Village, Panakkukang District). *J. Civ. Eng.* Hasanuddin University. Macassa
[3] Lubis and Ningsih E W 2018 Analysis of generation, composition and characteristics of household waste in Medan City Region I (case study: Medan Johor District and Medan Tembung District). http://repositori.usu.ac.id/handle/123456789/9222
[4] Ministry of Environment. 2008 on Waste Management. Jakarta: Bureau of Law and Public Relations of the Ministry of Environment. Law of the Republic of Indonesia Number 18 of 2008
[5] Jambeck J 2016 *Indonesia is the second-largest contributor to plastic waste in the world*
[6] Central Bureau of Statistics 2016 *Statistical Yearbook of Indonesia 2016.* © Central Bureau of Statistics/BPS-Statistics Indonesia. CV. Dharmaputera
[7] Ministry of Environment and Forestry 2015 *Environmental and Forestry Statistics 2014. Jakarta.*
[8] Regional Environmental Agency of South Sulawesi Province 2014 *Regional Environmental Status. Makassar*
[9] Regional Development Planning Agency 2016 *Makassar City Profile 2016.* https://makassarkota.bps.go.id
[10] Jumran S 2017 *Efektivitas Pola Penanganan Sampah di Kota Makassar*
[11] Setyono J S, et al 2018 Characteristics of community scale household waste management in Semarang City. Riptek Vol. **12**, No. 2, Year 2018 pp. 119-130
[12] Aswadi M and Hendra 2011 *Waste management planning at Tavanjuka Mas Housing*. Mektek Year XIII NO. 2, MAY 2011. https://media.neliti.com
[13] Sulistiyorini S G, Darwis N R R S and A 2015 *Community Participation in Waste Management in Margaluyu Neighborhood, Cicurung Village*," no. **5** (1) pp 71–80
[14] Maulina A S 2012 Identification of community participation in waste sorting in North Cimahi Subdistrict and its influencing factors *J. Urban Reg. Plan.* **23** (3) pp 177–196
[15] Government Regulation (PP) No. 81 of 2012 concerning the Management of Household Waste and Types of Household Waste
[16] Dwiyanto B M 2011 Model of Increasing Community Participation and Strengthening Synergy in Urban Waste Management. *J. Dev Eco.* **12** pp 239-256
[17] Dahlan and Sopiyuddin M 2017 Statistics For Medicine and Health. Jakarta : *Indonesian Epidemiology*
[18] Dewi S P 2015 The effect of extension with demonstration methods on increasing knowledge and actions on waste management of women in Cetan Village, Ceper Subdistrict, Klaten Regency *J. Environ. Health*, **3** (3) pp 120–126
[19] Posmaningsih D A A 2016 Factors affecting community participation in solid waste management in East Denpasar, *Husada's J.* **1** (13) pp 59–71
[20] Sulistiyorini N R, et al 2015 Community Participation in waste management in the Margaluyu Environment, Cicurug Village *Soc. Wor. J.* **5** https://jurnal.unpad.ac.id/share/article/view/13120/0
[21] Armadi N M 2021 Community participation in waste management as the key to success in managing waste *J. Soc. Political Sci.* **35** https://e-journal.stispolwb.ac.id
[22] Wardi I N 2011 Socio-cultural based waste management: efforts to overcome environmental problems in Bali *Bumi Lestari J.* **11** pp 167–177.
[23] Muljono P 2007 Learning society, explanation and national development. *Extension J.* **3** ISSN: 1858-2664
[24] Dewi S P, et al 2014 The effect of counseling using demonstration methods on increasing

women's knowledge and actions on waste management in Cetan Village, Ceper District, Klaten Regency in 2014 *Sanitation, Environmental Health Journal* **6** p 120–126

[25] Prasetyaningsih A Sudargo T Susilo J 2005 Knowledge, attitudes and skills of food handlers who were given food safety training at the nutrition installation of the heart hospital of Harapan Kita *Indonesian Journal of Clinical Nutrition* **2** p 71

[26] Gaspers V 1997 Total business management in the age of globalization. *Jakarta: Gramedia*

Hazard risk identification from the used masks in Jabodetabek

M M Simatupang[1] and E Veronika[2]

[1] Public Health Study Program, Faculty of Health Science, Respati University of Indonesia, East Jakarta, Special Capital Region of Jakarta, 13890, Indonesia
[2] Public Health Study Program, Faculty of Health Science, Esa Unggul University, West Jakarta, Special Capital Region of Jakarta, 11510, Indonesia

Corresponding Author: meithyra@urindo.ac.id

Abstract. COVID-19 has brought many changes of lifestyle to people around the world. The obligation to use masks as an effort to prevent the transmission of COVID-19 is such an example. This policy led to the increase of masks using in the general population. Some particular procedures of masks usage and used masks handling must be applied to prevent negative impacts. Many of these guidelines are not generally known. This study is carried out to determine the handling, cleaning, and disposal method of used masks at beginning of the pandemic. The research data were obtained from 152 random respondents who voluntarily filled out online questionnaires. Participants came from some areas in Jabodetabek that implemented Large-Scale Social Restrictions (PSBB) and were required to wear masks. Most of the respondents kept the used mask temporarily then reused it. Handling of the used masks, both surgical masks and cloth masks, was also showed poor sanitation for some participants. The waste generation from used surgical masks at the beginning of the pandemic was also quite alarming. Education on the handling method of used masks is important to be widely publicized to prevent potential hazards, such as the transmission of respiratory infections and environmental contamination risk.

1. Introduction

One of the preventive efforts of COVID-19 transmission was carried out by using personal protective equipment (PPE). During the pandemic, face masks are the most common PPE used to combat this respiratory infection, in the community or health care [1], [2]. Mask needed to suppress transmission, however the use of masks alone is not sufficient enough for adequate protection [3]. It is required face mask usage as well, combined with social distancing to increase infection prevention effectiveness [4].

World Health Organization declared COVID-19 as a pandemic in March 2020, afterward COVID-19 Response Acceleration Task Force, as a representative of the Indonesian government, issued the policy for the community to use masks while activity outside, along with providing wash hands facility, as an effort to prevent transmission of the virus. This regulation is applied in all regions of Indonesia [5].

The Public Behavior Survey during the COVID-19 pandemic organized by Statistics Indonesia (Badan Pusat Statistik) found that the level of respondent's compliance for wearing masks in September 2020 was quite high (91.98%) [6]. Although the conclusion of this survey has a positive outcome, some issues quite worrying regarding the handling of used masks, such as disease transmission or the problem of solid waste generated by disposable masks have become worrisome, as the handling procedure of used masks are not widely known in the community [2].

Cloth face masks, surgical masks, N95 mask, and facepiece respirators are the available type of mask in Indonesia. Medical masks (surgical masks, N95 masks, and facepiece respirators) that are classified as medical waste must be disposed of in a specific way to prevent disease transmission from used masks. As for the cloth masks user, they must apply a clean and healthy behavior, even if cloth masks can be washed and used repeatedly [7].

This research is conducted to observe the behavior of used mask handling in the community, reuse of masks, as well as perceive the method of single use masks disposal. The results of this study are expected to provide a recommendation regarding the importance of increasing public understanding and awareness of the procedures for handling used masks, to clean and dispose of waste safely. The goal is to reduce pathogens transmission from used masks and prevent pollutants contained in masks contaminate the environment.

2. Methods

This descriptive research used a cross sectional design and was conducted on residents in the Province of DKI Jakarta, Bogor City, Bogor Regency, Depok City, Bekasi City, Bekasi Regency, Tangerang City, Tangerang Regency, and South Tangerang City (Jabodatabek area) which implemented Large-Scale Social Restrictions (PSBB). This policy is obligated to wear a mask when doing the activity outside.

The sample calculation using 95% of confidence level ($Z\alpha$= 1,96), the proportion from literature is 0.1 [8], and the absolute precision is 0.05. The formula is:

$$n = \frac{Z^2 \alpha\, p\,(1-p)}{d^2}$$

$$n = \frac{1{,}96^2\ 0{,}1\ (1-0{,}1)}{0{,}05^2} = 138{,}3 \approx 138$$

To prevent the possibility of drop-out samples, the number of samples is increased by 10%. So that the total sample in this study is 152 respondents. The used mask handling, cleaning, and disposal behavior data for 3 months since the pandemic (March to May 2020) were collected using an online questionnaire which respondents filled out voluntarily (purposive sampling technique). The sample in this study was at least 15 years old. Working in hospitals, Puskesmas, laboratories, medical practices, and midwives were the exclusion criteria, considered the standard procedure of usage and handling of masks must be known. Data collection was held after passing the ethical test Number: 024/SK.KEPK/UNR/V/2020 issued by the Research and Community Service Institution (LPPM) of the Universitas Respati Indonesia.

3. Result and Discussion

Data collection was held in the end of May 2020. Most of the participants are woman (70.4%) and have completed high school (98%). Respondent characteristics by age group is shown in Figure 1 and 2:

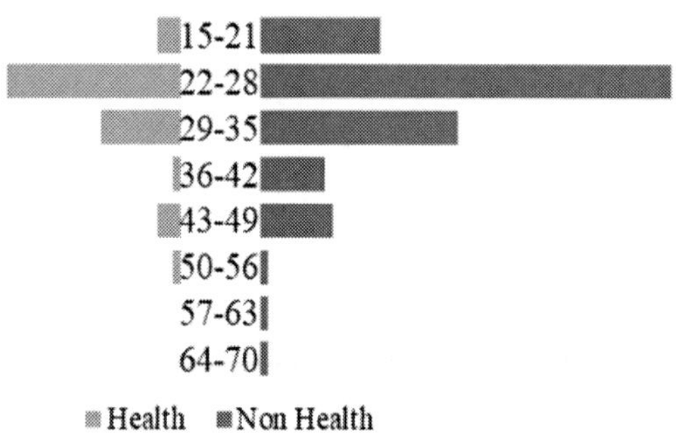

Figure 1. Frequency Distribution of Educational Background by Age Group.

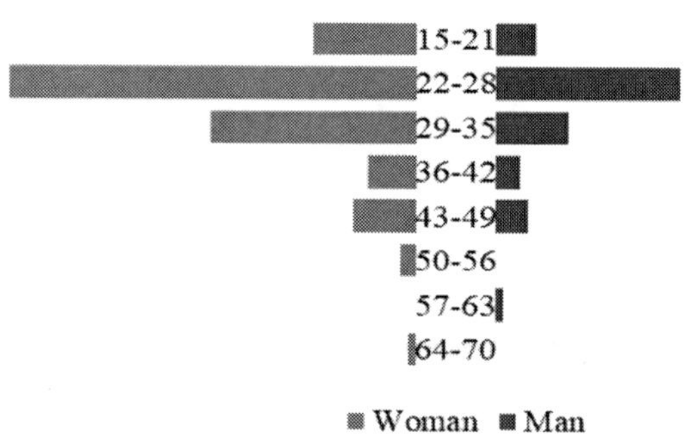

Figure 2. Frequency Distribution of Sex by Age Group.

At the beginning of the pandemic, scarcity of medical masks led to the increase of price, so that the people prefer to use cloth masks as a protection against COVID-19 infection. It can be seen from the data based on the type of mask used, 93.4% (142/152) of participants had used cloth masks in the last 3 months, while the number is lower at the respondents who had used surgical masks (74.3%; 113/152). Most of the respondents used both types, surgical masks and cloth masks (68.4%; 104/152), only 30.9% (47/152) of respondents used only one type.

3.1. Used Mask Handling

The handling of used masks could be concluded as less sanitary. Proven by only 39.5% (60/152) of participants who immediately threw away or washed used masks. The majority of respondents would reuse masks by storing them (51.3%; 78/152) or hanging masks around their necks temporarily (9.2%; 14/152). Among the respondents who stored the masks for reuse, some had folded the masks inward then rolled with mask straps (48.7%; 38/78). The rest did not fold the mask and put it on the surface of the object (28.2%; 22/78), or did not fold it and store it in various storage places such as in pockets, bags, drawers, or cupboards (23.1%; 18/78) (Figure 3 and 4).

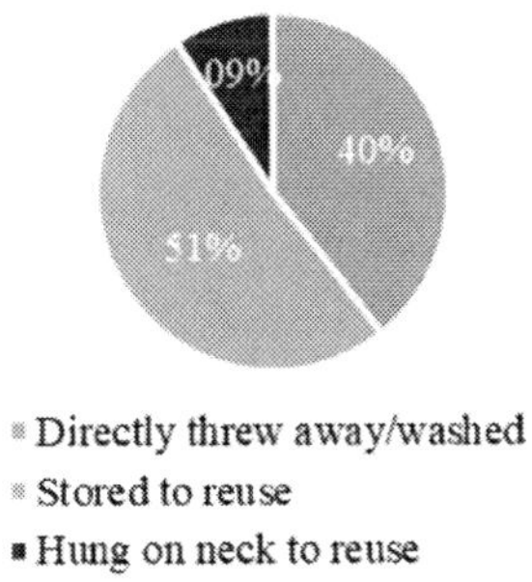

Figure 3. Respondent's used mask handling.

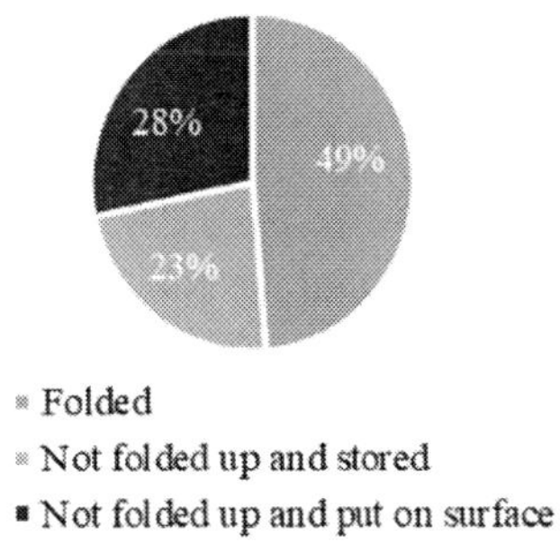

Figure 4. Respondent's habit: mask stored for reuse.

Scalvenzi et al mentioned that costs and the difficulty of obtaining masks during the pandemic affected the repeated use of masks [2]. However, some parts of the community did not know how to use, reuse and dispose the masks properly. Between uses, the mask was put on the wrist or stored in a trousers pocket. Only a small proportion kept the folded masks in a specific plastic bag before being reused. This non-sterile method was common and might increase the risk of mask contamination. Even the incorrect method of reuse masks over a long period time repeatedly can reduce filtration efficacy [2].

Most of the participants kept used masks for reuse the next day (46.2%; 36/78). Some reused masks after being stored for less than 1 hour (30.8%; 24/78), others kept it for 1-6 hours (23.1%; 18/78). Long-term use or reuse of single-use surgical masks (with or without reprocessing) should only be considered in critical lack situations. Guidance should be provided to enable the safest practice for reuse and prolonged mask usage. Prohibited to long-term use or reuse mask if soiled, damaged, moist, or hard to breathe through. The evidence for the impact of prolonged use on masks is still limited, but practice extended use is preferable since the risk of contact transmission is lower than reuse [9].

The result for the safety and effectiveness of surgical masks decontamination for reuse or extended use is insufficient, although some studies have been conducted. Further, decontamination intervention studies on surgical masks possibly carried out by methods that show satisfactory results on N95 [10]. As a face filtering respirator (FFR), N95 can be decontaminated with ultraviolet germicidal irradiation, sterilization by steam, ethylene oxide and vaporous hydrogen peroxide for reuse. The equipment for most of the methods is available in hospitals. These methods can reduce the source of infection in N95, although the level of particle penetration efficiency varies depending on the procedure performed [11].

Contamination has to be prevented during storage by fold the outer surface inward to decrease contact with the inner surface during storage, then kept in a clean and sealable bag or box [12] [9]. Potential hazards can be sourced from masks that were not treated sanitary when stored. Occurred because the respondent has noted coughs or sneezes when wearing a mask (64.5%; 98/152) or around people who cough/sneeze when wearing a mask (65.1%; 99/152) (Table 1).

Table 1. A potential source of contamination.

	Hung on Neck n (%)	Stored between Uses n (%)	Immediately Disposed/Washed n (%)
Coughs or sneezes during mask usage			
Ever	13 (92.9)	48 (61.5)	37 (61.7)
Never	1 (7.1)	30 (38.5)	23 (38.3)
Around people who cough/sneeze during mask usage			
Ever	12 (85.7)	50 (64.1)	37 (61.7)
Never	2 (14.3)	28 (35.9)	23 (38.3)

Coughing and sneezing expel droplets from the respiratory tract that can contain pathogens which can lead to infection [13]. Droplet nuclei can be dispersed in the environment and suspended in indoor air, then inhaled by the host and enter the respiratory system. In fact, SARS-CoV-2 is also estimated to be able to be transmitted via aerosols under certain circumstances [14].

A literature study summarized the resistance of pathogens that were reported to have infectivity on the surface of various objects. The study concluded some types of viruses can survive up to several days on surfaces. For example, Influenza A-virus survived on surfaces made of plastic for <2 hours-4 days and on cloth for <2 hours-1 week; influenza B-virus lasted for 24 hours on plastic and 6-8 hours on cloth, whereas SARS -CoV-2 was found to stay on cloth for 1 day and in surgical masks for 4-7 days [15].

Nosocomial bacteria have longer persistence on any kind of inanimate surface. *Mycobacterium tuberculosis* was known to be able to survive for 1 day - 4 months on dry surfaces, while *Streptococcus pneumoniae* was able to survive for 1 - 20 days [16]. These findings are the basis for the importance of sanitary and immediately handling of used masks before causing contamination or transmission of infection.

3.2. Cloth mask cleaning

Most of the respondents soaked the cloth masks in hot water (83.1%; 118/142) or in disinfectant (78.2%; 111/142) and separated the used masks from other clothes (68.3%; 97/142) before washed. Although dry clothes in the sun after washing was a habit of the community, some respondents did not expose their used masks to the sunlight (7.7%).

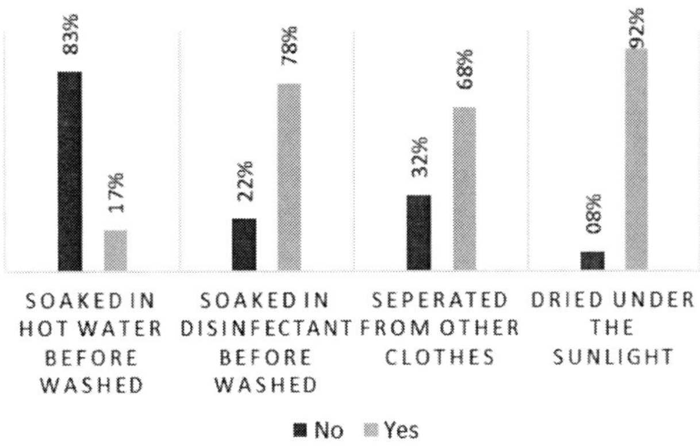

Figure 5. Respondent's used mask handling.

Cloth masks may have lower filtration abilities due to the larger average size of pores than surgical masks, allowing more particles and microorganisms to penetrate. However, the use of this type of face mask can also reduce the excretion of microorganisms, better than not using any masks [17]. Nevertheless, the use of cloth masks is not recommended for health care workers [7].

The material of cloth face masks affected the filtering efficiency. Polyester is considered to be the best fabric material for mask [17]. A face mask made of fabric is washable, has a low density and permeable filter. The material was able to transfer heat and moisture, so not easily dampen and deteriorate, make it more comfortable to use [18]. Cloth masks should be wash in detergent but prohibited from rubbing too much to prevent the pores of the fabric from widening. Cloth masks are suggested not to be used for more than 4 hours to prevent the development of bacteria or fungi due to damped masks [7].

Used masks from a patient are included in infectious waste. If the same treatment must be applied, it's recommended to place the used mask in a certain bag or container, then washing it in the washing machine with hot water at 60-90°C with detergent. Masks can also be soaked in hot water and washed with detergent in certain containers if a washing machine is not available. After that, soak it in 0.05% chlorine for about 30 minutes. After rinsing with clean water, dry the mask in the sun till completely dry [19]. According to those procedures, in case to categorized the result as "sanitary" if 3 or more of the methods was practiced, most of the respondents (92.3%; 131/142) are concluded had sanitary handling of cloth mask, while some others practice poor sanitation (7.7%; 11/142).

3.3. Surgical mask disposal

Used surgical masks washed to be reused by 8% (9/113) of respondents. On the other hand, some others did not apply sanitary behavior in throwing away used medical masks too. Only a small proportion of used mask that was torn off, folded up, wrapped, and soaked in disinfectant before disposal. If categorized into the sanitary and non-sanitary handling groups (sanitary in case performs 3 or more of the procedures) then 98.2% (111/113) of respondents do not dispose of surgical masks in a sanitary practice.

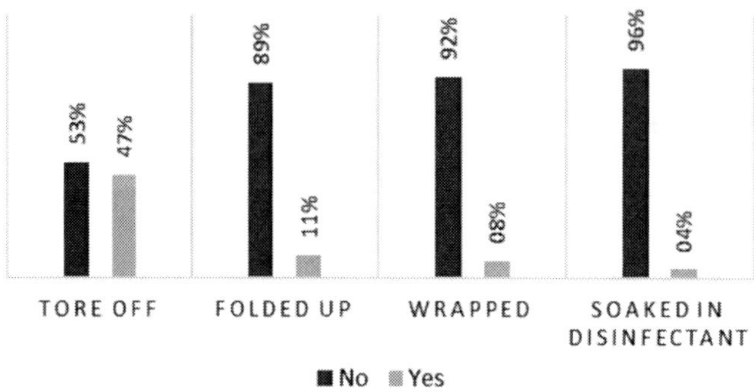

Figure 6. Disposal methods of surgical masks.

Most of the respondents disposed of masks in public trash bins or trash containers at home (93.8%; 106/113), only 5.3% (6/113) in the hazardous bin, some even threw masks in any place (0.9%; 1/113). Meanwhile, trash cans at home are in the form of plastic bags (39/113; 34,5%), trash bins without lids (35/113; 31,0%), or types that have a lid (39/113; 34,5%).

The guidelines for managing community mask waste issued by the Ministry of Health Republic of Indonesia states that masks used in the household are classified as domestic waste since not used in health services or patients in health care facilities. Disposal of waste have through the process of disinfection, store masks in a safe container, cut the strap and tear off the mask, throw masks into domestic trash, and maintain hand hygiene after touch masks [20]. These measures are effective to remove potential health risks from pathogens that may be present in masks and prevent the reuse of used masks by irresponsible people. However, it cannot overcome single-use masks waste generation.

Disposable surgical masks have become the standard for protection against the COVID-19 virus for reasons that are light, flexible, adequate filtering efficiency, affordable, comfortable to wear, and easy to breathing [21]. This medical mask also has better filtration of bacteria and air permeability than cloth masks [22]. The bad thing is the significant increase in the number of mask wearers globally. This phenomenon is a potential environmental risk due to improper disposal [21].

Several surveys found that surgical mask is the most frequent type of medical mask that use as personal protective equipment during the COVID-19 pandemic [2] [22]. This study reveals that the average respondent used 4 masks in a week. Considering the data submitted regarded behavior at the beginning of the pandemic, March - May 2020, this figure possibly increases since masks are more widely produced and easy to obtain in this present time.

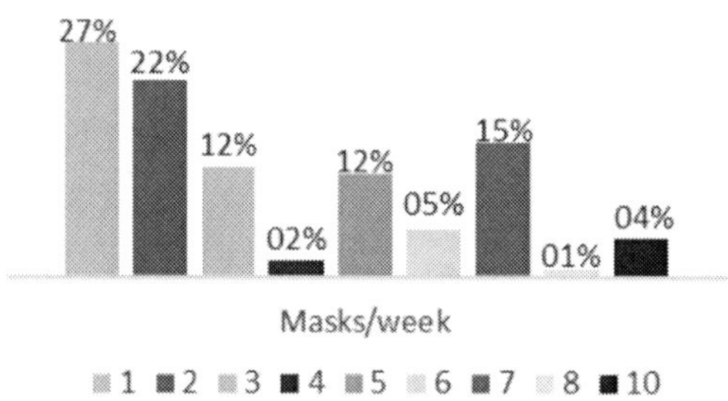

Figure 7. The amount of surgical masks that participants used in a week.

Individually, the number of masks waste generated per week depending on the duration of usage, type of mask, personal hygiene, or place visited. A survey by Selvaranjan et al took notes that more than 25% of participants generate above 5 masks waste a week, which might be influenced by the lockdown policy which decreased community activities outside, thus reducing the use of masks as well. Necessity of wearing masks around the world can significantly increase the number of masks disposed into the environment [22].

Indonesia ranks 3rd in the largest number of daily mask users after China and India, based on observations in the Pandemic database on July 31, 2020. This total daily face mask use was calculated based on 273,753,080 of population, 106,336 cases of COVID-19, 73% of the urban population, 80% of face masks acceptance rate, and the estimate of 1 mask is needed by general population each day. This calculation estimates that 159,214,791 masks are disposed of with a total weight of 420.03 tons/day [23].

The increase in disposable masks consumption contributes to a large amount of plastic waste and plastic particles to the environment. Disposable masks that are disposed of in landfills, dumpsites, freshwater, oceans or littering in public places create new sources of microplastic fibers due to the degradation process or break down into smaller sizes (<5 mm) [24]. The solid waste open dumping method, which is the most used landfill method in Indonesia, causes environmental and health risks due to irresponsible disposal, the release of various types of waste, particularly plastics. Some of these materials enter water bodies, contaminating rivers, lakes, right down to marine environments adding to the presence of plastics in the aquatic medium [24] [25].

Surgical masks can be produced from various polymer materials as well, such as polypropylene, polyurethane, polyacrylonitrile, polystyrene, polycarbonate, polyethylene, or polyester, which have been used as raw materials for various plastic products. A single-use mask consists of three layers; the inner layer which is a fibrous material, the middle layer (as filter part), and the outer layer (non-woven, waterproof and colored). The three layers of surgical face masks are confirmed to be a kind of plastic polymers [26]. The major material to produce surgical masks is polypropylene, as a polymer for the fabrication of surgical mask fiber [22], [21]. Polypropylene as a polymeric material has been recognized as a significant source of macroplastics and microplastic contamination in the environment [27].

Disposable masks develop a new source of environmental pollution if not disposed of with appropriate methods. Artificial weathering experiments were carried out to predict the release of surgical mask microfibers disposed to the marine environment. The study found that a surgical mask that was stirred vigorously in seawater and exposed to UV for 180 hours released 173,000 fibers/day. These results were corroborated by observations on surgical masks found on an Italian beach that showed the signature of morphological and chemical degradation similar to this experiment [21]. The degradation process may depend on various environmental conditions such as temperature, humidity, and salinity [27].

Microplastics can be ingested by microorganisms, and even higher organisms in aquatic life, such as fish, which will have an impact on the food chain and cause health problems for humans [26]. Urgent action is needed to limit the number of surgical masks disposed to marine ecosystems, namely a campaign to promote proper disposal and improve the management of waste. It is advisable to provide alternative masks that are biodegradable or contain materials that can be used for a long time [21].

Some countries in Asia have not implemented appropriate infectious waste management methods. Another problem is the lack of medical waste containers that impact the potential contamination from medical waste, infection disease transmission or environmental pollution, to general people. An appropriate medical waste management strategy (standardization, procedures, guidelines, and strict implementation) is necessary within the hospital, households, and public places, in line with the increase of mask disposal in the pandemic periods, particularly [23].

The law regarding solid waste or Environmental Health Standards for Hospitals has drawbacks in its practical application. These policies do not cover the entire region, occasionally. Indeed, it tends to be applied only to urban areas because of the limited infrastructure in rural areas. Several local governments in Indonesia stated directly burn the waste each day or collected and, in Padang, transported to a cement factory incinerator for a burning process is how they treat and dispose of healthcare waste source from households or quarantine locations [28].

Medical waste of household collection method has some consideration. The study Nowakowski et al showed the additional costs required to change the collection method, route, and schedule. Special vehicles for medical waste need to be used for collection, which is difficult to provide. Specific trash quarantine areas must also be prepared. Medical waste products produced by households have to be predictable, along with efforts to increase public awareness to separate medical waste from other waste before garbage collection [29].

All PPE (used masks, used gloves, and personal clothes related to COVID-19 infection) in households must be incinerated or landfilled. It's recommended to store used masks in a plastic bag or another container in order to separate them from another household waste, as this medical waste must be separated from other types of waste before being taken to the incinerator facility, without recycling process [30]. Although the research that has tried several strategies to recycle single-use masks by conducting thermal, morphological, and chemical analysis occurs, this experiment still requires other investigations [31].

The proper management of medical waste is an effort to prevent environmental pollution and disease transmission, incineration is either of the potential methods. Incineration is able to destroy pathogenic bacteria and reduce up to 99.95% volume of waste. The disadvantages of this process are ash residue and waste which may contain heavy metals, which require further process so that it does not cause health impacts and contaminate the environment [32].

The number of incinerator facilities in Indonesia can be categorized as insufficient when compared to medical waste generation, within hospitals and households. Out of 646 hospitals that treat COVID-19 patients, only 20 have their licensed incinerators in general, 110 incinerators with licenses in regular hospitals, and 9 incinerators located in urban areas. Some of the incinerators were in a damaged condition or did not have a license from the Ministry of Environment and Forestry [28].

4. Conclusion

The reuse of masks indicated a non-sterile method since respondents did not store used masks folded and stored in specific closed containers, such as plastic bags, to prevent contamination and avoid pathogens from masks to infect people or contaminate other objects. The majority of the respondents have practiced appropriate methods of cleaning cloth masks, considered the habits of respondents who mostly. Disposal of used surgical masks needs more attention, as becoming the potential environmental pollutants and sources of infection. Biodegradable masks and technological innovations to sterilize or recycle used masks can be an option. But for the first step, mask waste management must be generally known and implemented, from source to treatment location, to prevent the negative impacts. Promotion can be done by utilizing ever-growing technology and various media. The use of or work with social influencer is highly recommended to spread the necessary information effectively to the entire community. It is also important to use local language effectively on any medium including keywords, taglines, mottos, texts, typography, photos, graphics, etc. for the general public.

Funding

This research was funded by a research grant from the Research and Community Service Institution (LPPM) of the Universitas Respati Indonesia.

Authors' Contributions

MMS compiled the structure, data, and results of this study. EV and MMS searched and selected references for the discussion section. MMS arranged the draft of the manuscript and both authors collaborated to accomplish the final manuscript.

References

[1] Chughtai AA and Khan W 2019 Use of personal protective equipment to protect against

respiratory infections in Pakistan: A Systematic Review *J. Infect. Public Health* **12** 522–7
[2] Scalvenzi M, Villani A and Ruggiero A 2020 Community knowledge about the use, reuse, disinfection and disposal of masks and filtering facepiece respirators: results of a study conducted in a dermatology clinic at the University of Naples in Italy *J. Community Health* **46** pp 786-793
[3] World Health Organization 2020 *Updated guidance on the use of masks (COVID-19) update no. 30*
[4] Li T, Liu Y, Li M, Qian X and Dai SY 2020 Mask or no mask for COVID-19: a public health and market study *PLoS One* **15** :e0237691
[5] Kementerian Kesehatan Republik Indonesia 2020 *Circular letter HK.02.02 / I / 385/2020 about the use of asks and supply facility handwashing (CTPS) to prevent COVID-19* (in Indonesian)
[6] Badan Pusat Statistik 2020 Community behavior during COVID-19 pandemic (in Indonesian)
[7] Yudhastuti R 2020 The use of masks during the pandemic period in Indonesian people *Kesmas J. Kesehat. Masy. Nas. (National Public Health Journal)* Special Issue 1 32–6. (https://journal.fkm.ui.ac.id/kesmas/article/view/3945)
[8] Chughtai AA *et al* 2019 Contamination by respiratory viruses on outer surface of medical masks used by hospital healthcare workers *BMC Infect. Dis.* **19** 491
[9] Toomey EC *et al* 2021 Extended use or re-use of single-use surgical masks and filtering facepiece respirators: a rapid evidence review *Infect Control Hosp. Epidemiol.* **42** pp 75–83 (https://pubmed.ncbi.nlm.nih.gov/33028441/)
[10] Zorko DJ *et al* 2020 Decontaminationa interventions for the reuse of surgical mask personal protective equipment: a systematic review *J. Hosp. Infect.* **106** 283–94
[11] Polkinghorne A and Branley J 2020 Evidence for decontamination of single-use filtering facepiece respirators *J. Hosp. Infect.* **105** 663–9
[12] Centers for Disease Control and Prevention (CDC) 2020 *Strategies for optimizing the supply of facemasks* (https://www.cdc.gov/coronavirus/2019-ncov/hcp/ppe-strategy/face-masks.html)
[13] Bahl P, de Silva CM, Chughtai AA, MacIntyre CR and Doolan C 2020 An experimental framework to capture the flow dynamics of droplets expelled by a sneeze *Exp. Fluids* **61**(176)
[14] Dhand R and Li J 2020 Coughs and sneezes: their role in transmission of respiratory viral infections, including SARS-CoV-2 *Am. J. Respir. Crit. Care Med.* **202** 651–9 (https://doi.org/10.1164/rccm.202004-1263PP)
[15] Wißmann JE, Kirchhoff L, Brüggemann Y, Todt D, Steinmann J and Steinmann E 2021 Persistence of pathogens on inanimate surfaces: a narrative review *Microorganisms* **9**(2)
[16] Kramer A, Schwebke I and Kampf G 2006 How long do nosocomial pathogens persist on inanimate surfaces? a systematic review *BMC Infect. Dis.* **6** 130
[17] Rizki SA and Kurniawan 2020 A efficacy of cloth face mask in reducing covid-19 transmission: a literature review *Kesmas J. Kesehat. Masy. Nas. (National Public Health Journal)* Special Issue **1** pp 43–48
[18] Lee K-P, Yip J, Kan C-W, Chiou J-C and Yung K-F 2020 Reusable face masks as alternative for disposable medical masks: factors that affect their wear-comfort *Int. J. Environ. Res. Public Health* **17** 6623
[19] World Health Organization 2020 *Water, Sanitation, Hygiene, and Waste Management for the COVID-19 Virus*
[20] Kementerian Kesehatan Republik Indonesia 2020 *Guidelines for Community Mask Waste Management* (in Indonesian)
[21] Saliu F, Veronelli M, Raguso C, Barana D, Galli P and Lasagni M 2021 The release process of microfibers: from surgical face masks into the marine environment *Environ. Adv.* **4** p 100042
[22] Selvaranjan K, Navaratnam S, Rajeev P and Ravintherakumaran N 2021 Environmental challenges induced by extensive use of face masks during covid-19: a review and potential solutions *Environ. Challenges* **3** pp 1-11
[23] Sangkham S. 2020 Face mask and medical waste disposal during the novel covid-19 pandemic in

Asia. *Case Stud. Chem. Environ. Eng.*
[24] Fadare OO and Okoffo ED 2020 Covid-19 face masks: a potential source of microplastic fibers in the environment *Sci. Total Environ.* **737** 140279 (https://pubmed.ncbi.nlm.nih.gov/32563114)
[25] Ferronato N and Torretta V 2019 Waste mismanagement in developing countries: a review of global issues *International Journal of Environmental Research and Public Health* **16**
[26] Aragaw TA 2020 Surgical face masks as a potential source for microplastic pollution in the COVID-19 scenario *Mar. Pollut. Bull* **159** 111517
[27] Schmidt N, Thibault D, Galgani F, Paluselli A and Sempéré R 2018 Occurrence of microplastics in surface waters of the Gulf of Lion (NW Mediterranean Sea) *Prog. Oceanogr.* **163** 214–20 (https://www.sciencedirect.com/science/article/pii/S0079661117300484)
[28] UNEP (United Nation Environment Programme) 2020 *Waste management during the COVID-19 pandemic* (http://wedocs.unep.org/bitstream/handle/20.500.11822/33416/WMC-19.pdf?%0Asequence=1&isAllowed=y)
[29] Nowakowski P, Kuśnierz S, Sosna P, Mauer J and Maj D 2020 Disposal of personal protective equipment during the COVID-19 pandemic is a challenge for waste collection companies and society: a case study in Poland *Resources* **9** 116
[30] Rhee S-W 2020 Management of used personal protective equipment and wastes related to COVID-19 in South Korea *Waste. Manag. Res* **38** 820–4
[31] Battegazzore D, Cravero F and Frache A 2020 Is it possible to mechanical recycle the materials of the disposable filtering masks? *Polymers (Basel)* **12** 2726
[32] Sutrisno H and Meilasari F 2020 review: medical waste management for COVID19 *J. Kesehat.Lingkung.* **12** (Special Issue No. 1) 104–20

Metals in green mussels: is there any effect on their reproduction? A preliminary study of Muara Angke, Jakarta Bay

W Koagouw[1,3], N L P R Phadmacanty[2], M R Iskandar[1], D Hindarti[1] and Z Arifin[1]

[1] Research Center for Oceanography, National Research and Innovation Agency, Jl. Pasir Putih I, Ancol Timur, Jakarta 14430, Indonesia
[2] Museum Zoologicum Bogoriense, National Research and Innovation Agency, Cibinong 16911, Indonesia
[3] Centre for Aquatic Environments, University of Brighton, Lewes Road, Brighton BN2 4GJ, United Kingdom

Corresponding Author: wulan.koagouw@brin.go.id; W.Koagouw2@brighton.ac.uk

Abstract. Water pollution and its impacts on ecosystems in Jakarta Bay have been extensively studied over the years. Muara Angke is known anecdotally as one of the most heavily polluted parts of Jakarta Bay. In the context of the ongoing debate regarding contamination in the area, we present data on the concentration of metals in green mussels *Perna viridis*, one of the most popular edible commodities in Jakarta Bay. Histological examination of *P. viridis* gonads was also performed to demonstrate the direct effects of possible contamination on reproduction in this organism. The concentration of arsenic (As), copper (Cu), nickel (Ni), zinc (Zn), and mercury (Hg) exceeded the upper limits of heavy metal contents in marine biota, whilst only As exceeded the maximum limits of heavy metal contents in processed food, both limits set by Indonesian government. Although the histological observation of gonad tissue after being stained with hematoxylin and eosin revealed some pathologies, high contents of metals in mussels collected in Muara Angke seemed not to affect their reproduction. Ecologically, direct impacts on the tissue of *P. viridis* will provide vital information needed to predict the health of the species in Muara Angke and the sustainability of the population.

1. Introduction

Jakarta Bay is situated to the north of the capital city of Indonesia, Jakarta, between the Tanjung Pasir cape on the western side and the Karawang cape on the eastern side. It is shallow, with an average depth of 15 m and an area of sea approximately 490 km2 [1]. In the early 1970s, fisheries were one of the main sectors of the economy; however, the decreasing water quality of the bay has impacted both fisheries and the tourism industry [2].

Numerous investigations regarding water pollution and its impacts on ecosystems in Jakarta Bay have been recorded since 1952 [3]. A growing body of literature has highlighted the relationship with anthropogenic activities, while also directing attention towards ecosystem health and food safety in this area [4, 5]. With regards to tracing metal contamination, the accumulation in coastal areas of Jakarta has raised concerns since being reported by Palupi et al. [6], Williams et al.[7] and Rees et al. [8].

Williams et al. [7] highlighted that the concentrations of lead (Pb) and copper (Cu) were detected in Jakarta Bay stations at concentrations 5-10% above USEPA and CEC guidelines, although in general the median concentrations of all dissolved metals in seawater complied to these thresholds. Moreover, there is a possible anomaly of anthropogenic deviations for chromium (Cr) in the west of Muara Angke estuary. Results presented by van der Meij et al. [9] conclude that the deterioration of mollusc populations in Jakarta Bay is related to human activities. This study showed that predacious gastropods and plenty of molluscan fauna in association with reef are no longer extant in Jakarta Bay, including numerous comestible species. A study by Hosono et al. [4] on sediments collected from Jakarta Bay suggested that anthropogenic metals started to accumulate in the 1920s, with a notable increase in the 1970s to the end of the 1990s. However, this study also highlighted the constant decline in accumulation rates for zinc (Zn) and Pb indicates a positive outcome accredited to more stringent environmental policies implemented in the 1990s.

Mussels have been deployed as an indicator in the majority of pollution monitoring programmes worldwide, including the detection of trace metals such as Cu, cadmium (Cd), and Zn [10, 11]. Under controlled laboratory conditions, exposures to Cu, Cd, and Zn caused an enlargement in the digestive lysosomes of mussels as reported by Etxeberria et al. [12]. A publication by Cuevas et al. [13] investigated the histopathology in the digestive glands and gonads of mussels in relation to the level of contaminants including heavy metals. The study indicated that histopathological alteration is consistent with the contamination level in highly polluted sites, while it is less significant in sites with moderate or low pollution levels.

Focusing on *Perna viridis*, Vasanthi et al. [14] recorded histopathological conditions such as thickened epithelium in digestive glands, hemocyte infiltration in gills, and degeneration in adductor muscle tissue in *P. viridis* from Chennai, India. The study also quantified Cu, Pb, Zn, Cd, manganese (Mn) and iron (Fe) in mussel tissue, noting that some of these metals were above the standard limit. Another study showed that the highest biomagnification of heavy metals occurred in female gonads of *P. viridis*, with a positive correlation to the number of pathogenic bacteria in the female's digestive organs [15].

Muara Angke is known anecdotally as an area of Jakarta Bay that is heavily polluted by human activities. Within the ongoing debate regarding contamination, particularly of heavy metals in Jakarta Bay sites, there is a need to investigate the contamination levels in edible species and the impacts on the organism. This study investigates the concentration of heavy metals in the green mussel *Perna viridis*, one of the most popular edible commodities in Jakarta bay. This study also attempts to show the direct effects of contamination on the tissue of *P. viridis*, to show what effects these heavy metals may have on this organism. Ecologically, direct effects on the tissues of *P. viridis* will serve as important information to predict the health of the species in this area and the sustainability of the population as well.

2. Materials and Methods

2.1. Collection of mussels

Green mussels *Perna viridis* were collected from Muara Angke (6°03'48.2"S 106°45'04.5"E) (Figure 1) in September 2017. Mussels were placed in a cool box with ice to prevent degradation and brought immediately to the laboratory. All mussels were stored at -20°C prior to further analyses, which took place no more than a week after collection. The collection, storage and preparation of samples were conducted in accordance with good laboratory practice standards, following guidelines provided by US-EPA [16] and American Public Health Association (APHA) [17].

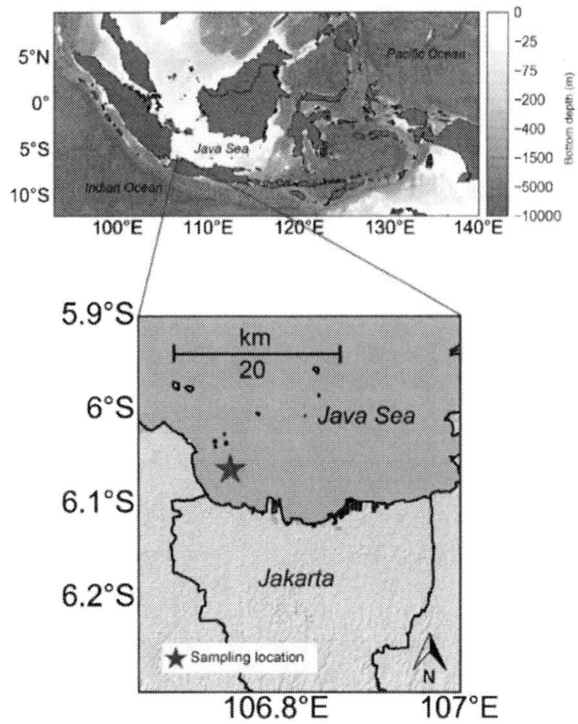

Figure 1. Map of the sampling site. An overview displaying the location of the study area within Indonesia (top), with a detailed map of Jakarta Bay (bottom).

2.2. *Metals analysis of mussels*

The analyses of metals were performed at Water Laboratory Nusantara, an accredited testing laboratory, and all methods comply with ISO/IEC 17025:2008. Analyses were performed following Standard Methods for Examination of Water and Wastewater [18], US-EPA, and standardised methods provided by instruments' protocols. US-EPA 245.7 [19] was adopted to quantify the level of mercury (Hg) using cold-vapor atomic fluorescence spectrometry (CVAFS), whilst the remaining metals were analysed using emission spectroscopy using inductively coupled plasma (ICP) following procedures in US-EPA 200.3 [20] and APHA-3120-B [18]. These method references describe the detailed procedures applied in this study.

2.3. *Histopathology*

About twenty mussels of various sizes were dissected, and their gonad tissues were fixed in 10% Neutral Buffered Formalin (NBF) and processed for histological analysis using the paraffin method [21]. The samples were washed with running tap water before being dehydrated with a series of concentrations of ethanol. All samples were embedded in paraffin and sectioned 7-micron using a rotary microtome (Yamato, Japan). Slides then underwent graded xylene washes to remove the paraffin. Following the deparaffinization, all samples were stained with hematoxylin and eosin, cleared by xylene, and then mounted with Entellan® (Merck Millipore). The histopathological condition of samples was examined and micrographs were recorded using compound microscopy (Nikon Optiphot-2, Japan) connected to a camera (Canon PowerShot S5IS) and PC with 4x/10x magnification.

3. Results

3.1. Stage of development of mussels

Figure 2 presents the stage of development of mussels collected in this study. Most individuals were in stages 3 and 4 of development, accounting for 40% and 35% of samples, respectively. Only very few mussels were at the other developmental stages (each 5% of samples). The stages of development identified here are according to Gosling [22].

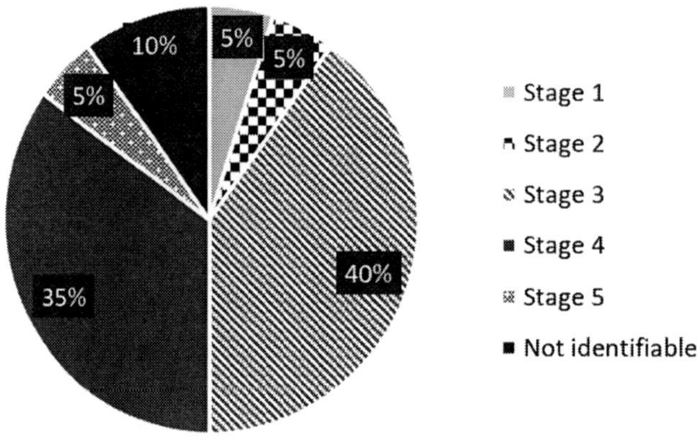

Figure 2. Stage of development of mussels collected in Muara Angke estuary during sample collection (September 2017).

3.2. Metals concentration in mussels

Chemistry analyses measured 29 metals in green mussels and their concentrations are shown in Table 1. We compared this result with the government regulation limit in the Decree of the State Minister of the Environment number 51 year 2004 regarding Standard Quality of Seawater for Marine Biota (GRL[1]) [23], as well as the limit set by the National Agency of Drug and Food Control Republic of Indonesia number 5 year 2018 regarding Maximum Limits of Heavy Metals Contents in Processed Food (GRL[2]) [24]. Among metals detected and quantified in the mussels, the levels of arsenic (As), Cu, nickel (Ni), Zn, and Hg exceeded the limits of heavy metals content in marine biota according to GRL[1]. Meanwhile, when we compared the results to the GRL[2], only As exceeded the upper limit of this standard regulation, with a concentration of 1.0 mg/kg wet weight. The other metal regulated in this standard, Hg, was also detected in mussels in Muara Angke, however, its levels were below the GRL[2]. Several metals, namely aluminum (Al), silver (Ag), and titanium (Ti) were also detected at levels that might require further concerns.

Table 1. Concentration of metals in green mussels (in mg/kg wet weight, ww; n = 1) compared to government regulation limit (GRL)

Test Description	Detection Limit	Result	GRL[1]	GRL[2]
Arsenic (As)	0.5	1.0	0.012	0.25
Aluminum (Al)	0.5	31.7	n/a	n/a
Silver (Ag)	0.01	0.93	n/a	n/a
Boron-dissolved (B)	0.5	2.6	n/a	n/a
Barium (Ba)	0.5	0.9	n/a	n/a
Beryllium (Be)	0.05	< 0.05	n/a	n/a
Bismuth (Bi)	0.5	< 0.5	n/a	n/a

Calcium (Ca)	10	60	n/a	n/a
Cadmium (Cd)	0.01	< 0.01	0.001	0.10
Cobalt (Co)	0.1	< 0.1	n/a	n/a
Chromium (Cr)	0.1	< 0.1	0.005	n/a
Copper (Cu)	0.5	0.8	0.008	n/a
Iron (Fe)	2	24	n/a	n/a
Potassium (K)	50	1200	n/a	n/a
Lithium (Li)	1	< 1	n/a	n/a
Magnesium (Mg)	5	470	n/a	n/a
Manganese (Mn)	0.5	7.3	n/a	n/a
Molybdenum (Mo)	0.1	< 0.1	n/a	n/a
Nickel (Ni)	0.1	0.2	0.05	n/a
Sodium (Na)	5	2820	n/a	n/a
Lead (Pb)	0.1	< 0.1	0.008	0.20
Antimony (Sb)	0.1	< 0.1	n/a	n/a
Selenium (Se)	0.5	< 0.5	n/a	n/a
Silicon (Si)	50	< 50	n/a	n/a
Tin (Sn)	0.5	< 0.5	n/a	n/a
Titanium (Ti)	0.5	0.5	n/a	n/a
Thallium (Tl)	0.5	< 0.5	n/a	n/a
Zinc (Zn)	0.5	5.5	0.05	n/a
Mercury (Hg)	0.004	0.007	0.001	0.50

[1]GRL: government regulation limit, refer to the Decree of the State Minister of the Environment number 51 year 2004 regarding Standard Quality of Seawater for Marine Biota
[2]GRL: government regulation limit, refer to the National Agency of Drug and Food Control Republic of Indonesia regulation number 5 year 2018 regarding Maximum Limits of Heavy Metals Contents in Processed Food

3.3. Histopathological condition of mussel gonads

Histological examination showed that mussels collected in this study consisted of 60% male and 40% female individuals. Moreover, it revealed that only one in twelve male mussels showed pathological conditions, while in female mussels these conditions were found in one in two mussels. Thus 55% of the sample size were healthy male mussels, whilst only 20% of the population were normal female individuals, as shown in Figure 3. In other words, one in every 4 mussels (or 25%) in Muara Angke exhibited pathologies in their gonads.

Hemocytic infiltration, inflammatory foci, and atresia were the pathologies recorded during the histopathological examination. Among these pathologies, atresia was the condition with the highest frequency, occurring in 15% of the samples overall, but in females only. The other two pathologies, hemocytic infiltration and inflammatory foci, occurred at a rate of only 5%, recorded in females and males respectively. Figure 4 shows the percentage of occurrence of these pathologies, while Figure 5 presents the photomicrographs associated with each histopathological condition.

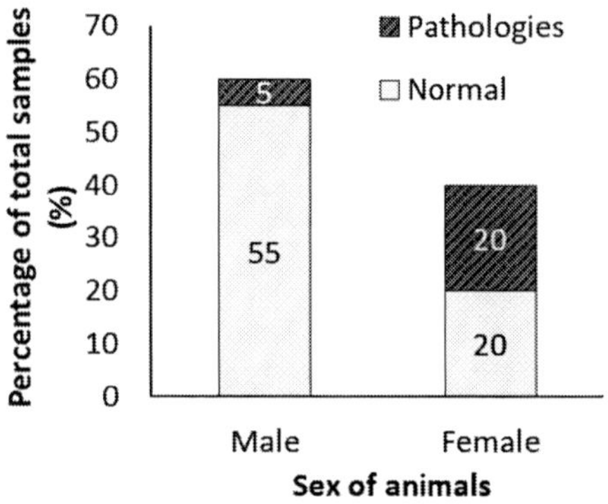

Figure 3. The proportion of healthy and pathological individuals in male and female mussels (n = 20).

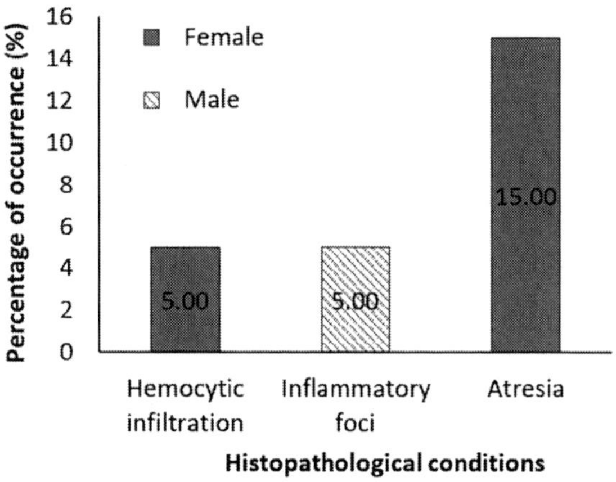

Figure 4. The occurrence of histopathological conditions observed in mussels (n = 20).

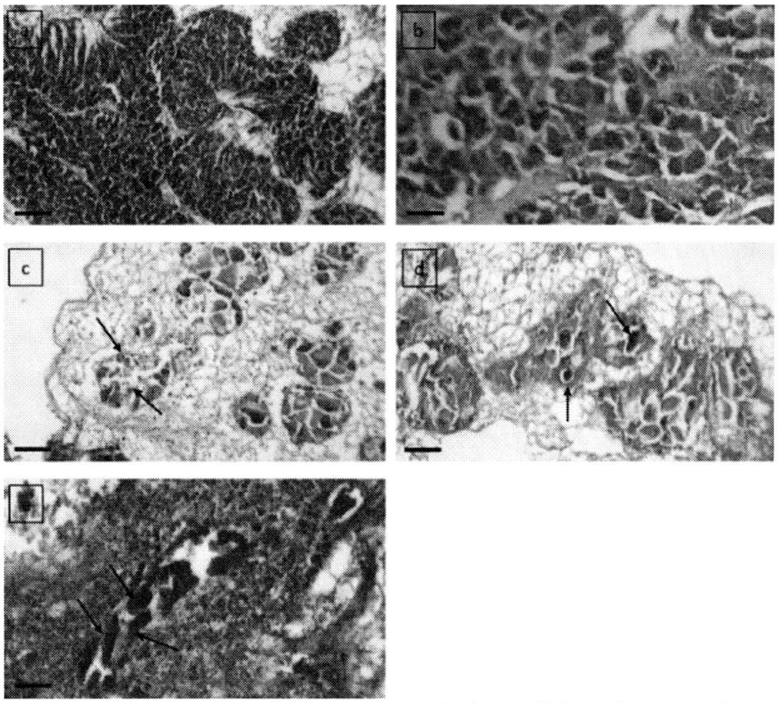

Figure 5. Micrographs of histopathological conditions in mussel gonads. Sections of 7 μm, stained with hematoxylin and eosin. (a) normal male, (b) normal female, (c) hemocytic infiltration (female), (d) atresia, (e) inflammatory foci (male). Arrows point to each pathological condition. Scale bar = 100 μm.

4. Discussion

4.1. Maturity of gonads at the time of collection

Most of the mussels collected (up to 75%) were at stages 3 and 4, which are considered the maturing stages, during collection in September 2017 (Figure 2). The results presented here provide information on the life cycle of green mussels *Perna viridis* in Muara Angke. Different species of mussel adopt different strategies for gametogenesis, spawning and fertilisation depending on their geographical location [25]. For example, another mussel species *Mytilus edulis* has a spawning season in September [26]. Hence, it would be interesting to see those strategies adopted by *P. viridis* in this area for their reproduction, and more research into this topic is therefore encouraged.

The data on the maturity of gonads also constitutes important information in the ecotoxicological context. Mounting research shows that contamination, especially heavy metals, produces more dramatic effects at the early stages of mussels life [27-29], while the effects are less prominent when contamination occurs in mature mussels. The stage of mussel maturation should therefore be taken into account when interpreting the effect of any contamination in ecotoxicological research, including in this investigation.

4.2. Levels of metals exceed the regulation limits

Among 29 metals analysed here, the levels of As, Cu, Ni, Zn and Hg exceeded the upper limits of heavy metals content in marine biota as set in GRL[1]. Previous studies have recorded heavy metal contamination in Muara Angke. A study on Hg and As content in seafood samples from the Jakarta fishing port recorded that in *P. viridis* these metals exceed the maximum limit of Indonesian government regulation [30]. Despite the temporal variation in the data, one of the results in Winarno et al. [31] showed that methyl mercury in *P. viridis* from Muara Angke was above the highest level permitted by WHO and FAO. An investigation by Inswiasri et al. [32] on the concentration of Cu, Pb, and Cd in fresh fish and shellfish from Muara Angke recorded that the mean concentration of Cd in shellfish surpassed the acceptable concentration, while others were within the range. The result here thus adds to the long list of the documentation of heavy metal pollution in this area.

The results here also recorded that As exceeded GRL[2]. Since *P. viridis* is one of the main fisheries commodities in this area, the results presented here should be of public health interest. Hg was also detected in mussels in Muara Angke, however, its levels were below the standard limit. Although these data are preliminary and only capture a snapshot of the area, it may be necessary to take more serious action to counter this issue, especially considering food safety in this popular aquafarm area.

Another result worth noting is the level of some metals (Al, Ag, Ti) detected in mussels from Muara Angke that might raise further concerns. Although these metals are not regulated in Indonesian government standards or limits, their levels here highlight the possibility of contamination as a result of runoff from the mainland, and may therefore be attributable to any industrial activity related to these metals. A recent report by Koagouw et al. [33] also documented the presence of some metals in Jakarta Bay seawater that can be related to runoff carrying contaminants from the land, as a prominent route of water contamination from terrestrial environments.

4.3. Histopathological assessment on the reproduction of mussels

Many studies have documented Jakarta Bay's transformation from a relatively clean and productive environment into one polluted by various contaminants, e.g. excessive nutrients [34] and trace metals [35]. Despite mounting evidence of contamination in Jakarta Bay, other than laboratory-based exposures few studies have documented the direct effects of these contaminants on the marine organisms in the area, especially on their reproduction.

Our data here show only 25% of mussels displaying pathologies in their gonads. Most ecotoxicological studies on mussels have successfully documented that the alterations by contaminants are primarily related to ontogeny. For example, shell formation in Mediterranean mussels [36] and impaired development of the embryo-larval stage of mussels *Perna perna* [37]. A recent study in the Jakarta Bay area also recorded deformed shells of *P. viridis* caused by metal pollution [38]. Although some previous studies reported that accumulated high concentrations of Pb, Cd, Cr and Hg in the gonads of both male and females of *P. viridis* influenced the development of secondary oocytes [39] and spermatogenesis process [40], our results here do not necessarily show the level of metal contamination to have a direct effect on the structure of their reproductive organs. Other explanations could be related to the plasticity of mussels in handling pollution [41], which is the reason why mussels are excellent species widely used in environmental monitoring (e.g. International Mussels Watch, etc.). However, as the data here are preliminary and show only a snapshot of the area at a single time point, further studies with more robust spatial and temporal approaches are recommended.

The female mussels in this study, however, are likely to be suffering the effects of metal contamination, as pathologies were observed in the gonads of up to 50% of female mussels here. This occurrence, where female mussels are more prone to effects of contamination than males, was also reported by Koagouw and Ciocan [42] in mussels exposed to metformin in a laboratory-based study. Atresia, a degenerative process with subsequent elimination and resorption of oocytes [43], was observed as the pathology with the highest occurrence in this study. This is similar to results from several recent studies [44, 45], which documented high occurrences of atresia in mussels even at low levels of paracetamol exposure. Also worth mentioning here is that the lower incidence of hemocytic infiltration and inflammatory foci compared to atresia might indicate that metal contamination is more directly related to degenerative effects than to inflammatory conditions in the mussels' reproductive organs. Although this can be related to the results from previous studies [39, 40] that show alteration in gametogenesis caused by metal accumulation in mussels, the results here together with said studies have not specifically pinpointed the presence of metals alongside the histopathological alterations. Hence, further studies showing the presence of these metals in gonads together with the direct effects of these contaminants are highly encouraged to form an explicit link between contamination and the observed effects in marine species.

5. Conclusion

Most mussels in this study were in the mature stage when collected in Muara Angke during September 2017. The concentration of As, Cu, Ni, Zn, and Hg exceed the upper limits of heavy metals content in marine biota, whilst only As exceed the maximum limits of heavy metals contents in processed food, both limits set by Indonesian government. The reproduction of the mussels, however, did not show severe effects, although there is prevalence toward degenerative effects in female gonads. The high content of metals in mussels collected in Muara Angke seemed not to really affect their reproduction. Further investigations are encouraged to establish or confirm the links between contamination and the observed effects in marine organisms.

References

[1] Arifin Z 2008 *Assessment of heavy metals contaminations in Jakarta Bay* (Assessment of ecological change of Jakarta Bay) ed Ruyitno et al (Jakarta: LIPI Press) pp 211 – 228

[2] Arifin Z 2004 *Trend of coastal pollution in Jakarta bay, Indonesia: its implication for fishery and recreational activities* (Bilateral workshop on Coastal Resources Exploitation and Conservation) (Bali: Indo-German Experiences) pp 61-66

[3] Zaneveld JS 1952 *A recent investigation on the geomorphology and the flora of some coral islands in the Bay of Djakarta* (Jakarta)

[4] Hosono T, Su C-C, Delinom R, Umezawa Y, Toyota T, Kaneko S and Taniguchi M 2011 *Estuarine, Coastal and Shelf Science* **92** pp 297-306

[5] Baum G, Kusumanti I, Breckwoldt A, Ferse SCA, Glaser M, Dwiyitno, Adrianto L, van der Wulp S and Kunzmann A 2016 *Marine Pollution Bulletin* **110** pp 778-789

[6] Palupi K, Sumengen S, Inswiasri S, Agustina L, Nunik SA, Sunarya W and Quraisyn A 1995 *Water Science and Technology* **31** pp 17-25

[7] Williams TM, Rees JG and Setiapermana D 2000 *Marine Pollution Bulletin* **40** pp 277-285

[8] Rees JG, Setiapermana D, Sharp VA, Weeks JM and Williams TM 1999 *Oceanologica Acta* **22** pp 627-640

[9] van der Meij SET, Moolenbeek RG and Hoeksema BW 2009 *Marine Pollution Bulletin* **59** pp 101-107

[10] Besada V, Fumega J and Vaamonde A 2002 *Science of The Total Environment* **288** 239-253

[11] Benali I, Boutiba Z, Grandjean D, de Alencastro LF, Rouane-Hacene O and Chèvre N 2017 *Marine Pollution Bulletin* **115** pp 539-550

[12] Etxeberria M, Sastre I, Cajaraville MP and Marigomez I 1994 *Archives of Environmental Contamination and Toxicology* **27** pp 338-345

[13] Cuevas N, Zorita I, Costa PM, Franco J and Larreta J 2015 *Aquatic Toxicology* **162** pp 152-164

[14] Arockia Vasanthi L, Revathi P, Arulvasu C and Munuswamy N 2012 *Ecotoxicology and Environmental Safety* **84** pp 92-98

[15] Tewari A, Joshi HV, Raghunathan C, Kumar VGS and Khambhaty Y 2001 *Current Science* **81** pp 819-828

[16] Decker C and Simmons K 2013 *Surface Water Sampling* vol R3 (Athens, Georgia: SESD US-EPA) pp 1-22

[17] Baird R, Eaton AD, Rice EW and Bridgewater L 2017 *Standard Methods for the Examination of Water and Wastewater* 23rd edn (Washington, D.C.: American Public Health Association)

[18] Rice EW and Bridgewater L 2012 *Standard Methods for the Examination of Water and Wastewater* 22nd edn (Washington, D.C.: American Public Health Association)

[19] US-EPA 2005 *Mercury in Water by Cold Vapor Atomic Fluorescence Spectrometry - Method 245.7* 2nd revision (Washington, D.C.: US-EPA)

[20] US-EPA 1991 *Methods for the Determination of Metals in Environmental Samples - Method 200.3* (Washington, D.C.: US-EPA)
[21] Kiernan J 1990 *Histological and Histochemical Methods: Theory and Practice* (Oxford: Pergamon Press)
[22] Gosling E 2015 *Marine Bivalve Molluscs* (United States: Wiley-Blackwell)
[23] Ministry of Environment Republic of Indonesia 2004 *Standard Quality of Seawater for Marine Biota, the Decree of the State Minister of the Environment, number 51* (Jakarta)
[24] National Agency of Drug and Food Control Republic of Indonesia 2018 *Maximum Limits of Heavy Metals Contents in Processed Food, number 5* (Jakarta)
[25] Hassan MM, Qin JG and Li X 2017 *Journal of Molluscan Studies* **84** pp 38-45
[26] Tyler-Walters H 2008 *Mytilus edulis Common mussel* (Marine Life Information Network: Biology and Sensitivity Key Information Reviews, online) (Plymouth: Marine Biological Association of the United Kingdom)
[27] Geffard A, Geffard O, His E and Amiard J-C 2002 *Marine Ecology Progress Series* **233** pp 131-142
[28] Wang Q, Liu B, Yang H, Wang X and Lin Z 2009 *Ecotoxicology* **18** pp 829-837
[29] Fathallah S, Mohamed Nejib M, Medhioub A and Kraïem MM 2011 *Journal of Environmental Chemistry and Ecotoxicology* **3** 277-285
[30] Koesmawati TA and Arifin Z 2015 *Marine Research Indonesia* **40** pp 9-16
[31] Winarno EK, Andayani W and Sumartono A 2009 *Indo J Chem* **9** pp 77-83
[32] Inswiasri I, Tugaswati AT and Lubis A 2012 *Buletin Penelitian Kesehatan* **25** 1
[33] Koagouw W, Arifin Z, Olivier GWJ and Ciocan C 2021 *Marine Pollution Bulletin* **169** 112558
[34] Sidabutar T, Srimariana ES and Wouthuyzen S 2020 *IOP Conf Series: Earth and Environmental Science* **429** 01202
[35] Arifin Z and Fitriati M 2006 Green mussels cultured in highly polluted area of Jakarta Bay, Indonesia *Int. Conf. on Hubs, Harbours and Deltas in Southeast Asia: Multidisciplinary and Intercultural Perspectives (Phnom-Penh, Cambodia)* pp 525-536
[36] Balbi T, Montagna M, Fabbri R, Carbone C, Franzellitti S, Fabbri E and Canesi L 2018 *Science of the Total Environment* **642** pp 601-609
[37] Fontes MK, Gusso-Choueri PK, Maranho LA, Abessa DMdS, Mazur WA, de Campos BG, Guimarães LL, de Toledo MS, Lebre D and Marques JR 2018 *Water Research* **13** pp 361-370
[38] Riani E, Cordova MR and Arifin Z 2018 *Marine Pollution Bulletin* **133** pp 664-670
[39] Julius, Setiyanto D, Sumantadinata K, Riani E and Ernawati Y 2008 *J Ris Akuakultur* **3** pp 43-52
[40] Julius, Setiyanto D, Sumantadinata K, Riani E and Ernawati Y 2008 *Jurnal Ilmu-ilmu Perairan dan Perikanan Indonesia* **15** pp 77-83
[41] Beyer J, Green NW, Brooks S, Allan IJ, Ruus A, Gomes T, Bråte ILN and Schøyen M 2017 *Marine Environmental Research* **130** pp 338–365
[42] Koagouw W and Ciocan C 2018 *Journal of Shellfish Research* **37** pp 467-474
[43] García-Gasca A, Leal-Tarin B, Ríos-Sicairos J, Hernández-Cornejo R, Aguilar-Zárate G and Betancourt-Lozano M 2010 *Journal of Environmental Science and Health Part A* **45** pp 56-61
[44] Koagouw W and Ciocan C 2020 *Environmental Science and Pollution Research* **27** pp 30933-30944
[45] Koagouw W, Stewart NA and Ciocan C 2021 *Environmental Science and Pollution Research* **28** pp 48823-48836

Characteristics of *moringa* leaf powder as fortification and consumer acceptance

Y Handayani[1], S Aminah[2], M Yanis[1] and Waryat[3]

[1] Jakarta Assessment Institute of Agriculture Technology, Jl. Raya Ragunan No. 30, Pasar Minggu, South Jakarta, Indonesia.
[2] Indonesian Center for Food Crops Research and Development, Jl. Merdeka No. 147, Bogor, West Java, Indonesia.
[3] Indonesian Center for Horticulture Research and Development, Jl. Tentara Pelajar No. 3C Kampus Penelitian Pertanian, Bogor, West Java, Indonesia.

Corresponding Author: mifa71@yahoo.com

Abstract: *Moringa Oleifera* is not only rich in nutrients but also has functional properties with efficacies and benefits for human health. Moringa leaf powder as an intermediate product can be used as a fortification to enrich nutrition in food products in an acceptable concentrate for consumers. This study aims to determine the physical and chemical characteristics of Moringa leaf powder and the level of consumer acceptance of the fortification product. The research included the process of making Moringa leaf powder until the preeminent temperature and drying time were obtained and the use of Moringa leaf powder as a fortification in ice cream. Fortification of ice cream used four treatments of the powder concentration in 0.5%, 1%, 1.5% and 2%. The results showed the yield of the powder was 25%, with a 7.80 % water content at a temperature of 47° – 50 °C for five hours. The level of consumer acceptance of fortified ice cream shows the same acceptance of color, aroma, texture, except taste. Consumers prefer formulas with a concentration of 1% Moringa powder, 5% creamer, 15% sugar, 0.5 % CMC, and 78.5 % milk. Moringa aroma and taste have an impact on consumer acceptance of Moringa ice cream.

1. Introduction

Moringa oleifera is a tropical plant that is relatively easy to cultivate in tropical areas, such as Indonesia. The nutrient content is sufficient for Moringa to have functional health properties and to overcome the problem of nutritional deficiencies or malnutrition. Moringa has earned the nicknames "Miracle Tree" and "Mother's Best Friend" due to its high nutritional value, as well as its properties and benefits. Moringa has begun to be recognized and developed in Indonesia as a product with economic value and industrial potential. Moringa powder, utilized to fortify various food processed products.

Moringa leaves were discovered to contain several essential nutrients, such as beta-carotene, vitamin C, protein, iron, potassium, and large amounts of phenol which is regarded as an antidote to free radical compounds [1]. Moringa leaves appear to contain vitamin C, vitamin A, calcium, potassium, and protein [2]. According to another study, moringa leaves contain 10 times the vitamin A of carrots, 17 times the calcium of milk, 15 times the potassium of bananas, 25 times the iron of spinach, and 9 times the protein of yogurt [3,4]. Moringa also has a high antioxidant content, as well as pharma logical and antimicrobial goods [5]. This is because ascorbic acid, flavonoids, phenolics, and carotenoids are present in moringa [6, 7, 8, 9].

Most people are aware that moringa is still limited to vegetables, and that the diversification of processed moringa is not yet widely known, so efforts are required to introduce the advantages and benefits of moringa. This type of form appears to be more flexible when applied to food products, in addition to serving as fortification it can also be a food additive (BTP) that serves as a natural dye. As fortifying material, by considering consumer acceptance factors such as taste, color, aroma and other physical properties that influence the level of consumer preference for fortified products. According to the findings of McLellan et al., [10], moringa leaf powder has been added to grits as a nutritious food supplement to meet the protein and micronutrient needs of children. The purpose of this study is to determine the process of producing moringa powder as well as consumer acceptance of ice cream containing moringa powder.

2. Methodology

2.1. Material

Ingredients used are fresh Moringa leaves, Moringa powder, liquid milk, CMC (Carboxymethyl Cellulose) sugar, and vegetable cream. The dryer, ice cream machine, and ice cream packaging are research aids.

2.2. Production of Moringa leaf powder

Moringa leaf drying begins with determining the temperature and drying time maximum moisture content standard of 8%, and the Moringa leaves used are 500 g per shelf. The drying process begins with washing the fresh Moringa leaves and arranging them on a drying machine rack with a thickness of 1 cm. The size of the drying rack is approximately 60 cm x 50 cm. Moringa leaves are dried for five hours at a temperature of 47-50 °C before being powdered with a blender and sieved through an 80 mesh sieve. Figure 1 depicts the process of producing Moringa powder.

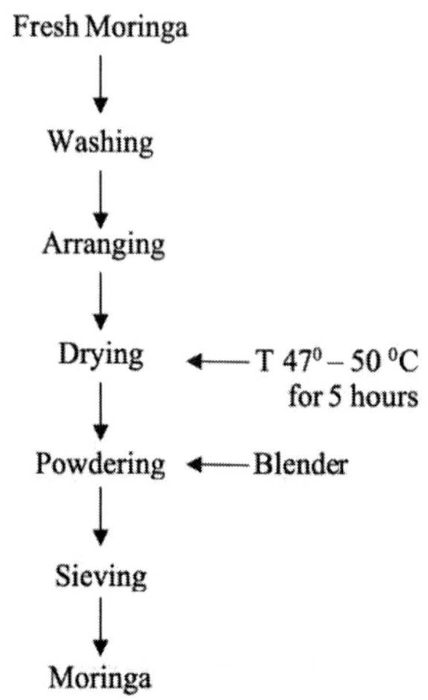

Figure 1. Flow chart for the production of moringa powder.

2.3. Formulation of ice cream moringa fortification

The treatment of Moringa powder to ice cream used the concentration of Moringa powder, namely 1) 0.5% Moringa powder (F1), 2) 1% Moringa powder (F2), 3) 1.5% Moringa powder (F3), and 4) 2% Moringa powder (F4). The ice cream formula with the addition of Moringa powder is presented in Table 1.

Table 1. Moringa ice cream formula.

Ingredients	F1	F2	F3	F4
Main Ingredient				
Moringa leaf powder (%)	0,5	1	1,5	2
Fresh Milk (%)	81,5	78,5	75,5	72,5
Additives (based on main ingredient volume)				
CMC (%)	0,5	0,5	0,5	0,5
Sugar (%)	15	15	15	15
Vegetable creamer (%)	2,5	5	7,5	10

2.4. Production of ice cream moringa leaf powder

Ice cream is made by heating milk for 2 minutes, then adding dry ingredients (sugar, CMC, vegetable cream), stirring evenly, and returning to the heat for another 2 minutes. After that, mix the solution for 2 minutes before adding the Moringa powder and mixing for another 2 to 3 minutes. It was determined that the solution had cooled for 6-7 minutes. Finally, pour the Moringa ice cream solution into the ice cream machine and let it run for 60 minutes, or until the ice cream is formed. The process of making Moringa ice cream is depicted in Figure 2.

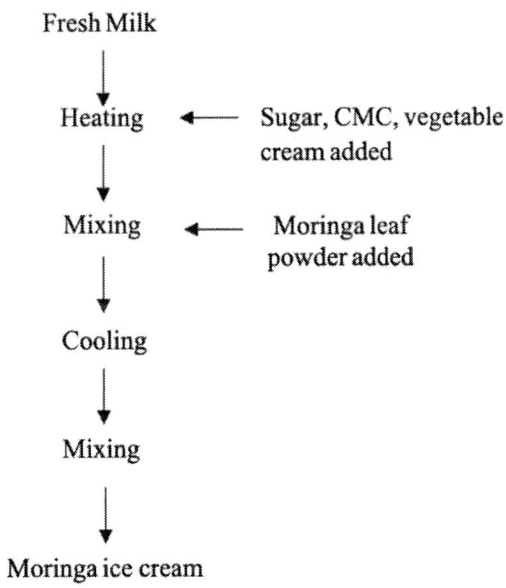

Figure. 2. Flow chart of the processing moringa ice cream.

2.5. Chemical and Physical Analysis
Physical and chemical characteristics were carried out on moisture content, yield, chroma color (L *, a *, b * values) and proximate analysis of moringa powder. Overrun analysis was performed on Moringa ice cream.

2.6. Preference Test
A consumer preference test for moringa ice cream was conducted based on liking and hedonic quality. The color, aroma, taste, and texture level of preference tests were performed using 1-5 hedonic scales, whereas the hedonic quality of moringa ice cream was performed using 1 to 5 on the hedonic quality scale.

2.7. Statistical Analysis
The study design used a completely randomized design with three replications. The data from the observations were then analyzed using ANOVA with further tests using Duncan with a significance <0.05.

3. Result and discussion
Table 2 shows the results of the yield and chemical characteristics measurements of fresh moringa leaves and moringa powder. The yield of fresh moringa leaves after they have been separated from the stems and are ready to be dried is 50%. After the drying process, a 25% yield of moringa powder was obtained with a depreciation of 50%. According to the standard, drying for 5 hours at a temperature of 50 °C can produce moringa powder with a moisture content of 7.80 %.

3.1. Physicochemical Analysis of Moringa leaf powder

Table 2. Moringa leaf powder physicochemical analysis.

Parameters	Moringa leaves	Moringa leaf powder
Rendement (%)	50	25
Moisture content	80.37	7.80
Colour		
- **L* value**	39.23	60.48
- **a* value**	-5.71	-14.69
- **b* value**	10.07	26.20

The drying process changed the brightness level, increasing the L* value from 39.23 to 60.48. Although the color of moringa powder lightens, the color remains green, as indicated by the value of a* (negative), which is -14.69.

3.2. Proximate composition of Moringa leaf powder
The proximate analysis results are shown in Table 3, which shows that the proximate content of moringa powder changes after drying. Moringa powder has a moisture content of 7.80%, which is in accordance with the standard for powder moisture content in order to extend the shelf life of Moringa powder and prevent the growth of mold and mildew. According to another study, the moisture content of Moringa powder was 7.55-8.65 percent [11]. According to Salma Sultana [11], the crude fiber content of Moringa powder ranged from 6.00 to 9.60 percent and could potentially be used as a dietary ingredient for humans and animals.

Table 3. Moringa leaf powder proximate composition.

Parameters	Moringa leaf powder
Moisture (%)	7.80
Ash (%)	9.32
Protein (%)	29.6
Fat (%)	6.98
Carbohydrate (%)	39.4
Crude fibre (%)	6.91

3.3. Mineral, β-Carotene and Vitamin C contents of Moringa leaf powder

Table 4. Mineral, β-Carotene and Vitamin C contents of moringa leaf powder.

Parameters	Moringa leaf powder
Potassium (%)	1,860
Calcium (%)	2.210
Iron (mg/kg)	91,2
Β-carotene (mg/kg)	271
Vitamin C (%)	498

Similarly, the vitamin C content of Moringa powder is 498 mg/100g, which is higher than the vitamin C content of other plant sources of vitamin C.

3.4. Overrun of ice cream with moringa leaf powder

Table 5 displays the results of overrun measurements of an ice cream formulation treatment with Moringa powder added. The greater the amount of Moringa powder added, the lower the overrun value. Overrun is a degree of development or increase in the volume of ice cream dough caused by trapped air in the ice cream formula as a result of the agitation process. Because it affects texture and density level, overrun becomes one of the determining parameters of ice cream quality.

Table 5. Overrun of moringa leaf powder ice cream.

Sample code	Overrun (%)
F1	70^{c}
F2	68,57^{c}
F3	63,46^{b}
F4	53,85^{a}

Formula F1 has the highest overrun value, which is 70%, and the value is decreasing as more Moringa powder is used. This is most likely due to the ice cream formula thickening, making it difficult to expand. The value of the overrun will be reduced if the formula or solution has a higher viscosity.

3.5. Sensory characteristics of moringa leaf powder ice cream

Table 6 shows the results of an organoleptic test of ice cream fortified with Moringa powder against the level of preference and hedonic quality. The addition of different moringa powders to an ice cream formula has no effect on consumer acceptance. The addition of Moringa powder in concentrations ranging from 0.5 to 2% has no effect on consumer preference for color, aroma, or texture. Colors 4.04 – 4.57 (liked moderately), aroma 3.70 – 3.90 (liked moderately), texture 3.96 – 4.35 (liked moderately). The use of Moringa powder above 1.5 percent demonstrates consumer acceptance of different flavors.

Table 6. Sensory properties of moringa leaf powder ice cream.

Sample code	Colour	Flavour	Taste	Texture
F1	4.57^{a}	3.96^{a}	4.70^{b}	4.35^{a}
F2	4.48^{a}	3.83^{a}	4.26^{ab}	4.17^{a}
F3	4.35^{a}	3.70^{a}	3.96^{a}	4.13^{a}
F4	4.04^{a}	3.70^{a}	3.61^{a}	3.96^{a}

Table 7 displays the results of the consumer assessment of hedonic quality tests of Moringa aroma and taste. The results of hedonic quality tests revealed that different concentrations of the powder had an effect on consumer acceptance. The higher level of Moringa leaf powder concentration, the lower consumer acceptance on Moringa powder.

Table 7. Sensory evaluation of Moringa leave powder ice cream.

Sample code	Aroma	Taste
F1	4.17^{c}	3.65^{c}
F2	3.39^{c}	2.87^{b}
F3	2.57^{b}	1.96^{a}
F4	2.26^{a}	1.65^{a}

Consumers prefer the aroma and taste of the ice cream formula with 1% Moringa, with aroma on a hedonic scale of 4.17 (liked moderately) and taste on a hedonic scale of 3.65 (liked moderately).

4. Conclusion

Making Moringa powder at a drying temperature of 47 - 50 ^{0}C for 5 hours yields a 25% yield and a moisture content of 7.80% - 8.14% with a green color. Except for taste, the application of Moringa powder to ice cream yielded the same consumer preference test results for color, aroma, and texture. Consumers prefer formulas with a concentration of 1% Moringa powder, 5% creamer, 15% sugar, 0.5 percent CMC, and 78.5 percent milk. Moringa aroma and taste have an impact on consumer acceptance of Moringa ice cream.

5. References

[1] A. Verma, M. Vijayakumar, C. Mathela, Chandra S, Rao, Chandana V. 2009 In vitro and in vivo antioxidant properties of different fractions of Moringa oleifera leaves. *Food and Chemical Taxicology,* **47** 2196-22-1

[2] Mahmood KT, Tahira Mugal, Ikram Ul Haq. 2011 Moringa oleifera: a natural gift-A review. *Journal of Pharmaceutical Sciences and Research* **2** 11 pp 775-781

[3] Fuglie, Lowell J., ed. 2001 The Miracle Tree: The multiple attributes of moringa. Dakar, Senegal: Church World Service.

[4] Hekmat, S., Morgan K., Soltani M., Gough R. 2015 Sensory evaluation of locally-grown fruit purees and inulin fibre on probiotic yogurt in Mwanza, Tanzania and the microbial analysis of probiotic yogurt fortified with Moringa oleifera. *J Healt Popul Nutr.* **33** 1 pp 60-67

[5] Das, A. K., Rajkumar, V., Verma, A. K., & Swarup, D. 2012 Moringa oleifera leaves extract: A natural antioxidant for retarding lipid peroxidation in cooked goat meat patties. *International Journal of Food Science and Technology* **47** pp 585–591

[6] Anwar, F., Latif, S., Ashraf, M., & Gilani, A. H. 2007b. *Moringa oleifera*: A food plant with multiple medicinal uses. *Phytotherapy Research* **21** pp 17–25

[7] Makkar, H. P. S., and K. Becker. 1997 "Nutrients and antiquality factors in different morphological parts of the Moringa oleifera tree." *Journal of Agricultural Science* **128** pp 311-322.
[8] Moyo, B. 2012 Antimicrobial activities of Moringa oleifera Lam leaf extracts. *African Journal of Biotechnology* **11**(11) pp 2797-2802.
[9] Dahot, M.U. 1998 Vitamin contents of flowers and seeds of moringa oleifera *Biochemistry* pp 2122 – 2124
[10] McLellan, L., Mckenzie, J. and Clapham, M.E. 2010 A study to determine if dried moringa leaf powder is an acceptable supplement to combine with maize meal for Malawian children. *Proceedings of the Nutrition Society*, 28 June–1 July 2010. Health Sciences, Queen Margaret University, Edinburgh EH21 6UU, UK.
[11] Salma Sultana. 2020 Nutritional and functional properties of Moringa oleifera. Departement of Biochemitry and Moleculer Biology. Bangladesh Agriculture University, Bangladesh. Metabolisme Open **8** pp 1-6

IOP Conf. Series: Earth and Environmental Science 1027 (2022) 012006 doi:10.1088/1755-1315/1027/1/012006

The effect of land cover forest on fluctuations in availability of water in the Batutegi Dam, Lampung, Indonesia

E Artika[1], S B Yuwono[2], I S Banuwa[3], A Setiawan[3], S Bakri[1], and E P Wahono[4]

[1] Faculty of Pascasarjana, Universitas Lampung, Lampung, Indonesia,
emi.artika10280@students.unila.ac.id. , samsul.bakri@fp.unila.ac.id
[2] Postgraduate Agriculture Doctoral Program, Faculty of Agriculture, Universitas Lampung, Lampung, Indonesia, slamet.budi@fp.unila.ac.id
[3] Faculty of Agriculture, Universitas Lampung, Lampung, Indonesia,
irwanbanuwa@yahoo.com. , aslulila@yahoo.com
[4] Faculty of Engineering, Universitas Lampung, Lampung, Indonesia,
epwahono@eng.unila.ac.id

* Corresponding Author: emi.artika28@gmail.com

Abstract: The Sekampung Hulu watershed has an important role in Lampung Province, as a catchment area for the Batutegi Dam with the main function as a supplier of irrigation water therefore it is very dependent on the availability of water (inflow). The inflow fluctuation is one of the indicators to determine the hydrological condition. The relationship between forest cover and inflow fluctuations was obtained from forest cover analysis on Landsat images using the object-oriented classification (OOC) method and inflow fluctuation analysis, followed by regression analysis between them. The results showed that the total inflow value in 2005 - 2020 is fluctuated with the lowest total inflow value of 6862.243 m^3/S in 2015 and the highest value of 9296.141m^3/S in 2020. Inflow fluctuation from 2005-2020 was included in the low and moderate category. The condition of forest cover in the Sekampung Hulu watershed in 2005-2020 has increased and decreased in each period. Analysis of the relationship between forest cover and inflow fluctuations is a significant with a p-value of 0.092 with a 90% confidence level. The results showed that the condition of forest cover greatly affects the stability of the water availability of the Batutegi Dam.

1. Introduction

The Sekampung Hulu watershed is a water catchment area for the Batutegi dam which has the main function as a provider of irrigation water in Lampung Province. Currently, based on data from the Lampung XX forest area consolidation center, land cover conditions in the catchment area of the Batutegi Dam, is the Sekampung Hulu watershed, experienced a 5.6% decrease in forest land area for the period 2005-2018. Conditions of land cover and use in an area reflect the natural and socio-economic conditions of the region and their use in terms of time and space [1]. The dynamics of land cover and use is a process that is widespread, rapid and significant by human activities related to space and time. Based on research, deforestation is the main factor causing land cover change in Indonesia [2]. Human activities drive the dynamics of land cover and use, which is a comprehensive, rapid and

IOP Conf. Series: Earth and Environmental Science 1027 (2022) 012006 doi:10.1088/1755-1315/1027/1/012006

significant process. LULC change analysis is one of the most appropriate techniques to understand how land was used in the past, what kinds of changes are expected in the future [3]. Conditions and characteristics of land cover and use in the watershed have a direct influence on downstream conditions and the availability of dam water resource potential [4].

Forests have an important role in maintaining the balance of the watershed ecosystem, because it will affect the hydrological processes that occur in a watershed. Forests have the ability to act as sponges, which absorb rainwater in the rainy season and release it slowly during the dry season. Land use conditions in the Sekampung Hulu watershed will affect the hydrological conditions of the Batutegi dam, one of which can be known by looking at the value of inflow fluctuations. Forests play a role in preventing flooding during the rainy season so that if there is forest conversion, it will increase the value of inflow fluctuations. The purpose of this study is to see how forest cover affects inflow fluctuations in the Batutegi watershed from 2005 to 2020. In the context of making decisions on the rehabilitation and management of the Sekampung Hulu watershed, this research is expected to provide input and advice for policymakers and other stakeholders.

2. Materials and Methods

2.1. Area Stud

This research was conducted in February 2021. The location of this research is in the Sekampung Hulu watershed and the Batutegi Dam in the Tanggamus Regency area. Geographically, the Sekampung Hulu watershed is located at 05°05'50"S and 104°30'34"E to 05°16'33"S and 104°49'14"E with an altitude between 175 m to 1,775 m above sea level, with an area of 42,400 ha. (Figure 1).

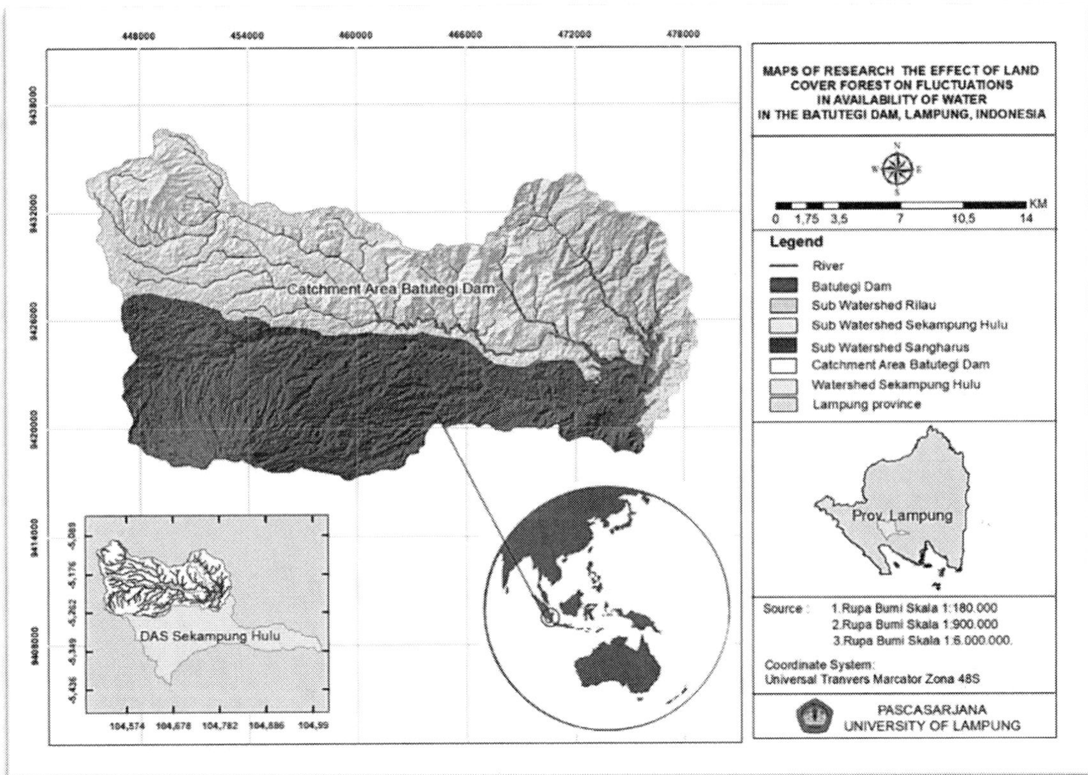

Figure 1. The Map of Research.

2.2. Research Tools and Materials
The study uses tools such as the Global Positioning System (GPS), cameras and supporting software including ArcGIS 10.4 and Microsoft Excel. The research materials used are rainfall data, inflow data and land use maps in the Sekampung Hulu watershed in 2005 and 2020 from the classification of Landsat ETM +5, ETM +7 and Lansat ETM +8 satellite imagery images 124 64

2.3. Forest cover and use analysis
Forest land cover analysis using eCognation Developer software and Arc GIS 10.4. land cover and land use classification on Landsat ETM +5, ETM +7 and Lansat ETM +8 path row 124 64 satellite images using the Object Oriented Classification (OOC) method to produce data on forest land cover in the Batutegi dam water catchment area for the period 2005 to 2020 i.e. 2005,2010,2015 and 2020.

2.4. Inflow Fluctuation Analysis
Analysis of inflow fluctuations in 2005-2020 by comparing the maximum inflow value with the minimum inflow for one year. The formula for calculating inflow fluctuations and categorizing inflow fluctuations is based on Permenhut P.61/Menhut-II/2014 [5].

$$\text{Inflow Fluctuation (KRA)}=\frac{Inflow\ Maximum}{Inflow\ Minimum}$$

Tabel 1. Categorization of Inflow Fluctuation Value (KRA)

Grade	Category	Score
KRA≤ 20	Very low	0,5
20 < KRA ≤ 50	Low	0,75
50 < KRA ≤ 80	Medium	1
80 < KRA ≤ 110	High	1,25
KRA > 110	Very high	1,5

2.5. Analysis of the Relationship between Forest Cover and Inflow Fluctuations (KRA)
Simple linear regression analysis was used in the study to see the relationship between forest cover and inflow fluctuations so that the relationship between the two was obtained. Simple statistical analysis between forest land cover area and fluctuations in water availability of the Batutegi dam (inflow) in 2005-2020 using Microsoft Excel software.Y= a+bx

3. Results and Discussion

3.1. Forest Cover Analysis
Based on the interpretation of Landsat ETM +5, ETM +7 and Lansat ETM +8 satellite images, the catchment area of the Batutegi dam is 42,400 ha consisting of 3 sub-watersheds, namely the rilau sub-watershed, the sangharus sub-watershed, and the sekampung Hulu sub-watershed. The condition of forest land in the catchment area of Batutegi Dam based on BPDASHL WSS data (2020) the forest area in that area always tends to decrease due to land conversion. The results of the classification of forest cover in the Batutegi dam catchment area from 2005 to 2020 obtained through land cover analysis using Landsat imagery can be seen in table 2.

Table 2. Forest Coverage Area of Batutegi Dam Catchment Area 2005-2020

Year	Forest Area	
	Hectares (Ha)	Percent (%)
2005	17777,22	41,93
2010	12647,12	29,83
2015	14014,51	33,05
2020	11034,61	26,03

The forest cover area in the catchent area of the Batutegi Dam experienced a significant change from 2005 to 2020, which decreased by 15.9%. Based on previous research in the Sekampung Hulu watershed, there was a change in forest land into agricultural land, and settlements [6]. The condition of forest cover in the Batutegi Dam water catchment area for the period 2005 - 2020 by classifying Landsat imagery in 2005, 2010, 2015 and 2020 experienced a decrease and increase in forest area from 2005 to 2010 the forest area decreased by 12.1%, in 2010 -2015 the forest area has increased by 3.22% and in 2015-2020 it has decreased again by 7%. Changes in land cover from natural cover to agrosystems are related to increasing human and livestock pressure, in this case deforestation due to the expansion of agricultural land which needs attention to maintain ecosystem stability and sustainability [7].Visually, a map of forest cover in the catchment area of the Batutegi dam can be seen in (Figure 2)

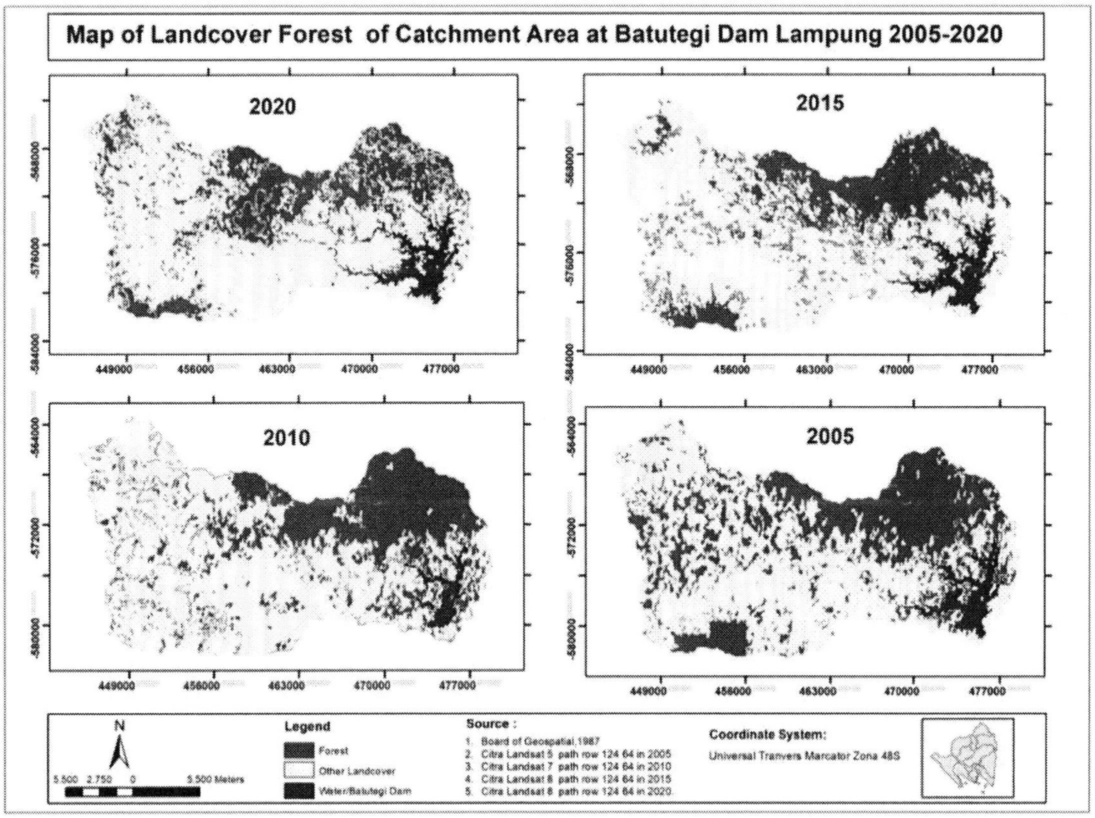

Figure 2. Map of Landcover Forest of Catchment Area Batutegi Dam.

3.2. Inflow Analysis

Inflow is the amount of water entering/availability of water from river water, rainwater, and springs [8]. The level of water availability (inflow) in a dam or dam is a very decisive factor in providing irrigation water for plant needs, especially in the dry season and as a guarantor of the availability of raw water [9]. The inflow data obtained from the Mesuji-Sekampung River Basin Center (BBWS-SM) can be seen in Table 3.

Table 3. Batutegi Dam Inflow 2005, 2010, 2015 and 2020.

Total Montly Inflow (m^3/S)				
Year	2005	2010	2015	2020
January	1179.730	631.190	1023.875	851.220
February	967.123	1073.179	1154.620	771.797
March	981.322	1134.909	1061.861	1170.594
April	742.284	725.762	878.675	1242.630
May	554.004	547.854	637.695	1421.398
June	679.309	473.206	526.196	948.150
July	465.026	492.122	313.060	659.470
August	326.336	586.725	248.166	489.514
September	417.808	692.752	217.101	385.178
October	263.466	702.137	143.590	333.366
November	394.282	312.618	201.575	493.042
December	259.231	824.226	455.829	529.783
Amount	7229.920	8196.681	6862.243	9296.141
Average	602.493	683.057	571.854	774.678

Based on the inflow data in table 3, the highest inflow occurred in May 2020, namely 1421.398 m3/S while the lowest inflow value occurred in October 2015 which was 143.590 m^3/S. Average monthly inflow in one year, the highest inflow occurred in 2020 at 774.678 m3/S and the lowest average monthly inflow in 2015 was 571.854 m^3/S.

3.3. Inflow Fluctuation

The calculation of the value of fluctuations in water availability (inflow) of the Batutegi Dam in 2005-2020 with a 5-year counting period, namely in 2005, 2010, 2015 and 2020, which is known by calculating the ratio of the maximum inflow and minimum inflow that occurs every year. The value of the Batutegi Dam inflow fluctuation can be seen in Table 4.

Table 4. Inflow Fluctuation Batutegi Dam in 2005, 2010, 2015 and 2020.

Year	Inflow (m^3/S)			Inflow Fluctuation (KRA)	
	Total	Maximum	Minimum	Value	Class Category
2005	7229.92	60.56	1.96	30.94	Low
2010	9380.72	101.07	2.07	48.87	Low
2015	6862.24	71.04	1.25	56.75	Medium
2020	8196.68	83.56	1.12	74.54	Medium

IOP Conf. Series: Earth and Environmental Science 1027 (2022) 012006 doi:10.1088/1755-1315/1027/1/012006

3.4. Regression analysis was carried out between forest land cover and fluctuations in debit

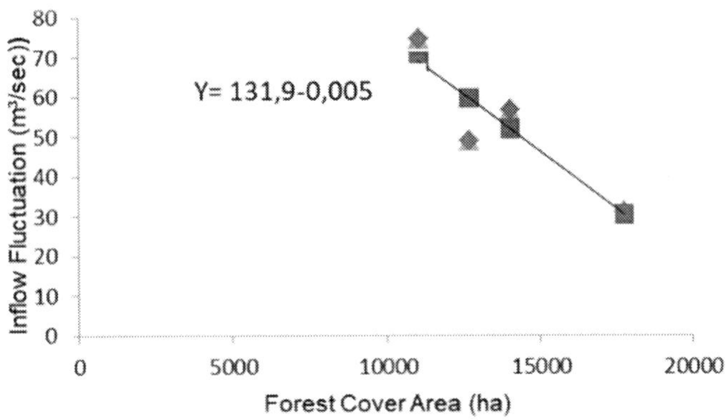

Figure 3. Relationship between Forest and Inflow fluctuation

Based on figure 3, it shows that if the forest area increases, it will decrease the value of fluctuations in the inflow that enters the dam. The condition of forest cover affects the watershed balance in maintaining the availability of water products which regulates fluctuations in discharge or in this case inflow, controls runoff, increases baseflow and reduces erosion and sedimentation. Forest conditions will affect inflow fluctuations in the rainy and dry seasons [10]. One of the indicators of watershed monitoring and evaluation is the value of fluctuations/KRA. The value of inflow fluctuation is obtained from the ratio between the maximum inflows to the minimum inflow, the ability to store water from an area is categorized as good if the inflow fluctuation value in the rainy and dry seasons is small. Discharge fluctuation as a key indicator of watershed system stability [11]. Land cover conditions from changes in land use will affect the ability of the land to accommodate rainwater that falls on the soil surface (infiltration) and surface runoff (runoff), so this will affect the inflow (inflow) as the volume of dam water availability.

Changes in land use, especially forest land, will greatly affect the runoff and infiltration conditions that occur. The amount of runoff and infiltration that occurs will affect inflow in the rainy season and dry season so that it will affect the Inflow Ratio in the rainy season and dry season. According to research, changes in forest land cover that turn into non-vegetative land cover will increase the maximum discharge in a watershed [12]. Deforestation leads to disruption of the hydrological cycle, resulting in a reduction in river flow once during the dry season, in the rainy season, the water discharge increases [13]. Changes in the use or cover of forest land will affect the environment so that it will have an impact on the contribution of the watershed for irrigation and power generation needs based on research changes in land cover patterns affect water quality [14].

4. Conclusions

Forest cover changes in the Catchment area of Batutegi Dam for the period 2005–2020 decreased from 41.93% to 26.03% and the value of inflow fluctuations increased from 30.94 in the low category to 74.54 in the medium category so that with a decrease in area the forest will increase the value of inflow fluctuations, which means that the hydrological conditions have disturbed the balance between maximum and minimum inflows. Based on the regression analysis, forest cover area has a significant effect on the value of inflow fluctuations with a p-value of 0.092.

Reference

[1] Rawat, JS, Kumar M 2015 Monitoring land use/cover change using remote sensing and GIS techniques: a case study of Hawalbagh block, district Almora, Uttarakhand, India. *Egypt. J. Rem. Sens. Space Sci.* **18** pp 77–84

[2] Tsujino R, Yumoto T, Kitamura S, Djamaluddin I, Darnaedi D. 2016 History of forest loss and degradation in Indonesia. Land Use Policy. **57** pp 335-347

[3] Alemayehu F, Tolera M dan Tesfaye G 2019 Land Use Land Cover Change Trend and Its Drivers innSomodo Watershed South Western, Ethiopia *African Journal of Agricultural Research* **14** pp 102-117

[4] Sumura H, Yuwono S B, Ismono H, Arifin B, Fitriani F dan Kada R 2018 Relationship between water quality variations and land use in the Batutegi Dam Watershed, Sekampung, Indonesia. *Lakes & Reserv* pp 1-9

[5] Menteri Kehutanan 2014 Regulation of the Minister of Forestry of the Republic of Indonesia No:P. 61/Menhut-II/2014. *Monitoring and EvaluationWatershed Management* p 36

[6] Spriyadi E, Banuwa IS dan Yuwono SB 2018 The Effect of Land Use Change on the Inflow Characteristics of Batutegi Dam. *Jurnal Hutan Tropis* **6** .1 pp 73-81

[7] Dinka OM and Chaka d D 2019 Analysis of land use / land cover change in Adei watershed . Central Highlands of Ethiopia *Journal Of Water And Land Development* 6.41, pp 146–153

[8] Hadryana IMAD, Arsana IGNK dan Suryantara IPG 2015 Analisis Keseimbangan Air/Water Balance di Das Tukad Sungi Kabupaten Tabanan. *Jurnal Ilmiah Teknik Sipil* . **19** 2 99-107

[9] Supriad H dan Rivai RS 2018 Small Irrigation Investment Development for Rice Production Enhancement to Support Rice Self-Sufficiency. *Jurnal Analisis Kebijakan Pertanian* **16** 1 43-57.

[10] Barkey RA dan Nursaoutra N 2019 The Detection of Forest Health Level as an Effort to Protecting Main Ecosystem in the term of Watershed Management in Maros Watershed, South Sulawesi. IOP Conference Series: Earth and Environmental Science. pp 1-15

[11] Arsyad S 2010 Soil and Water Conservation *IPB Press, Bogor* p 396

[12] Saraswati G F, Suprayogi A dan Amarrohman F J 2017 Analysis Of Change Of Land Cover In The Blorong Watershed To Increase The Maximum Debit Of The Blorong Kendal River. *Jurnal Geodesi Undip.* **6** 2 pp 90-98

[13] Mopangga S 2020 Water Balance Analysis Bolango Watershed. *RADIAL : Jurnal Peradaban Sains, Rekayasa Dan Teknologi* **7** 2 pp 162-171

[14] Rahmah D Y, Somura H, Yuwono S B, Arifin A, Ismono H and Masunaga T 2019 Assessment of soil erosion in social forest-dominated watersheds in Lampung, Indonesia *Environ Monit Assess*191, 726

IOP Conf. Series: Earth and Environmental Science 1027 (2022) 012007 doi:10.1088/1755-1315/1027/1/012007

Adaption strategy of fisherman to climate change: A case study from Limau Sub-District, Tanggamus Regency

A Mutolib[1], A Rahmat[2], I Listiana[3], H Yanfika[3], R A D Widyastuti[3] and M Riantini[3]

[1] Postgraduate Program, Siliwangi University, Jl. Siliwangi No.24, Kahuripan, Tawang, Tasikmalaya, West Java 46115, Indonesia
[2] Research Center for Limnology, Indonesian Institute of Sciences, Jl. Raya Bogor Jakarta, Cibinong, Bogor, Jawa Barat 16911 Indonesia
[3] Faculty of Agriculture, University of Lampung, Jl. Prof. Dr. Ir. Sumantri Brojonegoro, No: 1, Gedong Meneng, Bandar Lampung, Lampung 35141 Indonesia

Corresponding Author: amutolib24@yahoo.com and abdul.mutolib@unsil.ac.id

Abstract. Climate change is a natural phenomenon that occurs all over the world and has a negative impact on many sectors, one of which is the fishery sector. One of the communities that are most vulnerable to the impact of climate change is the fishing community in coastal areas. Fishermen depend on the availability of fish in the sea, as well as the weather and other aspects related to climate and nature. This study aims to identify the fishermen's efforts and adaptation strategies towards climate change. Data collection was carried out from July to August 2020 in Limau Sub-District, Tanggamus Regency with a total respondent number of 80 fishermen households. Respondents were determined through the approach of simple random sampling. The data collected consisted of both qualitative and quantitative data. Data collection was performed through the method of survey, Focus Group Discussion (FGD), and interviews with key informants. Data collected was analyzed using the Qualitative Descriptive approach. The results of the study show that fishermen in Limau Sub-district have made some adaptations to climate change in coastal areas through income diversification, diversification of fishing gear, migration, and social network development among fishermen

1. Introduction

Indonesia as a country possesses high fisheries potential [1]. The fishery sector is a great source of income and livelihood for fishermen on the coast. However, due to certain conditions, many of the workers are now forced to live below the poverty line. Some of the causes of poverty amongst fishermen could be due to poor education and inadequate access to the necessary knowledge and technology required [2], [3]. Apart from these two aspects, another cause of poverty involves the destruction of coastal marine resources, which are the major sources of livelihood for many fishermen [4]. The impacts of anthropogenic activities significantly damage aquatic resources which can lead to climate change.

Climate change refers to a change in global or regional climate patterns [5], [6], [7]. It can lead to changes in sea surface temperatures, which in turn lead to differences in the patterns of fish migration [8]. In addition, it leads to water column stratification which affects the upwelling process (this is positively correlated with fish schooling), thereby making the process more difficult for the fishermen. Furthermore, it leads to changes in the fishing grounds. This migration or damage to the habitats exacerbates the conditions of the workers, leading to increased demand on both time and transport facilities

Limau Sub-District Tanggamus Regency is one of the capture fisheries centers in Lampung Province [4]. Apart from the capture fisheries center, in Limau Sub-District, there is a home industry that processes fish into processed fish products. Climate change also has an effect on the social and economic conditions of coastal communities in Limau Sub-District, whereby it encourages the workers to adapt to new techniques regarding the profession. The fishermen carry out these adaptations to maintain catches that are simultaneously decreasing due to climate change. Furthermore, this study aims to analyze the adaptation patterns of fishermen towards climate change that threatens the sustainability of fishing businesses in Limau Sub-District, Tanggamus Regency.

2. Research methods

2.1. Location and time of research

Data collection was conducted from July to August 2020 in Pekon Tegineneng, Limau Sub-District, Tanggamus Regency. The location was chosen deliberately considering that Limau Sub-district is located on the coast of Tanggamus Regency, where the people mostly work as fishermen and fisheries processing centres.

2.2. Design and research methods

The study used a qualitative research design with a survey approach. The qualitative research format aims to describe various conditions, various situations, or various social reality phenomena and attempt to draw that reality to the surface as characteristics, characters, characteristics, models, or descriptions of certain conditions, situations, or phenomena. Alternatively, research collects and analyzes data in the form of words (spoken or written) and human actions and does not try to quantify the data obtained [9].

2.3. Research data and respondents

The types of data used in this study were primary and secondary data. The Primary data was obtained through household surveys, key informant interviews using a questionnaire, and direct observation; while the Secondary data was obtained from literature reviews and reports/documents from various agencies that were related to the research studies. The respondents in this study were fishermen in Limau Sub-District, who were from 80 different households.

2.4. Data analysis technique

In qualitative research, this analysis is carried out at any time during the research [9]. Data collection and analysis activities in this study are not separate from one another. Both take place simultaneously, and the process takes the form of a cycle [10]. Therefore, this study uses an interactive data analysis model consisting of three stages: data reduction, data display (presentation), and determination of conclusion/verification [11].

3. Results and discusion

3.1. Characteristics of respondents

The average fisherman's age is 39 years old and is said to be a productive age amongst fishermen [12], [13]. The study respondents were predominantly male, having a ratio of 61 male to 19 female. In general, as is the case with other agricultural sectors, men played a more prominent role than women [14], [15]. The average dependence of a fisherman's family was 2-3 members with a length of stay of about 27 years. Most of them were from the Sundanese ethnic group and as many as 49 respondents and junior high school graduates dominated the educational level of the workers. Furthermore, it has been studied that the ability to fish is influenced by the educational level of a worker [16]. Only a few of the workers in the Limau Sub-district had second jobs and they include laboring (*buruh*), small-time entrepreneurship and farming. The characteristics of the research respondents are shown in Table 1

Table 1. Characteristics of respondents.

No	Characteristics	Description
1	Number of respondents	80 Fishermen
2	Average of age	39 Years old
3	Proportion of sex	Male =61, Female 19
4	Duration of living in the area (average)	27 Years
5	Number of dependents	2-3 People
6	Ethnic	Sundanese=49, Lampungese=18, Javanese=13
7	Education level	SD=21, SMP=43, SMA=14, Not graduated from SD= 2
8	Primary job of respondent	Fisherman=68, Housewife =11, Pekon (Village official)=1
9	Side Job	None=43, Farmer=15, Labor= 6, Entrepreneur=10, Boat builder =2, Driver 2, Teacher=1, Pekon (Village official) =1
10	Fishing experience (average)	19 Years
11	Type of fishing tools	Net and fishing rod.
12	Types of catch (fish)	Mackerel tuna, cutlass, selar, mackerel scad, giant trevally, Spanish mackerel, blue marlin, chub mackerel, tuna, and lobster
13	Daily fishing frequency (average)	1.55 times
14	Fishing time (average/day)	7.1 hours

The average working experience of the respondents as fishermen is 19 years. Fishermen's experience is significant to see climate change and its impact on the capture fisheries sector [17]. Fishermen in Limau Sub-District use nets and fishing rods as fishing gear. The fish caught by fishermen in Limau Sub-District is dominated by Mackerel tuna, cutlass and other types (See Table 1). The frequency or number of fishermen fishing in the Limau Coast is 1 to 2 times per day, with the length of time going to sea for about 7.1 hours per day. The frequency and duration of fishing are influenced by climatic factors and the economic capacity of fishermen. If the weather is terrible, the fishermen do not fly for several days.

3.2. Fisherman adaptation strategy to climate change

Implementing appropriate adaptation strategies in dealing with climate change needs to be prepared so that people are not vulnerable to the impact of the rate of change occurring [18]. The adaptation strategy carried out by fishermen in Limau Sub-District to climate change consists of four strategies. The four adaptation strategies include income diversification, fishing gear diversification, migration and social networks. Every fisherman carries out a strategy according to his needs and abilities. Not all fishermen apply the four adaptation strategies simultaneously.

3.2.1. Diversification of income. Many fishermen are adopting adaptive strategies to address climate change through job diversification [19], [20]. In the Limau Sub-District, the fishermen's adaptation strategy through income diversification consisted of six strategies. The overall strategy and the most widely adopted by at least 50% of the population was of having a side job outside of fishing. Another strategy widely adopted was the involvement of their families in working to increase the household income. While the least strategy adopted was in livestock rearing. This adaptation strategy is shown in Table 2.

Table 2. Fisherman adaptation strategy to climate change through income diversification.

No	Fisherman adaptation strategy	Yes	No
1	Have another job	50.0%	50.0%
2	Bring family members to work	22.5%	77.5%
3	Work as a construction worker	8.75%	91.25%
4	Cultivate garden produce	20.0%	80.0%
5	Sell seafood in other forms (cooked, canned, etc.)	12.5%	87.5%
6	Buy and raise livestock	1.25%	98.75%
7	Opened a food stall	13.75%	86.25%

3.2.2. Diversification of fishing gear. Fishing gear diversification is one strategy that many fishermen do to deal with climate change and increase catches [21]. The fishermen in Limau Sub-District periodically purchased new fishing gear, to increase the volume of their catch by at least 40 percent. In addition, this strategy was carried out by producing the fishing gears independently, and the type commonly made was the fishing net. The strategies least adopted in this case were the adding and renting of various types of fishing gear. The workers however, still made use of traditional fishing gears with a little variety. This adaptation strategy is shown in Table 3.

Table 3. Fisherman adaptation strategy to climate change through diversification of fishing gear.

No	Fisherman adaptation strategy	Yes	No
1	Renting other fishing gear	3.75%	96.25%
2	Buy other fishing gear	40.0%	60.0%
3	Making other fishing gear	32.5%	67.5%
4	Adding various types of fishing gear	7.50%	92.5%

3.2.3. Migration. This study is divided into two aspects: the Migration from fishermen to non-fishermen (including living locations) and the Migration of fishing locations. Migration was the strategy most widely and as many as 85% had expanded their fishing areas. This was to increase the number of catches, as well as their income [22], [23]. Meanwhile, the strategy least adopted was the day-to-day movement involving their entire family. This strategy can be seen in Table 4.

Table 4. Fisherman adaptation strategy to climate change through migration.

No	Fisherman adaptation strategy	Yes	No
1	Expanding fishing areas	85.0%	15.0%
2	Work to migrate and settle in the destination	20.0%	80.0%
3	Work overseas and come home every day	3.75%	96.25%
4	Migrate and settle in their destination with family	16.25%	83.75%
5	Migrate and come home every day with family	2.50%	97.5%

3.2.4. Social network. Another effort made towards dealing with climate change was through the development of social networks [24]. These networks could help the fishermen increase their knowledge and information as regards the ever changing patterns of the weather, their economic potentials and also their family networks [25]. If there was any problem encountered, as many as 77.5% of the fishermen joined in attending the village meetings on the social network platforms while the remaining percent resrted to asking family members for help. The strategy that was least adopted in handling an issue was in seeking the boss' (*tauke*) help. This adaptation strategy is shown in Table 5.

Table 5. Fisherman adaptation strategy to climate change through social network.

No	Fisherman adaptation strategy	Yes	No
1	Ask neighbors for help if there is a problem	32.5%	67.5%
2	Asking siblings for help if there is a problem	50.0%	50.0%
3	Ask village officials for help if there is a problem	23.75%	76.25%
4	Ask the boss (*tauke*) for help if there is a problem	5.0%	95.0%
5	Maintain good relations with officials at national borders	30.0%	70.0%
6	Join in attending village meetings to add to my social network	77.5%	22.5%

4. Conclusion

Fishermen on the Limau coast are implementing several adaptation strategies to deal with climate change. The adaptation strategy adopted by fishermen includes four approaches: income diversification, diversification of fishing gear, migration, and social network development among fishermen divided into 22 types of adaptation strategies more precisely. The types of adaptation most widely adopted by fishermen are expanding fishing areas (85 percent), joining in attending village meetings to add to my social network (77.50 percent), Asking siblings for help if there is a problem (50 percent), have another job (50 percent), and buy other fishing gear (40 percent). The adaptation strategy for each fisherman is different depending on the level of climate change and the socio-economic capacity of the fishermen.

References

[1] Pursetyo K T, Tjahjaningsih W and Pramono H 2015 Perbandingan morfologi kerang darah di perairan Kenjeran dan periran Sedati *Jurnal Ilmiah Perikanan dan Kelautan*. **7** (1) pp 31-33

[2] Ansaar 2019 Pola adaptasi nelayan terhadap perubahan iklim di Desa Bambu Kecamatan Mamuju Kabupaten Mauju Provinsi Sulawesi Barat *Pangadereng*. **5** (2) pp 349-364

[3] Yanfika H, Rangga K K, Viantimala B, Listiana I, Mutolib A and Rahmat A 2020 Evaluation of the Success of Programs and Strategy for Sustainable Coastal Community Development in Tanggamus Regency *Journal of Physics: Conference Series*. **1467** 012026

[4] Mutolib A, Rahmat A, Yanfika Y, Listiana I, Rudy and Haryanto Y 2021. Level of income, knowledge, and impact of climate change on fishing household in Limau Subdistrict, Tanggamus Regency *IOP Conference Series: Earth and Environmental Science*. **739** 012041

[5] Rudy R, Yonariza Y, Yanfika H, Rahmat A, Ramadhani W S and Mutolib A 2021 Forest Cover Change and Legal Pluralism in Forest Management: A Review and Evidence from West Sumatra, Indonesia *Indonesian Journal of Science and Technology*. **6** (2) pp 299-314

[6] Murniati K and Mutolib A 2020. The impact of climate change on the household food security of upland rice farmers in Sidomulyo, Lampung Province, Indonesia *Biodiversitas*. **21** (8) pp 3487-3493

[7] Rangga K K,Yonariza,Yanfika H, Mutolib A 2020 Perception, attitude, and motive of local community towards forest conversion to plantation in Dharmasraya District, West Sumatra, Indonesia *Biodiversitas*. **21** (10) pp 4903-4910

[8] Diposaptono, S., Budiman, dan F. Agung. 2009. Menyiasati Perubahan Iklim di Wilayah Pesisir dan Pulau-Pulau Kecil. Bogor: PT. Sarana Komunikasi Utama

[9] Afrizal, 2015 *Metode Penelitian Kualitatif: Sebuah Upaya Mendukung Penggunaan Penelitian Kualitatif dalam Berbagai Disiplin Ilmu* Jakarta: Raja Grafindo Persada.

[10] Creswell, J. W, 2014 *Research design Qualitative quantitative and mixed methods approaches (Fourth)* Thousand Oaks, California: SAGE Publications.

[11] Miles M B and Huberman A M, 1984 *Qualitative Data Analysis: A Sourcebook of New Methods*. California: SAGE Publications.

[12] Nsilapa E S, Budiyanto and Siang R S 2017 Faktor-Faktor yang Mempengaruhi Produktivitas Nelayan Pancing Cumi di Kelurahan Petoaha Kecamatan Abeli Kota Kendari *J. Sosial Ekonomi Perikanan FPIK UHO.* **2** (1) pp 10-19

[13] Listiana I, Efendi I, Mutolib A, Rahmat A 2019 The behavior of Extension Agents in Utilizing Information and Technology to Improve the Performance of Extension Agents in Lampung Province *J Phys: Conf Ser.* **1155**, 012004.

[14] Mutolib A, Yonariza, Mahdi, Ismono H 2017 Gender Inequality and the Oppression of Women within Minangkabau Matrilineal Society: A Case Study of the Management of Ulayat Forest Land in Nagari Bonjol, Dharmasraya District, West Sumatra Province, Indonesia *Asian Women.* **32** (3) pp 23–49

[15] Mutolib A, Yonariza, Rahmat A and Yanfika H 2019 Competition among actors and challenges of production forest management in Dharmasraya, West Sumatera *IOP Conference Series: Earth and Environmental Science.* **399** 012074.

[16] Rahim A, Hastuti D R D, Firmansyah A S 2018 Pengaruh lama melaut, kekuatan mesin tempel, dan karakteristik Responden terhadap pendapatan nelayan tangkap tradisional di Kabupaten Takalar *Jurnal Agrisocionomics.* **2** (1) pp 50-57

[17] Cintra A K A, Setyobudiandi I and Fahrudin A 2017 Analisis kerentanan perikanan tangkap akibat perubahan iklim pada skala provinsi *Marine Fisheries.* **8** (2) pp 223-233

[18] Wibowo A and Satria A 2015 Strategi adaptasi nelayan di pulau-pulau kecil terhadap dampak perubahan iklim (kasus: Desa Pulau Panjang, Kecamatan Subi, Kabupaten Natuna, Kepulauan Riau) *Sodality: Jurnal Soliologi Pedesaan.* **3** (2) pp 107-124

[19] Hasmah 2018 Stretegi adaptasi nelayan tradisional di Desa Sumare Kabupaten Mamuju Provinsi Sulawesi Barat *Walasuji.* **9** (2) pp 403-413

[20] Liando V Y, Aling D R R and Wasak M P 2020 Strategi adaptasi sosial dan ekonomi nelayanalat tangkap sero di Desa Jayakarsa Kecamatan Likupang Barat Kaupaten Minahasa Utara *AKULTURASI_Jurnal Ilmiah Agrobisnis Perikanan.* **8** (1) pp 61-66

[21] Dewiyanti S, Ma'ruf A and Indriyani L 2019 Adaptasi nelayan Bajau terhadap dampak perubahan iklim di peisisr Soropia Kabupaten Konawe Sulawesi Tenggara *Ecogreen.* **5** (1) pp 23–29

[22] Limbong I, Wiyono E S and Yusfiandayani R 2017 Factor affecing the production of purse seine unit in fishing base Sibolga Fishing Port, North Sumatera. *Albacore.* **1** (1) pp 89-97

[23] Nurmeiana D A, Wiyono E S and Riyanto M 2020 Strategi adaptasi nelayan Eretan Kulon, Indramayu terhadap kebijakan pengoperasian arad *Jurnal IPTEKS PSP.* **7** pp 136-150

[24] Vibriyanti D 2014 Kondisi sosial ekonomi dan pemberdayaan nelayan tangkap Kota Tegal, Jawa Tengah *Jurnal Kependudukan Indonesia.* **9** (1) pp 45-58

[25] Iriani 2019 Sekuritas sosial pada nelayan tradisional di Penggoli Kota Palopo *Walasuji.* **10** (1) pp 69-83

The estimated total area of forest fire in Siak Regency, Riau Province during the early period of COVID-19 outbreak

M H Mustofa [1], L Syaufina [1] and N Puspaningsih [1]

[1] IPB University, Jl. Raya Dramaga, Kampus IPB Dramaga, Bogor 16680, West Java, Indonesia

Corresponding Author: mhmustofa@apps.ipb.ac.id

Abstract. Forest fires in Indonesia are more prominently caused by human factors rather than natural factors. In 2020, the COVID-19 outbreak resulted in social distancing policies which leads to the restriction and limitation of human activities. This research aims to compare the forest fire areas and the number of cases between the early period of the COVID-19 outbreak and the normal period before the outbreak. In this research, we used Sentinel-2 images for further analysis through visual method and Normalized Burn Ratio method. Additionally, the ground check was also carried out in the burnt area. The estimation based on the visual analysis resulted in a total burnt area of 459.71 ha with 4 counts of cases. While based on the NBR result, it was 302.79 ha. For comparison, during the same period in 2019 in the same focus research area, there are 7 forest fire cases with a total burnt area of 1,236.80 ha. These results indicate that in terms of total burnt area and count of cases, there are fewer burnt areas in 2020 compared to the burnt areas in 2019. In conclusion, restrictions on human activity are suspected to be the cause of the reduced area of the fire.

1. Introduction

Forest fire is a problem every year in Indonesia during the annual dry season, especially in Sumatra and Kalimantan [1]. In Riau, one of the provinces in Sumatra, the forest and peat fires mostly happened because of land clearing activities for plantations [2]. Forest fires occur in Indonesia more frequently during El Nino years due to a deficit in rainfall [3]. The losses areas caused by forest and land fires are enormous, in the form of economic, health, and environmental damage [4].

Forest fires themselves are influenced by four factors: fuel, weather conditions, ignition agents, and people. The type of fuel, continuity, structure, humidity, and quantity are important elements of a wildfire [5]. One example of a large disaster due to forest fires was in 2015 during El-Nino, where conditions were drier than normal, causing forest fires that occurred by humans to be out of control in fire-prone areas in Indonesia, including Sumatra, Kalimantan, and Papua. Millions of people were exposed to hazardous air quality for weeks and at its peak, dense pollution from wildfires stretches across the globe on the equator [6]. The frequency of forest fires is influenced by a complex combination of natural and human factors [7].

Mostly, the motive for the forest fire caused by humans is economic motive [8]. Burning peatlands is considered a fast way to conduct a land preparation [9]. As a way of preventing, controlling, and monitoring forest fires in areas with high risk such as peatlands, the researchers used hotspots which are indicators of fire in land and forests, even though hotspots does not mean an actual wildfire, they are useful to identifying transient wildfires patterns by detecting seasonal and annual variations in the number of wildfires occurrences [10]. The relationship between hotspots and burned areas obtained from Landsat imagery can be quantified in different ecosystems to find information on the occurrence of fires [11].

Remote sensing is the study of information about the earth's land and water surface using images obtained from an overhead perspective, using electromagnetic radiation in several parts of the electromagnetic spectrum that is reflected or emitted from the earth's surface [12]. In this study, sentinel-2 imagery is used as remote sensing that has a function for making thematic designs that are related to spatial planning, ecosystems, and changes in the appearance of the earth. The use of sentinel-2 is also related to the Google Earth Engine (GEE) application which functions to archive all data sets and link them to the cloud computing engine for open source use. Data archives may include those from other satellites, as well as vector data sets based on Geographical Information Systems (GIS), social, demographic, weather, digital elevation models, and climate data layers [13].

Riau Province was chosen as the location of this study because it has 2 fire seasons, which are January-March and June-September. Based on research data by Mubekti [14], the districts that have a large distribution of peatlands are Siak Regency, Indragiri Hilir Regency, Bengkalis and Dumai, Palalawan, Rokan Hilir, and Meranti. Based on data from the Riau Province Forestry Service, the total area of forest and land burned in 2019 is 90,550 ha, greater than in 2018 which amounted to 37,236.27 ha.

COVID-19 outbreak caused the enactment of social distancing policies. As a result, several activities related to forest and land fire control have also been disrupted. It is known that some types of infectious diseases are more easily spread in areas with high temperature [15], and to reduce the spread of the COVID-19 disease, various types of social distancing and lockdown policies have been implemented in various countries. In Indonesia, the implementation of social distancing is manifested in the form of large-scale social restrictions. Social distancing that was done to slow the spread of the coronavirus is already hampering fire prevention programs and could do the same for firefighting efforts when the intensity of the dry season increase. Then, regarding the incidence of COVID-19, in Riau Province, starting from March 2020 to October 30, 2020, a total of 14,671 positive cases were found. The objective of the present study is to compare the forest fire areas and the number of cases between the early period of the COVID-19 outbreak and the normal period before the outbreak.

2. Material and Methods

2.1 Location and material preparation

This study was done at Siak Regency, Riau Province. One of the reasons for Siak Regency to be chosen as a research area is because the majority of its land is peatland. The vast peatlands in Siak Regency increases the possibility that the forest and land fires are caused by human factors. The climatic conditions in Riau Province are bimodal, which means there are two dry season peaks. In this study, the data collection time was planned for July-November 2020, and the data processing and analysis were carried out in November 2020-May 2021.

The tools used in this research included several software such as ArcGIS, Google Earth Pro, Google Earth Engine, and Microsoft Office. The materials used are in the form of certain data from the European Space Agency (ESA), KLHK, BMKG, and Manggala Agni Daop Siak Patrol Report. The framework in this study can be seen in Figure 1.

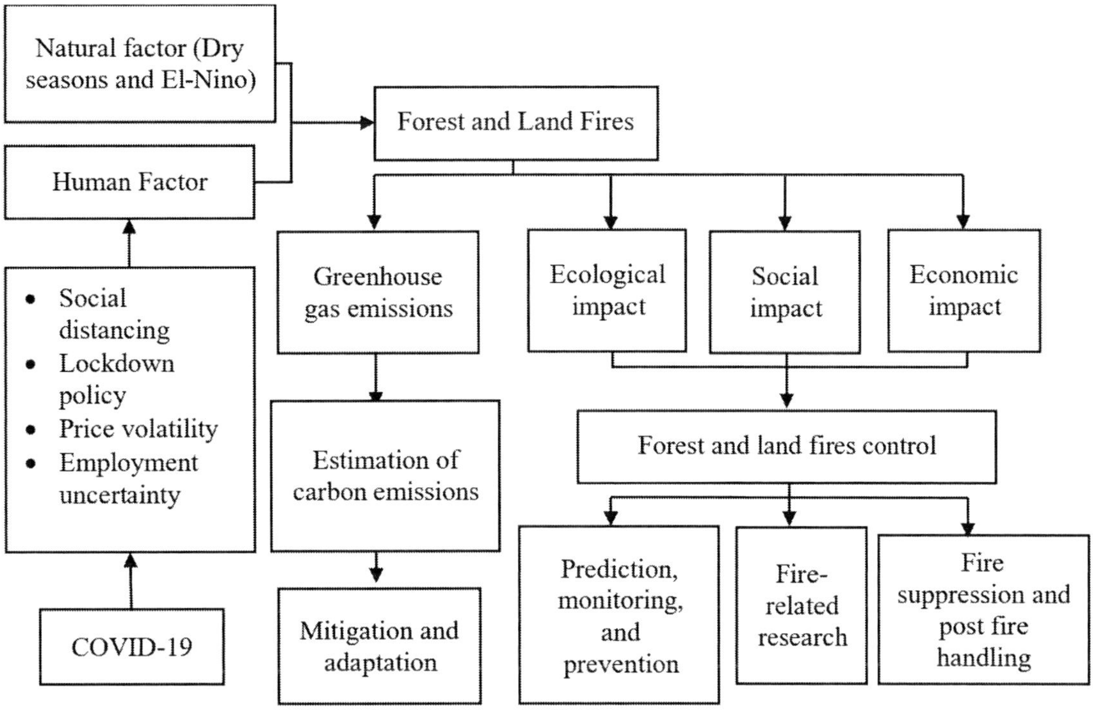

Figure 1. The framework of research.

2.2 Research Method

This research method contains several parameters or variables, data sources, and analysis methods from the previous research conducted in the same regency [16]. The purpose of this study was to estimate the area of forest and peatland fires in Siak Regency, Riau Province, with one of the measured parameters being the burned area in hectares (ha) and the data needed is the burned area data from Sentinel-2 imagery sourced from ESA. The analysis method used is to analyze the Sentinel-2 image to identify the burned area using a visual method based on hotspot data and fire suppression data, and a digital method based on the NBR (Normalized Burn Ratio) model. The research stages of the estimated burned area can be seen in Figure 2.

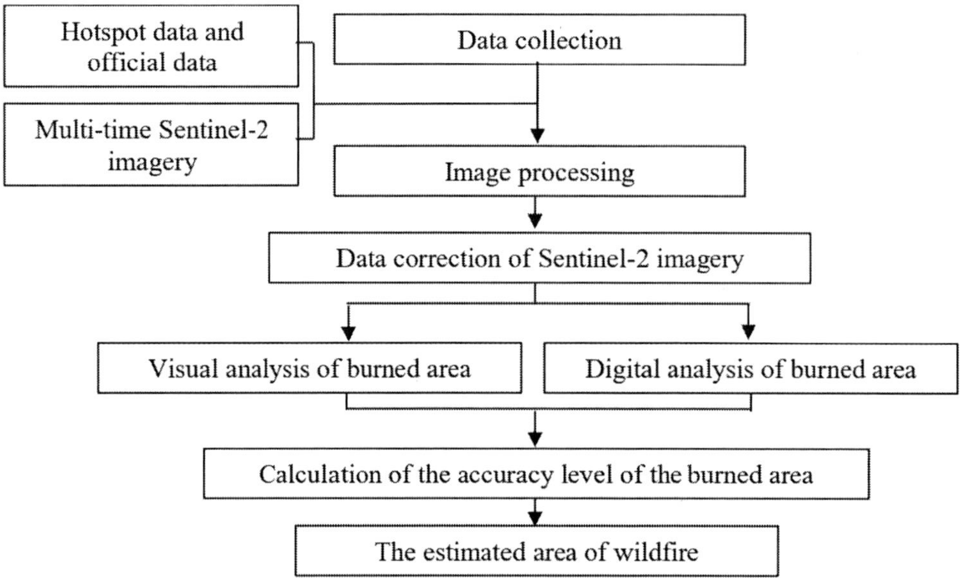

Figure 2. Research stages of estimated burned area.

2.3 Collecting data

Data collection was implemented by estimating the area of burned area based on hotspot data obtained through the data of active MODIS fire that generated based on a contextual algorithm in which a threshold is applied to the middle-infrared temperature and the brightness of thermal infrared were observed by the wildfires that detected by the Terra and Aqua MODIS satellite channels [17]. The active fire was identified from wildfire data sensed remotely by satellites at a spatial resolution of 1 km pixel which was reprocessed by a series of tests, closings, additional corrections, and false alarm rejections to improve prediction accuracy [18]. Hotspot data used is monthly MODIS hotspot data sourced from the website: http://earthdata.nasa.gov. This hotspot data is used to assist in identifying the period of fire occurrence which is then used in selecting Sentinel-2 imagery.

The data of remote sensing satellite image used is Sentinel-2. The channels used are channels 2, 3, 4, 8, and 12 from the Multispectral Instrument (MSI). Channels 2, 3, 4, and 8 have a spatial resolution of 10 meters, while channel 12 has a spatial resolution of 20 meters. In one fire episode, it takes one scene of the recording date image before the fire period and one scene of the recording date image after the fire period. Sentinel-2 measures different frequency ranges along the electromagnetic spectrum in a color, although not always the color visible to the human eye. Each range is called a band, and Sentinel-2 has 12 bands. The Sentinel-2 indicated the red, green, and blue sensors as the number of 4, 3, and 2, and when it is combined, you get true color images.

The data collection technique to estimate carbon emissions is by sampling the random layered referring to SNI 7724 of 2011 that concerning the measurement and calculation of carbon stocks. The largest plot size is 20 m x 20 m, and for each level of vegetation growth the plot sizes are as follows: (a) seedlings with a minimum area of 4 m^2, (b) saplings with a minimum area of 25 m^2, (c) piles with a minimum area of 100 m^2, and (d) trees with a minimum area of 400 m^2.

2.4 Processing and data analysis

The data analysis was performed to estimate the area of the burned area by correcting Sentinel-2 image data which includes geometric and radiometric corrections. In this processing, the radiometric correction that is carried out is a correction of TOA (Top of Atmosphere) reflectance or peak atmospheric reflectance, which converts the digital value into a peak atmospheric reflectance value. This correction is useful for producing surface reflectance, reflectance correction to the direction and the slope (slope correction), or atmospheric correction to minimize disturbance by the atmosphere.

The next analysis is to detect burned areas visually which is done by combining several separate images into one data set file (dataset). The data files that are combined are channel data 2, 3, 4, 8, and 12. After that is image sharpening and color composite image creation, image sharpening that included contrast enhancement and spatial enhancement, then is the determination of the selected Sentinel-2 imagery period to identify areas of ex-forest and land fires. Fire hotspot data (vector format) which has been adjusted to the projection system (into the UTM datum WGS84) is compiled with Sentinel-2 imagery, on the date before and after the forest fires. Then, delineation of the burned area visually is done by drawing a line vector which is the boundary between the image pixels which are the burned area, and the unburned area. The method used to identify burned areas is by determining the changes of land cover conditions before and after the burning of vegetated land. The image layer on-off technique is used to make it easier to identify these changes, the identification process is only carried out in areas where there are hotspots, smoke, and burnout data. Visually, the burned area can be identified using 7 characteristics such as tone or color, shape, size, pattern, texture, location, and association.

The next step is detecting the burned areas digitally by combining channel data 8 and 12, then creating an NBR image for all image data, so that the NBRpre (NBR before the fire) and NBRpost (NBR after the fire) images are obtained. The index variables extracted from the Sentinel-2 image are the vegetation index variable from channel 8 and the fire index from channel 12. Channel 8 is a channel that displays data with Near Infrared (NIR) techniques, while channel 12 can display data related to Short-wave Infrared (SWIR). The equation for calculating the two variables [19] is:

$$NBR = \left(\frac{NIR - SWIR}{NIR + SWIR}\right) = \left(\frac{Band\ 8 - Band\ 12}{Band\ 8 + Band\ 12}\right) \qquad \text{........(1)}$$

Explanation:
NBR = Normalized Burn Ratio
Band 8 = Reflection of channel 8 Sentinel-2 (NIR; 785-799nm)
Band 12 = Reflection of channel 12 Sentinel-2 (SWIR; 2100-2280nm)

After that, it was continued with NBR image fusion, which was combining several separate NBR image data files, namely NBRpre and NBRpost into one dataset, then extracting the burned area pixels. A pixel is stated as a burned area pixel if it meets 2 (two) requirements which are the burned area pixel threshold. Statistically, the threshold value can be measured from the mean value (μ) and standard deviation (σ) of the sample area that representing the burned area in the NBRpost and ΔNBR images. By using the assumption of a normal distribution, the threshold value chosen is the value $\mu \pm 2\sigma$. Furthermore, based on these assumptions and criteria ($\mu \pm 2\sigma$), the α value is obtained from the mean value plus two times the standard deviation of the burned area sample in the NBRpost image. The β value is obtained from the mean value minus two standard deviations from the burned area sample in the ΔNBR image. The estimation of the area of the fire is based on the best model.

3. Result and Discussion

3.1 The condition of burned area

The majority of land in the Siak Regency is peatland, so the forest fires that happened are caused by humans. In 2020, the COVID-19 outbreak resulted in the enactment of a social distancing policy, as a result, several activities related to forest and land fire control were disrupted. The data were collected by taking samples on predetermined area plots. Figure 4 is a sampling of plots in 3 different areas.

There are two major advantages of this research. First, a ground check has been done during the early period of COVID-19 outbreak. Second, this type of research which tries to find the correlation between pandemic and forest fire can be considered new. Meanwhile, this research also has some disadvantages, namely this research does not include the carbon emission calculation and epidemiologic analysis.

Field observations have been done by measuring the total weight of peat and the total weight of litter on three predetermined plots, then taking 2 samples at different depths and analyzing the physical and chemical properties of the soil. Peat samples were obtained directly from the burned area in Sungai Apit District, Siak Regency, then the peat samples were sent to Bogor for further analysis. In the chemical analysis of the three measurement plots, the average pH value was 3.48 and the average C-organic value was 47.69%, then in the physical analysis the average value of bulk density of peat soil with a depth of 30 cm was 0.06 while at a depth of 60 cm, the average value is 0.09 g/cm^3.

3.2 Area estimated of the burned area by visual method

In this study, Sentinel-2 imagery was used to monitor burned areas, Sentinel-2 imagery has been used in various studies [20, 21, 22]. The classification of all Sentinel-2 Bands that are common to Landsat-8 has resulted in overall accuracy, which is 5% and 4% better than Landsat-8 [23].

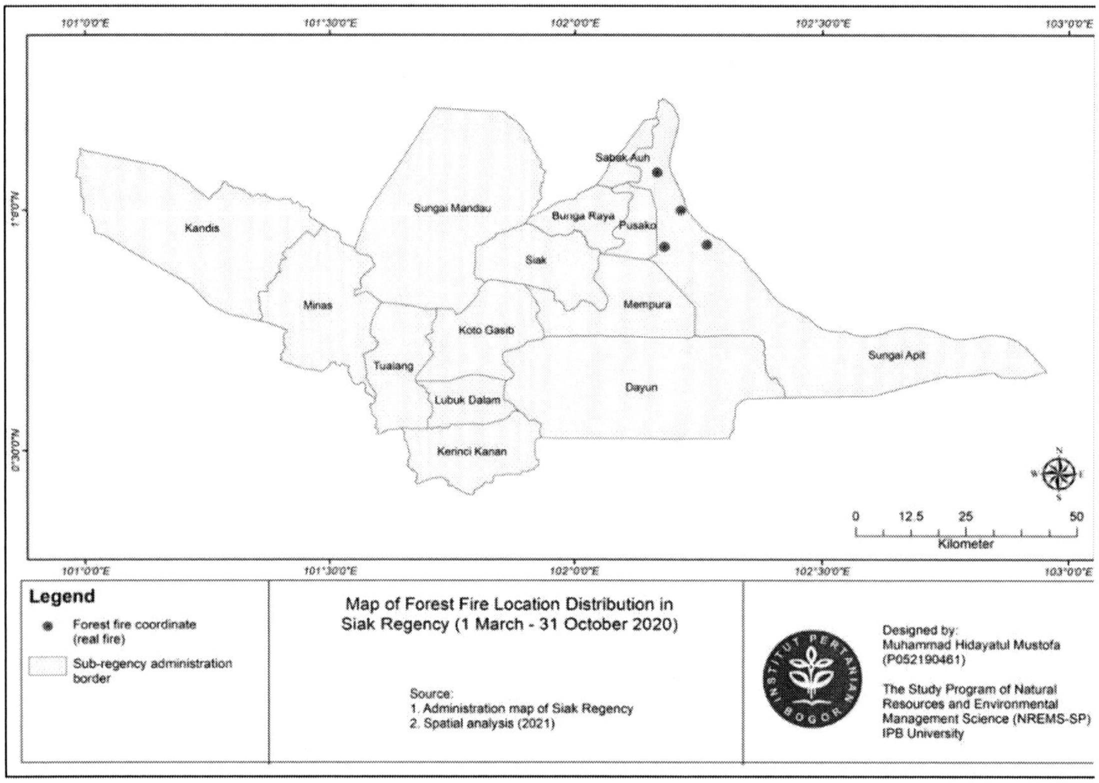

Figure 3. Map of distribution of the location of forest and land fires in Siak Regency, Riau for March-October 2020.

Figure 3 shows that the burned areas were in Mengkapan Village and Lalang Village that was in Sungai Apit District. Three burned areas were found in the Mengkapan Village and one burning area at Lalang Village. According to Manggala Agni, there were also other fire areas with an area of about one hectare in Dayun Village and Merempan Hilir Village, but after matching the coordinates there was no burned area on the satellite imagery. With this in mind, these two areas were excluded from the analysis due to a lack of data.

The period used in finding hotspots or the location of wildfires at a predetermined place is the dry season period that starts from March to October. From the five data of hotspots that have been determined, 4 locations of peat fires are located in Sungai Apit District. The pattern of hotspots associated with the rainy season can become a recommendation in the formulation of strategies for disaster management of forest and land fires in Indonesia as well as preparations for handling haze disasters caused by forests and land fires.

Sentinel-2 (MSI Level-1C) images were obtained from the ESA Copernicus portal via Google Earth Engine (GEE). As can be seen in Figure 4, the analyzed satellite images have cloud coverage of less than 15%. The determination of the satellite image period has been directly matched and confirmed with official data from Manggala Agni Siak related to forest and land fires during 2020 in Siak District. The Sentinel-2 satellite is similar to the Landsat satellite, especially in terms of making thematic designs that are related to spatial planning, ecosystems, and the changes of the earth sighting.

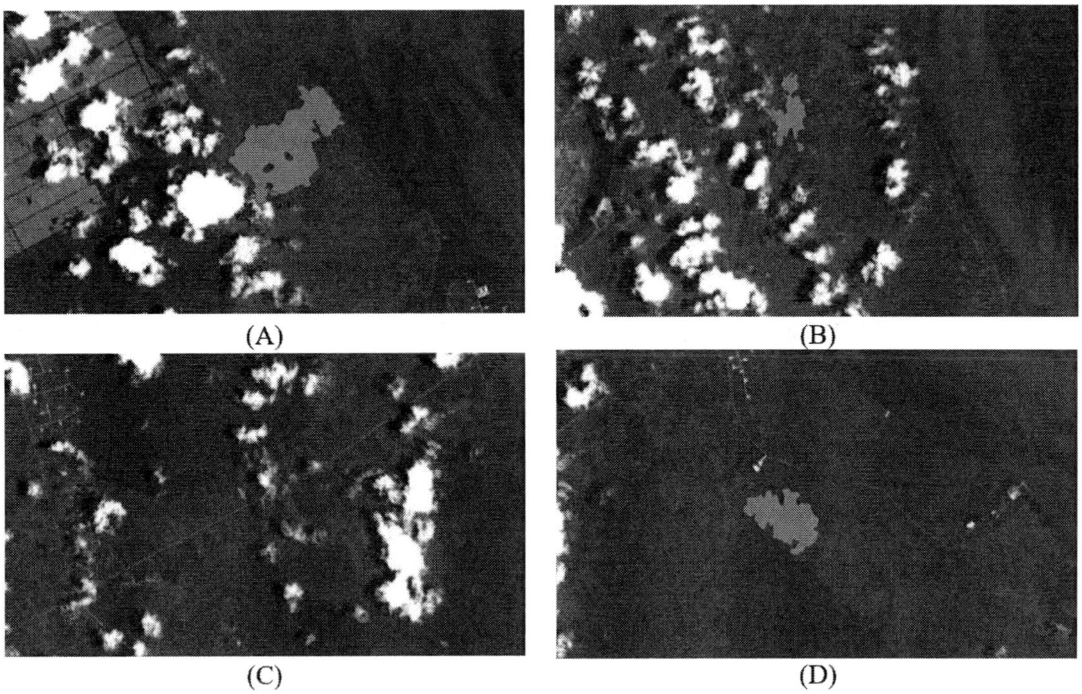

Figure 4. Sentinel-2 image and digitization of burned areas at locations A, B, C, and D.

3.3 Area estimated of the burned area by digital method

The use of the burned area index for mapping burned areas has been widely used. On one hand, by highlighting the charcoal signal in post-fire pictures, the Burn Area Index (BAI) identifies burned land in the red to NIR spectrum in which burned regions are indicated by brighter pixels [24]. On the other hand, NBR uses near-infrared (NIR) and shortwave-infrared (SWIR) wavelengths highlight burned areas in large fire zones greater than 500 acres. NBR and BAI are ways to measure the area burned. In this study, NBR was used to help estimate the burned area and estimate emissions due to forest and land fires. For the mapping of burned areas in this study, NBR is calculated based on equation 1 in sub-chapter 2.4.

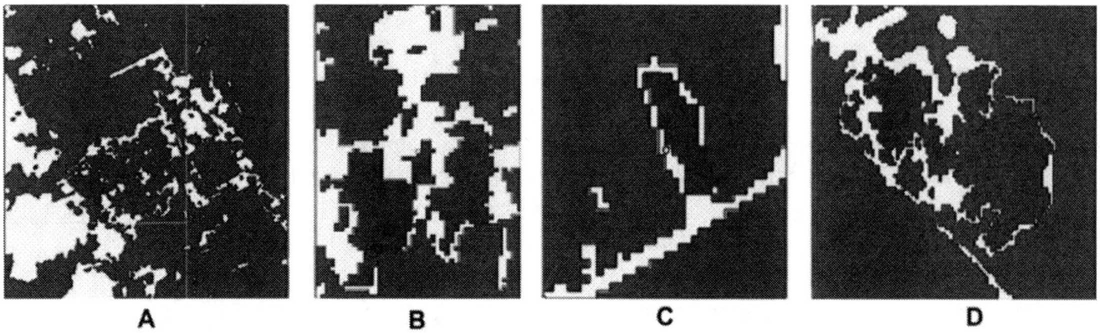

Figure 5. The results of determining the NBR (Normalized Burn Ratio) at locations A, B, C, and D.

The results of the Sentinel-2 image analysis at location A, which is at Lalang area, precisely at the coordinates 0.99936 °N 102.21062 °E, at this location, the area burned based on the NBR analysis was 187.00 ha. This is equivalent to 11.53% of the total area. Then in area B, which is located around the Kayu Ara River, to be precise at the coordinates 1.08044 °N 102.16792 °E. At this location, the burned area based on the NBR analysis is 27.00 ha, which is equivalent to 21.09% of the total area. Furthermore, at location C around the Buton area with location coordinates at 0.92233 °N 102.18787 °E, at this location, the burned area based on the NBR analysis was 2.79 ha, which is equivalent to 11.32% of the total area. And the last location is location D which is located in the area around Koempai with location coordinates at 0.92607°N 102.25860 °E. At this location, the area burned based on the NBR analysis was 86.00 ha, which is equivalent to 21.18% of the total area.

Table 1. Data of forest and land fires in Siak District for March-October 2020.

Code	Location	Hotspot Coordinate	Real Fire Coordinate	Burned Area (ha)	NBR (ha)	NBR (%)
A	Mengkapan Village Sei Apit District, Siak Regency	0.9995 102.2120	0.9986 102.2140	277.28	187.00	11.53
B	Lalang Village, Sungai Apit District, Siak Regency	1.0762 102.1671	1.0775 102.1663	59.68	86.00	21.18
C	Mengkapan village Sungai Apit District Siak Regency	0.9463 102.2318	0.9227 102.1812	2.66	2.79	11.32
D	Mengkapan Village Sungai Apit District Siak Regency	-	0.9269 102.2666	120.10	27.00	21.09
	Total			459.72	302.79	

The data in Table 1 are related to the locations of forest and land fires that have been processed. There are 4 cases in 4 locations. All hotspots mentioned have a high confidence level. The first location is Mengkapan Village, Sungai Apit District, Siak Regency, precisely at coordinates 0.9986°N 102.2140°E, the burned area there is 277.28 ha and NBR of 187.00 ha or 11.53% with the hotspot obtained at the coordinates 0.9995°N 102.2120°E which was found via the SNPP/VIIRS satellite and AQUA/MODIS. The second location is Lalang Village, Sungai Apit District, Siak Regency, precisely at the coordinate 1.0775°N 102.16638 °E, with a burned area of 59.68 ha and the NBR obtained was 86.00 ha or 21.18%, and the hotspot coordinates obtained are 1.0762°N 102.1671°E via SiPongi and

NOAA-20 satellite. Furthermore, the third location is in Mengkapan Village, Sungai Apit Subdistrict, Siak Regency with the coordinate 0.9227°N 102.1812°E, it was found that the burned area was 2.66 ha and the NBR obtained was 2.79 ha or 11.32%, and the hotspot was obtained at the coordinates 0.9463°N 102.2318°E which was obtained via SiPongi with a confidence level of 79% obtained through the SNPP/LAPAN satellite. The 4th location is Mengkapan Village, but precisely at the coordinate 0.9269°N 102.2666°E with the burned area of 120.10 ha and NBR obtained 27.00 ha or 21.09% with hotspots found through spatial analysis.

The analysis was carried out based on satellite images through the visual method and NBR method. The fire area based on visual analysis is 459.71 ha while the total area of fires based on the NBR results was 302.79 ha.

As a comparison data, the burnt area in 2019 was analyzed. The focus area of analysis used in estimating the burnt area in 2019 is the same as the focus area for the burnt area estimation in 2020. T time period used for the burnt area analysis was also the same, which is March to October. As a result, the estimated burnt area in 2019 is 1,236.80 ha with 7 cases.

The results showed that both of them are less compared to the burnt area during the same period in 2019 in the same focus area, which is 1,236.80 ha. Apart from the size of the burnt area, the forest and land fire can be compared based on its count of cases. In 2019 the number of forest and land cases was 7 cases, while in 2020 it was 4 cases. That means there is a fewer count of cases in 2020 than in 2019.

While it is clear that the burnt area in 2020 is less than in 2019, for the forest fire cases in 2020, the locations of the forest fire are largely concentrated in the western region of the Siak District. Specifically, the Village of Mengkapan and the Village of Lalang in the Sungai Apit District. Based on the ground check most of these locations are relatively difficult to access especially during the early COVID-19 outbreak.

Forest fires in Indonesia are mostly caused by human factors. Due to the partial lockdown, human activities are more limited. Subsequently, it can be determined that forest fire areas and cases during the early period of the COVID-19 outbreak are lower compared to the normal period prior to the outbreak because of the limited human activities. This is also in accordance with the findings during the ground check. Future studies should integrate carbon emission calculations and epidemiologic analysis to have a better understanding of the relationship between pandemic situations and forest fire cases in multiple forest fire prone areas.

4. Conclusions

Based on the results of the analysis, it was found that the total area of wildfires based on the NBR results was 302.79 ha. Meanwhile, the wildfire area based on visual analysis was 459.71 ha. The results showed that both of them are less compared to the burnt area during the same period in 2019 in the same focus area, which is 1,236.80 ha. Based on the count of cases, the number of forest fire cases of 2020 is less than that of 2019. Restrictions on human activities are highly suspected to be the cause of the reduced areas and cases of forest fire. Carbon emission calculation and epidemologic analysis should be included in future research to improve the acquire more insight into the correlation between pandemic situation and forest fire cases in various areas which are prone to forest fire.

References

[1] Kustiyo *et al* 2015 Detection of forest fire, smoke source locations in kalimantan during the dry season for the year 2015 using landsat 8 from the threshold of brightness temperature algorithm *Int. Journal of Remote Sensing and Earth Sciences* **12** 151

[2] Thah TH and Sitanggang IS 2016 Contextual outlier detection on hotspot data in Riau Province using k-means algorithm. *Procedia Environmental Sciences* **33** 258

[3] Yulianti N *et al* 2012 Recent forest and peat fire trends in Indonesia the latest decade by MODIS hotspot data *Global Environmental Research* **16** pp 105-116

[4] Konecny K *et al* 2016 Variable carbon losses from recurrent fires in drained tropical peatlands *Global Change Biology* **22** pp1469-1480

[5] Flannigan M D *et al* 2005 Future area burned in Canada *Climatic Change* **72** pp 1–16
[6] Shawki D *et al* 2017 Long-lead prediction of the 2015 fire and haze episode in Indonesia *Geographic Research L* **36** pp 9996-10005
[7] Young AM *et al* 2017 Climatic thresholds shape northern high-latitude fire regimes and imply vulnerability to future climate change *Ecography* **40** pp 606-617
[8] Japan International Cooperation Agency 2017 *Survei Pengumpulan Data Mengenai Pengendalian Hutan & Lahan Gambut dan Restorasi Lahan Gambut di Indonesia* (Jakarta: JICA)
[9] Febrie H *et al* 2017 Karakteristik Tanah Gambut yang Distabilisasi Terhadap Pembakaran *Jom Fteknik* **4** pp 1-8
[10] Takahata C *et al* 2010 Remotely-sensed active fire data for protected area management: eight-year patterns in the Manas National Park, India. *Environmental Management* **45** pp 414-423
[11] Hantson S *et al* 2013 Strengths and weaknesses of MODIS hotspots to characterize global fire occurrence *Remote Sensing of Environment* **131** pp 152-159
[12] Campbell J B and Wynne R H 2011 *Introduction to Remote Sensing (5th Edition)* (New York: The Guilford Press)
[13] Mutanga O and Kumar L 2019 Google Earth Engine Applications *Remote Sensing* **9** pp 1-4
[14] Mubekti 2011 Studi Pewilayahan dalam Rangka Pengelolaan Lahan Gambut Berkelanjutan di Provinsi Riau. *Jurnal Sains dan Teknologi Indonesia* **13** pp 88-94
[15] Zhou X *et al* 2013 High temperature as a risk factor for infectious diarrhea in Shanghai, China *J. Epidemiol* **23** pp 418-423
[16] Hafni D A F *et al* 2018 Estimation of carbon emission from peatland fires using Landsat-8 OLI imagery in Siak District, Riau Province *IOP Conf. Ser.: Earth Environ. Sci* **149** 012040
[17] Giglio L, Schroeder W, Justice CO 2016 The collection 6 MODIS active fire detection algorithm and fire products *Remote Sens Environ* **178** pp 31-41
[18] Fornacca D *et al* 2017 Performance of three MODIS fire products (MCD45A1, MCD64A1, MCD14ML), and ESA Fire_CCI in a mountainous area of northwest Yunnan, China, characterized by frequent small fires *Remote Sens* **9** 1131
[19] Kovács K D 2019 Evaluation of burned areas with Sentinel-2 using snap: the case of Kineta and Mati, Greece, July 2018 *Geographia Technica* **14** pp 20-38
[20] Westerling A L 2016 Increasing western US forest wildfire activity: sensitivity to changes in the timing of spring *Phil. Trans. R. Soc. B* **371** 20150178
[21] Quintiere J G 2017 *Principles of Fire Behavior* (New York: CRC Press)
[22] Sannigrahi S *et al* 2020 Examining the effect of forest fire on terrestrial carbon emission and ecosystem production in India using remote sensing approaches. *Science of the Total Environment* **725** pp 1-13
[23] Farda N M 2017 Multi-temporal land use mapping of coastal wetlands area using machine learning in Google Earth Engine *IOP Conf. Ser.: Earth Environ. Sci* **98** 012042
[24] Martín M P 2006 Burnt Area Index (BAIM) for burned area discrimination at regional scale using MODIS data *Forest Ecology and Management* **234** pp 193-197

Bird's species response to the ex-coal mining reclamation stages

T Atmoko[1], A W Nugroho[1] and Mukhlisi[1]

[1] Research and Development Institute for Natural Resources Conservation Technology, Jl. Soekarno Hatta Km 38 Samboja, Kutai Kartanegara, Kalimantan Timur 75271, Indonesia

Corresponding Author: mukhlisi.arkan@gmail.com

Abstract. Birds played an essential role in the life cycle. This study aimed to determine the bird diversity in coal mining reclamation areas with different ages of reclamation plants: one, four, and eight years old, located in East Kalimantan, Indonesia. The bird diversity was determined using two methods: direct observation and capture-released using mist nets. The birds captured by mist nets were identified, and morphometric measurements were conducted. The results showed 26 species of birds found in this study. Birds species were dominated by Blue-throated bee-eater (*Merops viridis*), Yellow-vented bulbul (*Pycnonotus goiavier*), Asian-glossy starling (*Aplonis panayensis*), and Javan myna (*Acridotheres javanicus*). The bird's diversity in the reclamation area with four years of reclamation plants was higher than the other areas. However, the number of birds in the reclamation area with eight years of reclamation plants was the highest. Differences in habitat conditions in the three research areas might cause a low similarity index of bird diversity. Javan myna and Yellow-vented bulbul were found in all research sites. In conclusion, pioneer bird species were observed during the early process of post-coal mining reclamation. The existence of cover crops was essential as a source of food for the birds and nests.

1. Introduction

East Kalimantan is one of the provinces that has the largest coal reserves in Indonesia. By the end of 2018, East Kalimantan's coal production reached ± 257 million tonnes, equivalent to 46% of total national production [1]. In general, the coal mining system in Indonesia is open-pit mining. On a large scale, this coal mining technique can affect soil structure, sedimentation, habitat loss, and a reduction in the population of biodiversity that lives above the soil surface [2-3]. One of the stages of efforts to restore environmental conditions after coal mining is through reclamation activities.

The reclamation of coal mining land was through land-use activities, revegetation, and the completion process in assessing canopy cover and plant maintenance [4]. Unfortunately, the implementation of coal mine reclamation in Indonesia does not include the criteria for restoring biodiversity, such as wildlife [5]. The presence of wild animals is an essential indicator of ecological restoration's success, especially for reclamation areas with the Borrowing and Use of Forest Area Permit (IPPKH= Izin Pinjam Pakai Kawasan Hutan) scheme.

Birds are wild animals that have an essential role in helping the succession process in coal mine reclamation areas. Bird communities have an ecological role as seed dispersers, thus triggering new plant's emergence in the natural succession process [6]. Birds are accelerating the ongoing revegetation process. This phenomenon causes bird community's presence in the reclamation area of coal mines to be an essential indicator in the process of restoring biodiversity [7]. Many studies have been conducted to record bird species in the coal mine reclamation area [6-8]. Environmental variations between coal mine reclamation sites affect the bird community's existence [7-8]. So, the presence of bird species is not always the same between reclamation locations, so conducting a more specific study for each location is necessary.

This study aims to determine the response to bird specie's presence at various ages of coal mine reclamation. Understanding the coal mine reclamation area's succession process is expected to help accelerate the ongoing environmental restoration efforts.

2. Materials and Methods

2.1. *Study site*

The research was conducted in September 2012 at coal mining company areas of PT Kideco Jaya Agung (KJA), located at Batu Kajang Subdistrict, Paser District, East Kalimantan Province of Indonesia (S0°51'- 1°58' E115°50'-115°53'). The company's coal mining concession area was about 23,000 hectares, which was previously a forest area. However, according to the regulations in Indonesia, the company could not occupy all the concession areas to extract coal. They were also obliged to reclaim and rehabilitate after the coal was extracted. The plants used for reclamation were usually fast-growing and harvested at the ages of 6-8 years.

Three reclamation areas with different ages of reclamation plants were selected as the observation areas: 1) Location A was planted with a year old *Enterolobium cyclocarpa* plantation area with 3 x 3 m spacing. The average height and diameter of the plants were about 3 m and 4 cm, respectively. This place was adjacent to a limestone hill with a diverse vegetation composition. 2) Location B was four years old of mixed *Paraserianthes falcataria* and *Acacia mangium* plantation area, with 3 x 3 m spacing. The plant's average height and diameter were about 12 m and 17 cm, respectively. 3) Location C, was an eight-year-old of *P. falcataria* plantation with the spacing of 4 x 4 m. The plant's average height and diameter were about 17 m and 30 cm, respectively.

2.2. *Procedure*

Bird biodiversity was estimated using direct observation using 10x50 binoculars and capture-release using mist nets for 10 days in a row ($\pm$120 hours). The observations were performed using the transect method in the morning (5-7 AM) and evening (5-6.30 PM) at each reclamation area: locations A, B, and C. There were two bird observers at each location. In the second method, a mist net-sized 6x2 m was installed at each observation area. The mist nets were placed in the area where the birds were most likely to cross. Mist nets were inspected every two hours, and the captured birds were identified and measured. Identification was done using a field guide of birds [9]; morphometric measurements were taken using calipers for beak length, head length, head width, total length, tail length, limb length, wing length, and weight by digital scale. Subsequently, the data were tabulated, and the bird's abundance was compared among the research locations.

2.3. *Data analysis*

The recorded bird species were further analyzed to determine the species dominance index and species diversity index. Species dominance index (*Di*) was calculated by (*ni/N*) x 100%, where *N* is the total individual of birds, and *ni* is the individual number of birds *i* [10]. A bird species is not dominant if the *Di* is less than 2%. When the *Di* is between 2-5%, the species is sub-dominant, and it is a dominant species when the *Di* is more than 5%. Meanwhile, the diversity index was calculated using Shannon index (H') = $-\sum_{i=1}^{s}(\mathrm{pi})\,(\log_2 \mathrm{pi})$, where *pi* is the proportion of species *i* to total species (*s*) [11]. Overlapping of species presence between locations were described using Venn diagram analysis. Inter-location clustering of bird communities is analyzed using hierarchical clustering with the "ward.D" method. Meanwhile, cluster distance was determined with Jaccard similarity index (*J*), which was calculated by *a/(a+b+c)*, where *a* is the total number of species in both location A and B, *b* is the number of species found only in location A, and c is the number of species found only in location B [12]. Hierarchical clustering analysis was conducted using software R version 3.4.4, the vegan package [13]. Additionally, the dietary preference of each bird species was categorized into six groups of feeding guilds, i.e., carnivorous (bird feed mainly non-insect animals, e.g., fish, lizard, small mammal, etc.), insectivorous (feeding insect, arthropods, etc.), frugivorous (fruit-eating species), nectarivorous (feeding floral nectar), granivorous (feeding seed/grains), and omnivorous (feeding both on animals and part of the plant) [14].

3. Result

3.1 Bird species

There were 26 species of birds found in all reclamation areas, which comprised 16 families. Of the total, 20 bird species were classified as the understorey species, recorded through direct observation using the transect method and the mist nets. There were differences in the composition of bird species between locations. Location B (4 years old reclamation plant) has the highest number of bird species with the highest species diversity. Meanwhile, location C (8 years old reclamation plants) has the highest number of individual birds than other locations (Table 1).

Table 1. Composition and diversity of birds in reclamation stage

Parameter	*Research location (Age of reclamation plants)*			Total number
	Location A (1 year)	Location B (4 years)	Location C (8 years)	
Number of species	7	14	10	20
Number of family	6	11	9	12
Number of individuals	18	18	20	56
Diversity indices (H')	2.17	3.73	2.70	3.47

3.2 Bird dominance

The under-canopy bird species found in this study were dominated by the Blue-throated Bee-eater (*Merops viridis*) (Di = 16.1%), Yellow-vented Bulbul (*Pycnonotus goiavier*) (Di = 16.1%), Asian Glossy Starling (*Aplonis panayensis*) (Di = 15.5%), and Javan Myna (*Acridotheres javanicus*) (Di = 6.5%). On the other hand, several non-under-canopy bird species were also observed flying around in the reclamation areas, such as Glossy Swiftlet (*Collocalia esculenta*), Silver-rumped Spinetail (*Rhaphidura leucopygialis*). There were also observed several raptor species of birds: *Haliastur indus, Pandion haliaetus, Macheiramphus alcinus*, which often flown around the mining area to search for some prey. The Black hornbill (*Anthracoceros malayanus*) had also observed in this study. This species often flew across mining areas, moving from and to the forest fragments, but these birds did not visit the reclamation areas.

3.3 Bird Species Community

The similarity of bird communities between locations A-B was 40%, higher than the similarity between locations A-C, and B-C which are 21% and 20%, respectively. Meanwhile, the habitat clustering based on the dissimilarity value with the threshold value of 50% indicated that locations A and B were in the same cluster (Figure 1).

Javan myna and Yellow-vented Bulbul were two species found at the three locations in this study. On the other hand, Asian Glossy Starling (*A. panayensis*), Short-toed Coucal (*Centropus rectunguis*), Eastern Spotted Dove (*Spilopelia chinensis*), White-breasted Waterhen (*Amaurornis phoenicurus*), and Blue-throated Bee-eater (*M. viridis*) were considered as the initial bird species because they were found in the coal mine reclamation area with 1-year-old reclamation plants (Figure 2).

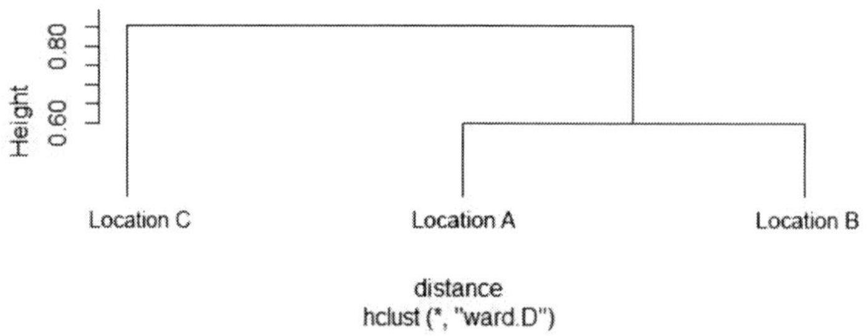

Figure 1. Dissimilarity-clustering birds species community among reclamation site

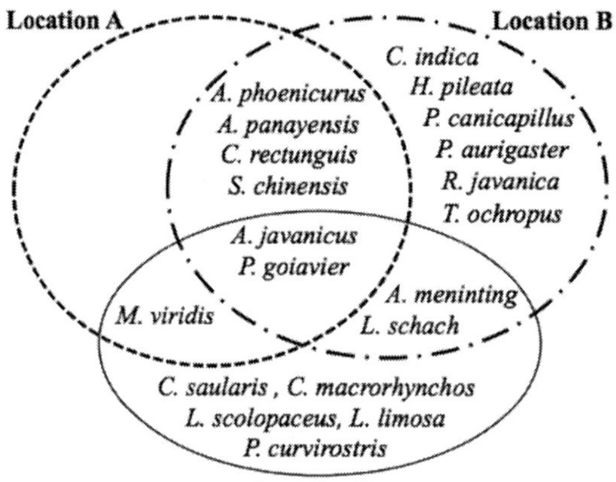

Figure 2. Distribution of birds in three reclamation stages of ex-coal mining

3.4 Birds Morphometry

Morphological measurements were carried out on all birds caught in nets in this study. However, analysis can only be conducted on those birds that were found in two or three locations. These birds were *A. panayensis, M. viridis and P. goiavier*. *Aplonis panayensis* and *Merops viridis* caught at locations A and B, while *P. goiavier* was captured in all locations. The average size of all morphometric parameters of *A. panayensis* found at Location A was larger than in Location B. Meanwhile, except for the sizes of wing length, the average size of all morphometric parameters of *M. viridis* found at Location A was also larger than Location B (Figure 3). The body weights of *A. panayensis* and *Merops viridis* at Location A were also heavier than Location B by 1.75% and 1.60%, respectively. In general, *P. goiavier* found at Location C was the largest compared to the other two locations (Figure 3), accounted for 1.75-1.90% heavier. However, our data is limited, so it cannot be supported by further statistical analysis.

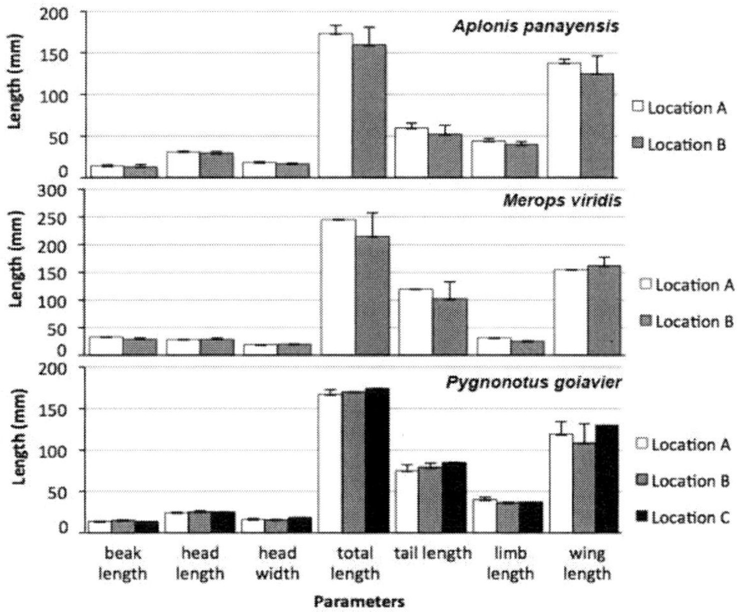

Figure 3. Morphometry of three bird species on the three stages reclamation stage

3.5 Guild Classification

All identified bird species can be grouped into five dietary categories. We found that insectivorous dominated each study site (Figure 4). In the location A and B, the dominance of insectivorous was followed by omnivorous. Meanwhile, carnivorous became dominant after insectivorous in Location C. There were no species in frugivorous and granivorous in location C.

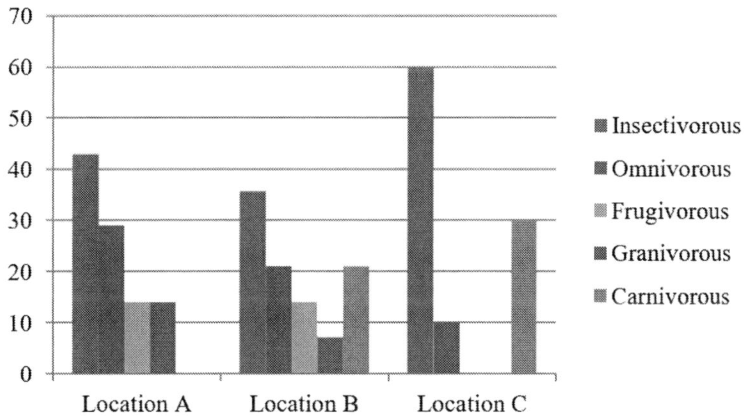

Figure 4. Percentage of bird trophic guilds in each location

4. Discussion

In this study, the dominant bird species, such as *M. viridis, P. goaivier,* and *A. panayensis*, had wide adaptability to various habitat conditions [6], [8] and [15]. The presence of these species of birds in the coal mining reclamation area was also reported in the other studies in Kalimantan [8]. Naturally, these bird species preferred to inhabit various types of habitat: secondary forests, plantation areas, and surrounding human settlements [9]. These habitat conditions might be comparable to the conditions in the research areas. The reclamation area provided food sources, especially for bird species with generalist feeding behavior, such as *P. goaivier* and *A. panayensis*. In other words, that both species have high adaptability compared to others.

Direct observations also found the species of raptor birds, which often flew over above the mining reclamation areas during the day. This flying behavior was generally carried out when the birds hunted for prey, such as rats or other small mammals on the ground. The presence of the raptor bird species as the top predator in this study might indicate that a local food chain had been established. Based on the analysis of the feeding guild, it can be seen that the occurrence of carnivorous birds was increasing along the natural succession process. Location C was starting to become a habitat for top predatory birds, e.g., eagles. It has denser vegetation than locations A and B. Frugivorous and granivorous play a role in helping the colonization of pioneer plants. It is found in younger reclamation areas (locations A and B). Furthermore, the dominance of insectivorous birds in all study sites is natural since it is a key indicator of the coal mining reclamation until the stage of 5 years old [7].

The *Centropus rectunguis* observed in the research area were categorized as vulnerable species according to the IUCN Red List (The International Union for Conservation of Nature's Red List of Threatened Species). Consequently, the vulnerable category required conservation efforts for this species so that the population could be maintained. Human access in coal mining areas was limited, reducing the potential threat to bird species, including protected and vulnerable species.

The different composition of bird species among locations A, B, and C in this study might be related to the different ages of reclamation plants. Among the three locations, dynamic changes affected bird preferences for selecting places to inhabit [8]. Various plants for post-mining reclamation and the pioneer plant species at various age levels have influenced the bird species composition. For example, it was generally preceded by frugivores/granivores bird species [3]. Furthermore, nectarivore birds could be a marker of the succession of reclamation areas after 5 years [7]. However, the bird similarity index among the reclamation areas in this study was low, indicating only a few species-specific birds as the habitat characteristic. Species-specific animals were those that were only found in specific habitats [16].

Species of plants usually planted at the reclamation areas were fast-growing species, such as *Paraserianthes falcataria, Enterolobium cyclocarpa*, and *Acacia mangium*. At the age of four, these species of plants had grown into a small tree, and its canopy began to form with shrubs dominating the floor at the reclamation sites. Shrubs usually can produce seeds and small fruit, which could be one of the factors these birds visit; some even created nests, such as *P. goiavier*. Shrubs could also provide adequate shelters for small birds against predators, strong winds, and cold air [17]. At the age of eight, these fast-growing plants could reach the mature age, ready for harvest with a height of 15 m and a diameter of 20 cm; The tree canopies were already formed and created a microclimate underneath.

The number of bird species and the diversity value at the reclamation site with eight years old of reclamation plants (location C) were lower than at the reclamation site with four years old (location B). There were still many shrubs in the reclamation area with 4-year-old reclamation plants that provided a food source for pioneer birds. On the other hand, in the reclamation area with 8-year-old plants, a 17-meter-hight tree canopy has been formed, reducing sunlight for undergrowth. However, the reclamation area with eight-year-old plants was often visited and crossed by birds on the canopy, such as swallows and three species of eagles.

The reclamation area with one-year-old reclamation plants was still an open area where the sunlight reached the floor. In addition, there were also found several pioneer plant species such as *Melastoma malabathricum, Trema cannabina, Trema tomentosa, Geunsia pentandra, Vitex pinnata*, and *Piper aduncum* [6]. At the beginning of the post-mining reclamation process, these plants were planted as cover crops to reduce soil erosion. These pioneer plant's fruits and seeds were generally food sources for various birds from the *Pycnonotidae* family.

Birds observed in the reclamation area with one-year-old reclamation plants could come from the limestone hills near that reclamation area. The limestone hills were a suitable habitat for these birds because they had various vegetation structures and compositions. The presence of birds in the reclamation areas might purposely for foraging for food or passing because there were no bird nests found in this reclamation area. This limestone hill was a small patch of natural forest remaining in the mining concession area. The remaining forest fragments could have benefits because they were a suitable habitat for birds and other animals. Furthermore, the birds could spread plant seeds from the remaining forest fragments traveling to the reclaimed land, accelerating the natural succession process.

5. Conclusion

At the early stage of post-coal mining reclamation, birds with wide adaptability appeared to the reclamation areas. Meanwhile, the existing cover crops planted before the reclamation plants had become the source of food for the birds. These cover crops created shrubs in the above-ground of reclamation areas. The growth of wild shrubs in the reclamation areas aged 1 and 4 years was needed to be maintained because they were to provide a seed and fruit feed source for pioneer birds and serve as a nesting site. Furthermore, the presence of raptor species suggested that a local food chain had been established, along with small terrestrial mammals.

References

[1] Badan Pusat Statistik [BPS] 2020 *Produksi barang tambang mineral 2017-2019.* https://www.bps.go.id. Accessed on Februari 24, 2021
[2] Sonter L J, Ali S H and Watson E M 2018 *Proc. Royal Soc. B* **285** 20181926
[3] Haigh M, Woodruffe P, D'Aucourt M, Alun E, Wilding G, Fitzpatrick S, Filcheva E and Noustorova M 2020 *Minerals* **10** pp 1-27
[4] [PermenESDM, 2014] Peraturan Menteri Energi dan Sumber Daya Mineral Nomor 7 Tahun 2017 *tentang Pelaksanaan Reklamasi dan Pascatambang pada Kegiatan Usaha Pertambangan Mineral dan Batubara* Jakarta: Kementerian ESDM.
[5] Nugroho A W and Yassir I 2017 *Jurnal Analisis Kebijakan Kehutanan* **14** pp 121-136
[6] Yassir I and Atmoko T 2014 *Burung dan kelelawar di lahan bekas tambang batubara* Samboja: Balai Penelitian Teknologi Konservasi Sumber Daya Alam.
[7] Soegiharto S, Zuhud E A M, Setiadi Y, dan Masyud B 2017 *Jurnal Biologi Indonesia* **13** pp 297-304
[8] Soendjoto M A, Riefani M K, Triwibowo D and Metasari D 2018 *Biodiversitas* **19** pp 323-329
[9] MacKinnon J, Phillips K and van Balen B 2010 *Burung-burung di Sumatera, Jawa, bali dan Kalimantan.* Bogor: LIPI-Burung Indonesia.
[10] Dewi R S, Mulyani Y and Santosa Y 2007 *Keanekaragaman jenis burung di beberapa tipe habitat Taman Nasional Gunung Ciremai* Bogor: Bogor Agricultural University.
[11] Chiarucci A, Bacaro G and Scheiner S M 2011 *Phil. Trans. R. Soc. B* **366** pp 2426-2437
[12] Bag S, Kumar S K and Tiwari M K 2019 *J. Inf. Sci.* **483** pp 53–64
[13] Oksanen J, Blanchet F G, Kindt R, Legendre P, Minchin P R, O'Hara R B, Simpson G L, Solymos P, Henry M, Stevens H, et al. 2019. *Vegan: Community Ecology Package. R Package Version 2.* 5-6.
[14] Sohil A and Sharma N 2020 *Ecol. Process.* **9** 49
[15] Mukhlisi M, Sayektiningsih T, and Yassir I 2021 *P. J. Zool* pp 1-10
[16] Rahayuningsih M, Purnomo F A and Priyono B 2010 *Biosaintifika* **2** pp 82-89
[17] Rusmendro H, Ruskomalasari, Khadafi A, Prayoga H B dan Apriyanti L 2009 *Vis Vitalis* **2** pp 50-64

IOP Conf. Series: Earth and Environmental Science 1027 (2022) 012010 doi:10.1088/1755-1315/1027/1/012010

Effectiveness of using of brown algae alginate to immobilize the indigenous bioremediation bacteria for reducing waste water from shrimp culture

A Setyawan[1], Supono[2], A Wijayanti[2] and UT Anti[2]

[1]Department of Fisheries and Marine Science, Faculty of Agriculture, Lampung University, Jl. Soemantri Brojonegoro No.1, Bandar Lampung, Lampung, Indonesia 35141

[2]Magister of Coastal and Marine Management, Postgraduate of Multidicipliner Program, Lampung University, Jl. Soemantri Brojonegoro No.1, Bandar LampungLampung, Indonesia 35141

Corresponding Author: agus.setyawan@fp.unila.c.id

Abstract. *Bacillus coagulans* T1.2, an indigenous bacterium from shrimp ponds in East Lampung, has been found to reduce total ammonia nitrogen (TAN). The purpose of this study was to know the effectivity of immobilization of bacterial by sodium alginate from marine brown algae to reduce wastewater pollution from shrimp culture. The brown algae, *Sargassum* sp. and *Padina* sp., were collected from Pesisir Barat and Ketapang beach, Lampung. Alginate was extracted from both *Sargassum* and *Padina* by alkali method. The bacterial immobilization beads were made by mixed *Bacillus coagulans* suspension with alginate (1:3 v/v) and formed beads by 1,5 ml syringe without needle. In the in vitro study, the immobilized bacteria were submerged in artificial wastewater (2 beads ml^{-1}) and incubated for 20 days. The control group includes the same bacteria without immobilization. The TAN was measured and the viability of bacteria was evaluated after the incubation period. The results showed that using alginate from *Sargassum* sp. and *Padina* sp. as a matrix of immobilization indigenous bacterial *Bacillus coagulans* effectively and significantly reduce the content of Total Ammonia Nitrogen (TAN) in wastewater. The viability of bacteria immobilized with the alginate of *Sargassum* sp. and *Padina* sp. better than the bacterial treatment without immobilization.

1. Introduction

In the last five years, the national shrimp production volume has shown a positive growth trend with an average growth of 15.7% [1]. Shrimp production that occurs continuously will cause problems of decreasing environmental carrying capacity. This happens because biochemical processes involving suspended particles, inorganic nitrogen, and phosphorus nutrients in the cultivation environment will have a direct impact on the content of ammonia, nitrite, hydrogen sulfide (H_2S) compounds, and carbon compounds that are toxic for cultivation system [2].

Bioremediation is one way to overcome these problems. The utilization of bioremediation bacteria derived from external products is not effective to reduce organic waste in shrimp ponds due to unequal environmental conditions. Indigenous bacteria *Bacillus coagulans* are expected to accelerate the degradation process of pond waste because the habitat of these bacteria is same as current environmental conditions. Based on previous research, *Bacillus coagulans* bacteria has the potential as a bioremediation agent. This bacterium is not pathogenic and can reduce levels of Total Ammonia Nitrogen (TAN) in vitro [3].

Storing isolates for a certain time can cause a decrease in bacterial bioactivity so that their potential is not optimal. Technology improvement is needed to increase the activity of these bacteria, a potential alternative is to improve its performance through immobilization using sodium alginate. The use of sodium alginate was chosen because its presence is easily obtained from the extraction of brown seaweed *Sargassum* sp. and *Padina* sp. which are plentiful in the coastal area of Lampung. Utilization of this type of brown seaweed as a source of alginate is expected to support the potential of *Bacillus coagulans* to optimally degrading TAN.

This study aimed to analyze the effectiveness of immobilization of indigenous *Bacillus coagulans* bacteria to reduce TAN using sodium alginate from *Sargassum* sp. and *Padina* sp.

2. Methods

2.1. Algae Collection

This study used the brown algae *Sargassum* sp. and *Padina* sp. obtained from coastal areas in Lampung Province. The *Sargassum* sp. was obtained from Pesisir Barat beach, and *Padina* sp. was obtained from Ketapang Beach.

2.2. Design Experimental. This study was conducted using four treatments of immobilized bacteria to degrade TAN in Sewage medium, both for sodium alginate from *Padina* and *Sargassum*. The four treatments were control group that was without adding sodium alginate and bacteria (A), adding sodium alginate without bacteria (B), adding immobilized bacteria using sodium alginate (C), and adding bacteria without immobilization (D).

2.3. Alginate Extraction and Characterization.

2.3.1. Alginate Extraction. Alginate extraction was carried out on *Sargassum* sp. and *Padina* sp. which has a moisture content of <15%. The alginate extraction process was preparing 100 g of *Sargassum* sp. and *Padina* sp. powder then soaked in distilled water (1:10). After that, added 5% Na_2CO_3/50μM EDTA for 24 hours. The immersion results were filtered and added with 0.13 M KCL. Then, precipitation was carried out with 96% cold ethanol (1:1) while vigorously stirring until homogeneous and allowed to stand for 24 hours. The precipitation results were centrifuged at 3,500 rpm for 5 minutes. The results of the alginate extraction were then collected and dried in an oven at 60 °C overnight [4].

2.3.2. FTIR Analysis. Extraction results of *Sargassum* sp. and *Padina* sp. the functional groups were analyzed using the FTIR (Fourier Transform Infra-Red) Spectroscopy method. The test compared extracted alginate from *Sargassum* sp. and *Padina* sp. with commercial alginate. This test was weighing of 1 mg samples, then mixed with 100 mg of KBr. Then pressed for ± 10 minutes in 8-10 psi, until a thin pellet is obtained. The pellet was inserted into the cell holder and the spectra were made for analysis [5].

2.4. Bioremediation Bacteria Culture and Immobilization

The immobilization followed an entrapment method using *Bacillus coagulans* bacteria. *Bacillus coagulans* is an indigenous bacterium from traditional tiger shrimp ponds in Mulyosari Village, Pasir Sakti District, East Lampung Regency, Lampung Province. These bacteria have the potential as biodegradation agents to reduce TAN [3].

Bacillus coagulans bacteria were cultured in 1000 mL liquid SWC (Sea Water Compete) medium. The SWC medium consists of 5 g bacto peptone, 1 g yeast extract, 3 ml glycerol, 75% distilled water, and 25% sterile seawater. And then the bacteria were incubated for 24 hours at 30 °C. These bacterial culture was mixed with Na-Alginate solution extracted from *Sargassum* sp. (5%) for *Padina* sp. (10 %) in a ratio (1:3) (bacteria:Na-alginate(v/v)).

The mixture was printed using a syringe 0.1 mL and dripped it drop by drop in a container that already contains a 0.2 M CaCl2 solution for the gelatinization process to form alginate beads (balls). The alginate beads were stored in the refrigerator for 6 hours. After that, the alginate beads were washed with physiological NaCl and put in 150 mL of sewage medium [6].

2.5. Sewage medium assay

Sewage medium is a TAN simulation medium consisting of 13.5 g K_2HPO_4; KH_2PO_4 0.7 g; $MgCl_2.6H_2O$ 0.1 g; $NaHCO_3$ 0.5 g; $FeCl_3.6H_2O$ 0.014 g; $CaCl_2.2H_2O$ 0.18 g; NH_4Cl 0.1 g; EDTA 0.2 g [7]. The using of immobilized *Bacillus coagulans* bacteria 1 – 2 balls/mL into sewage medium. Incubation was carried out at 24°C using a shaker (100 rpm) for 20 days [8].

2.6. Total ammonia nitrogen (TAN) reduction

Analysis of TAN content was carried out at the beginning and end of the study. The steps in the analysis of TAN content include: standard ammonia solution prepared with a concentration of 0.1; 0.2; 0.4; 0.6; 0.8; 1 mg/L. This standard solution is used to determine the standard curve for ammonia. From each filtered sample, 10 mL was taken, and then 0.5 mL of phenol solution (C_6H_5OH) was added and homogenized. After that, 0.5 ml of 0.5% sodium nitroprusside ($C_5FeN_6Na_2O$) solution and 1 ml of oxidizing solution were added. The oxidizing solution consisted of an alkaline citrate solution ($C_6H_5Na_3O_7$) and 5% sodium hypochlorite (NaClO). The samples were allowed to stand for 1 hour (28–31°C). The absorbance was measured using a spectrophotometer with 640 nm wavelength. The absorbance value was entered in standard curve formula to determine TAN content in sewage medium [9].

2.7. Statistical Analysis

Statistical analysis was tested with One Way ANOVA using the SPSS 22.0 program to determine the level of difference between treatments. If there was a difference between treatments, then further test of LSD (Least Significant Difference) with 95% confidence interval.

3. Results and Discussion

3.1. Yields

The yield is a percentage between the weight of the final product and the weight of dry samples used during extraction. Determination of yield is useful for knowing the results of Sargassum sp. and Padina sp extraction. The results showed that the yield of Na-Alginate from 100 g Sargassum sp. powder extraction was 19.80%. This value was lower than previous study that it reached 40.34% [10], but it higher than several previous studies 17.39% [11], and 12.88% [12].

Meanwhile, the yield of 100 g of *Padina* sp. extraction reached 14.44%. This value was lower than the results of previous studies, which reached 25% of the dry weight of Padina sp. extracted [13]. The factors that affect the yield of alginate include: type of seaweed used, conditions of seaweed grow place or habitat (light intensity, wave size or current, water-nutrition, etc.), climate, and the extraction method used, as well as how to handle it [14].

In the extraction process, using Na_2CO_3 solvent was considered to separate cellulose and alginate in brown algae cells. The concentration of Na_2CO_3 used during the extraction process also affects the yield produced. A high concentration of Na_2CO_3 can produce more yield [15].

3.2. FTIR (Fourier Transform Infra-Red) Spectra Analysis

The FTIR test was carried out at 4000 – 650 cm-1 wave length on 32 scanned samples and 4 cm-1 resolution. The test was conducted to determine the presence of chemical bonds in organic compounds in Sargassum sp. and Padina sp. extract. The results of functional group analysis were compared with data from various references to confirm the presence of functional groups representing a sodium alginate bond at a certain wavelength (Table 1).

Table 1. Functional group of alginate based on FTIR spectra.

Wavelength (cm^{-1})				
Na-Alginate *Sargassum* sp. Extract	Na-Alginate *Padina* sp. Extract	Na-Alginate Commercial	Functional Group Interpretation	Wavelength Reference (cm^{-1})
3332.2	3272.6	3227.9	Hydroxyl Group (O-H)	3200 – 3500[16]
2937.1	2929.7	2877.5	Aliphatic Group (C-H)	2850 – 3000[16]
1595.3	1595.3	1580.4	COO-asymmetric	1620[17]
1394.0	1401.5	1416.4	COO-symmetric	1410[17]
1028.7	1028.7	1319.5	Manuronic Acid and Carboxyl Group (C-O)	817,82 – 1315[18] and 1000 – 1300[16]
	857.3	1021.3	Guluronic Acid	817,82 – 1029[18]

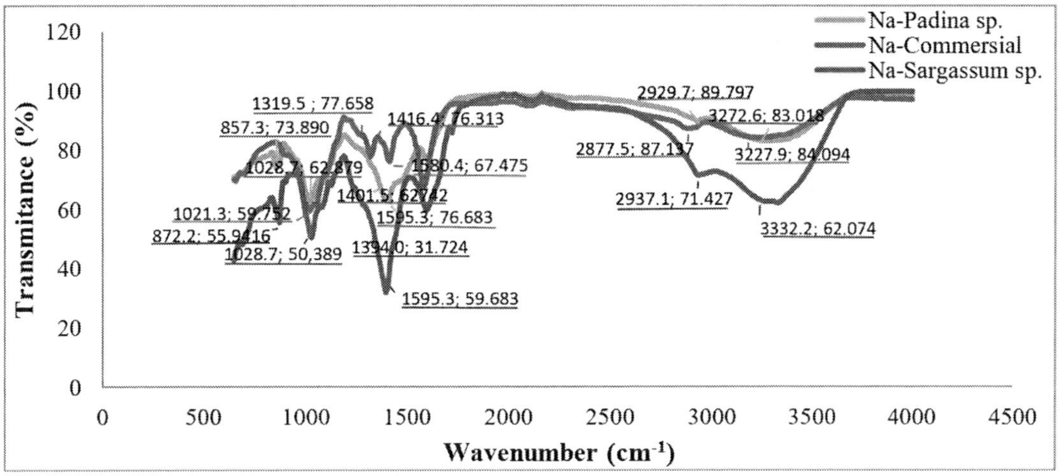

***Figure 1**. FTIR spectra of sodium alginate.*

Vibration spectra at 3200 – 3500 cm-1 indicate the presence of a hydroxyl group (O–H) bonded to hydrogen, vibrations at 2850 – 3000 cm-1 indicate the presence of aliphatic compounds, and at 1000 – 1300 cm-1 indicate the presence of a carboxyl group (C– O) [16]. Vibration spectra at 1620 cm-1 indicated the presence of COO-asymmetric and at 1410 cm-1 indicated the presence of COO-symmetric [10]. Vibration spectra at 817.82–1315 cm-1 indicated the presence of mannuronic acid, and at 817.82–1029 cm-1 indicated the presence of guluronic acid [18]. The presence of mannuronic acid and guluronic acid means that the sample contains alginate compounds [19].

*The FTIR test results on samples of Sargassum sp., Padina sp., and commercial alginates (**Figure 1.**) showed slightly different spectral patterns. The sodium alginate extracted from Padina sp. has a lower spectrum pattern than sodium alginate extracted from Sargassum sp. and commercial alginates (**Figure 1**). This happens because other functional groups are also filtered during the sample purification process. However, the extraction results from Sargassum sp. and Padina sp. are still identified as sodium alginate because it has met the standard by having 3 spectral peaks consisting of hydroxyl groups, COO-asymmetric, and COO-symmetric [17].*

3.3. Analysis TAN (Total Ammonia Nitrogen)

Analysis of TAN content was carried out to determine the effectiveness of immobilization of Bacillus coagulans bacteria using sodium alginate from Sargassum sp. and Padina sp. The TAN content at the beginning treatment was 0.45 – 0.48 mg/L, then decreased until the end of the study. The decrease TAN content occurs because Bacillus coagulans bacteria can utilize TAN compounds as metabolism for their growth [3].

The results showed the utilization of immobilized bacteria Bacillus coagulans with sodium alginate from Sargassum sp. and Padina sp. can increase the activity of bacteria better than free cells. This is known based on the decrease in TAN content in both treatments. Both treatments had a higher reduction in TAN levels than Bacillus coagulans without immobilization ***(Figure 2.)****. The utilization of immobilized bacteria can absorb nitrogen better than free cells [20].*

The highest decrease TAN content occurred in immobilization treatment of Bacillus coagulans bacteria using sodium alginate with Sargassum sp. The decrease TAN content in this treatment reached 60% ***(Figure 2)****. At the end of study, the results of One Way ANOVA test showed that there were differences in concentration of TAN in each treatment (α<0.05). The results of LSD test showed that each treatment had significantly different results. Thus, assumed that immobilization of Bacillus coagulans bacteria using sodium alginate matrix from Sargassum sp. and Padina sp. can to increase activity of these bacteria to reduce TAN significantly.*

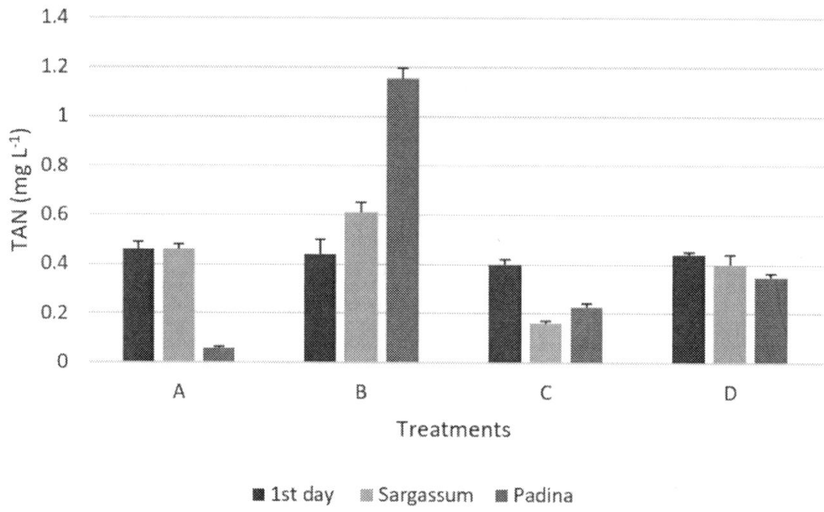

Figure 2. TAN Concentration in Sewage Medium: (A) Control (without adding sodium alginate and bacteria), (B) Adding sodium alginate (C) Adding immobilized bacterial using sodium aginate (D) Adding bacteria without immobilization. The values is average of TAN concentration ± standard deviation (SD).

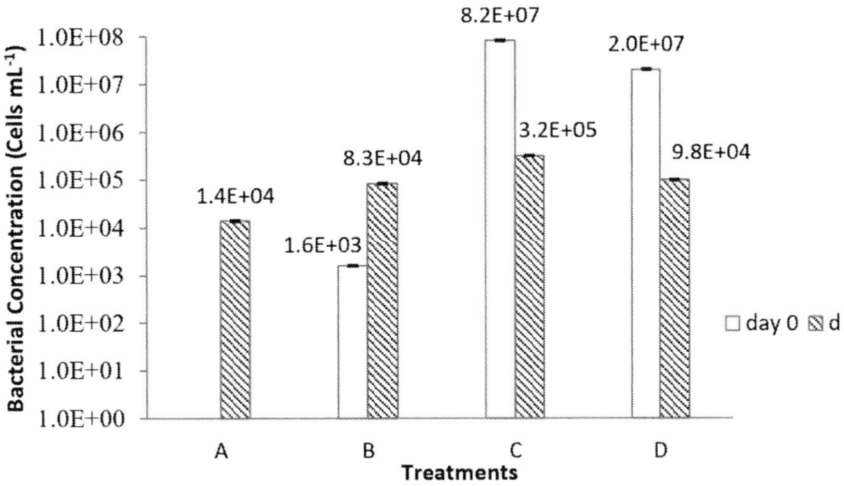

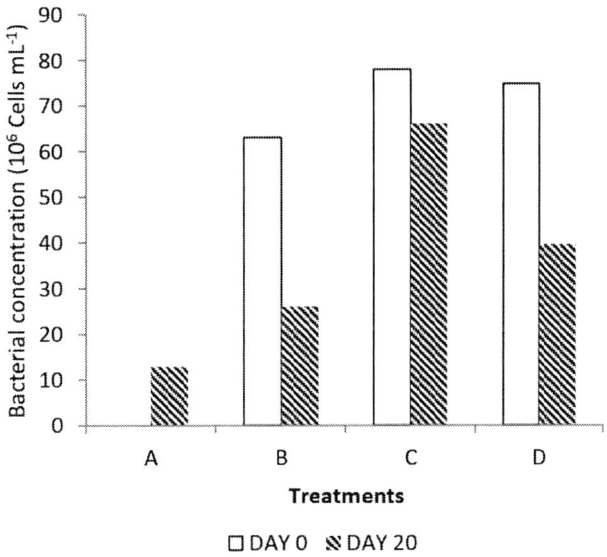

Figure 3. Viability of bacteria in Sewage medium which treated by sodium alginate immobilization matrix from *Sargassum* (A) and *Padina* (B); A: Control, B: Immobilization of *Bacillus coagulans* with Na-*Sargassum* sp., C: Immobilization of *Bacillus coagulans* with Na-*Padina* sp., D: *Bacillus coagulans* without immobilization.

3.4. Bacterial Viability

The bacterial viability test was performed to determine the density of bacteria that grew in sewage medium. This method has counted the colonies that grow on agar medium. The bacterial counts were performed on free cells detached from alginate beads. During the research process, bacteria were released from the alginate beads and counted as free cells. The cells released from the alginate beads occurred due to the diffusion of nutrients in alginate beads, causing a decrease in density and rigidity of the gel matrix. The use of sodium alginate concentration also affects the strength of matrix protection in holding cells out of the alginate bead [21].

Figure 3 showed that immobilization of *Bacillus coagulans* bacteria using sodium alginate from *Sargassum* sp. had the highest viability and was more stable than other treatments. Meanwhile, in the treatment of immobilized *Bacillus coagulans* using sodium alginate from an extract of *Padina* sp. and *Bacillus coagulans* without immobilization occurred decrease in bacterial viability until the end of the study. The decreased viability of bacteria in a medium can be caused by a reduction source of nutrients in the media, and saturation of the growth of bacteria is achieved [22].

Some factors cause decreased bacterial viability, including availability and amount of bacteria concentration in the samples. Bacterial growth biomass can also be affected by nutrients in the media. The lack of nutrients in the media for the needs of microorganisms causes the metabolism of *Bacillus coagulans* bacteria is not optimal so that the bacterial density decreases [23].

4. Conclusion

Based on the results, it was concluded that the immobilization of endogenous bacterium *Bacillus coagulans* using sodium alginate from *Sargassum* sp. and *Padina* sp. extract effective to reduce the content of Total Ammonia Nitrogen (TAN). The treatment showed that the viability of bacteria was better than Indigenous *Bacillus coagulans* without immobilization.

Acknowledgment

This research was made possible thanks to the support by the DIPA BLU Grant No. 1927/UN26/.21/PN/2019, University of Lampung.

References

[1] Ministry of Marine and Fisheries Affairs of Republic of Indonesia 2019 *Indonesian Fishery Productivity (Ann. Report)* (Jakarta: kementerian Kelautan dan Perikanan republic Indonesia). Available from: https://kkp.go.id/wp-content/uploads/2018/01/KKP-Dirjen-PDSPKP-FMB-Kominfo-19-Januari-2018.pdf

[2] Song X *et al* 2016 Integrated bioremediation techniques in a shrimp farming environment under controlled conditions *Act. Oce. Sin.* **35** 88

[3] Susanti E *et al* 2014 Penapisan bakteri pendegradasi total ammonia nitrogen dari sedimen tambak tradisional udang windu (*Penaeus monodon*) *Aquasains* **2** 145

[4] Yudiati E *et al* 2016 Innate Immune Stimulating and Immune Genes Up-Regulating Activities of Three Types of Alginate from *Sargassum siliquosum* in Pacific White Shrimp, *Litopenaeus vannamei Fish. Shellfish. Immunol.* **54** 46

[5] Herdianto R W *et al* 2019 Optimasi Suhu Ekstraksi Terhadap Kualitas Alginat yang Diperoleh dari Rumput Laut *Sargassum muticum Pengolahan Hasil Perikanan Indonesia* **22** 164

[6] Kiran M G *et al* 2018 Heavy metal removal from aqueous solution using sodium alginate immobilized sulfate reducing bacteria: Mechanism and process optimization *J. Env. Man.*. **218** 486

[7] Hastuti Y P 2011 Nitrifikasi dan denitrifikasi di tambak *J. Akua. Ind.* **10** 89

[8] Dong Y *et al* 2014 Immobilization of ammonia oxidizing bacteria by calcium alginate *Ecol. Eng.* **73** 809

[9] Indonesian national Standard 2005 *How to Test Ammonia Levels with a Spectrophotometer with Phenolic. SNI 06-6989.30-2005* (Jakarta: Badan Standarisasi Nasional)

[10] Yudiati E and Isnansetyo A 2017 Characterizing the three different alginate type of *Sargassum siliquosum Ilmu Kelautan* **22** 7

[11] Putriyana R S *et al* 2018 Sintesis natrium alginat dari *Sargassum* sp dengan proses leaching *Industrial Research Workshop and National Seminar vol 9* (Bandung: Politeknik Negeri Bandung) 934p
[12] Jayanudin *et al* 2014 Pengaruh suhu dan rasio pelarut ekstraksi terhadap rendemen dan viskositas natrium alginat dari rumput laut coklat *Sargassum* sp *Integrasi Proses* **5** 51
[13] Hamrun N *et al* 2018 Physical properties of irreversible hydrocolloid dental impression materials obtained from brown algae species *Padina* sp. *J. Phys.: Conf. Ser.* **1073** 052018
[14] Rasyid A 2003 Perbandingan kualitas natrium alginat beberapa jenis alga coklat *Oseana* **28** 33
[15] Mushollaeni W and Rusdiana E 2011 Karakterisasi natrium alginat dari *Sargassum* sp., *Turbinaria* sp. dan *Padina* sp. *J. Teknol. dan Industri Pangan* **22** 26
[16] Hamrun N and Rachman S A 2016 Measuring sodium alginate content of brown algae species *padina* sp. as the basic matter for making dental impression material (irreversible hydrocolloid impression material) *J. Dentomaxillofac. Sci.* **1** 129
[17] Ju H K *et al* 2002 pH/temperature-responsive semi-IPN hydrogels composed of alginate and poly(N-isopropyl acrylamide) *J. Appl. Polym. Sci.* **83** 1128
[18] Eriningsih *et al* 2014 Rumput laut coklat untuk proses pewarnaan kain sutera *Arena Tekstil* **29** 73
[19] Iriyanti *et al* 2018 Potential Na-alginate extract from brown algae *Sargassum* sp. *Indonesia Chimica Acta* **11** 17
[20] Erna *et al* 2013 Immobilized nitrifying bacterial consortium for improving water quality, survival and growth of *Penaeus monodon* Fabricius 1798 postlarvae in hatchery system *Asian Fish. Sci.* **26** 212
[21] Ratnasari N *et al* 2014 Pengaruh konsentrasi natrium alginat sebagai penjerat sel *Lactobacillus acidophilus* FNCC 0051 dan lama penyimpanan terhadap jumlah sel yang terlepas dan karakter carrier *Teknol. Pangan dan Gizi* **13** 81
[22] Advinda L *et al* 2015 Penambahan gliserol pada bahan pembawa alginat sebagai penstabil pertumbuhan bakteri *Pseudomonas* berfluoresen *Prosiding Semirata 2015 Bidang MIPA BKS-PTN Barat* (Pontianak: Universitas Tanjungpura)
[23] Nasrah S N *et al* 2012 Viabilitas dan keriap *Bacillus* sp. BK17 dan *Enterobacter* sp. BK15 pada sumber karbon dan nitrogen yang berbeda *Saintia Biologi* **1** 1

Survey of knowledge and practice related health risks of mercury exposure among gold craft workers in Rappocini area, Makassar city

Nanang Rahmadani [1] and Muhammad Syafri [2]

[1] Institut Teknologi dan Kesehatan Tri Tunas Nasional; Jl. Todopuli Raya Timur No. 4, Makassar, South Sulawesi, 90234, Indonesia

* Corresponding Author: nanang.rahmadani@tritunas.ac.id

Abstract. Mercury (Hg) can cause acute and chronic intoxications in humans, such as damage to the central nervous system, lungs, and kidneys. Goldsmiths pose the risk of exposure to Hg through direct skin contact or inhalation of Hg vapor during the amalgam burning process. This study aims to assess the level of workers' knowledge regarding the risk of exposure to Hg and investigate the practice of preventing health risks from exposure to Hg in household-level gold-crafting activities. The study used a descriptive observational approach. The population of the study was 110 gold craftsmen in the Rapokkaling Village, Makassar City. The number of samples was 52 people determined by the purposive sampling technique. The data collection instruments included a structured questionnaire, camera, recorder, and field note. The results of the data analysis showed that the proportion of respondents with a sufficient level of knowledge was greater than that with a low level of knowledge (65.4% and 34.6%, respectively). On the other hand, the results of the observation showed that 100% of gold craft workers had unsecure work practices. Sufficient knowledge about the risks of exposure to Hg does not guarantee the discipline of implementing secure work practices including using personal protective equipment.

1. Introduction

Indonesia is wide and stretches from Sabang to Merauke. The wealth of mining products is also abundant from the island of Sumatra to Papua[1]. Among all available natural resources, the most promising wealth to be managed is in the mining sector, considering that it only needs to take existing treasures from the earth, absorb a large number of labor, excite so many supporting sectors, and the selling value of the products taken are valuable [1]

Since the case of the mercury accident in Minamata Japan in 1953 which was intensively reported [2], the issue of heavy metal pollution has increased in line with the development of various studies which have begun to be directed at various technological applications to deal with environmental pollution caused by heavy metals at very low concentrations directly and accumulates in the food chain therefore there is a concern of having an impact on human health [3]. As with other sources of environmental pollution, these heavy metals can be transferred in a very far range in the environment, then have the potential to disrupt the life of environmental biota and ultimately affect human health even in the long term and far from the main pollution source [4]

Mercury (Hg) is one of the metals that is included in the teratogenic material [5]. Hg is a dangerous heavy metal, which in even small concentrations can be toxic[6], [7] . In Indonesia, there have been cases of Hgy poisoning in several places, such as the case of pollution in Buyat Bay as a result of pollution from gold mining company Perseroan Terbatas Newmont and the activities of Unlicensed Gold Miners (PETI) polluting rivers in Central Kalimantan [8]. Hg levels in the fish body reached 0.257 mg/l in the Rungan river and 0.676 mg/l in the Kahayan River. Meanwhile, Hg levels at the bottom of the Rungan river are 0.554 mg/l and at the bottom of the Kahayan River are 0.789 mg/l even though the threshold for sediment is only 0.005 mg/l [9]. Meanwhile, in West Kalimantan, Hg pollution cases were found in the Menyuke and Kapuas rivers. The Hg content in the two rivers has exceeded the threshold of 199.27 ppm [10]. In a study on Hg levels in gold miners in Renggas Tujuh Village, Titi District, Ketapang Regency, West Kalimantan, 44.4% of gold mining workers had Hg poisoning in their urine with an average content of 7.6 g/l.9 [11].

Research that has been conducted in Sulawesi, one of which was in Palu, Central Sulawesi, regarding the use of Hg in the gold processing process at the Poboya People's Mining has affected the land around the people's mining [12]. The level of Hg in the air was between 20 m^3 and 40,000 m^3 which threaten the health of the population residing in the region [13]. Meanwhile, research conducted in Bombana showed that the Hg content in the gold processing plant reached 0.0315 mg/l, far exceeding the quality standard at the processing site and at the mining site, which is only 0.005 mg/l [14].

Research that has been conducted in Makassar on the impact of Hg on marine sediments shows that the highest Hg concentration is in Cambayya Village at 1.871 mg/kg and the lowest Hg concentration is in Pulau Lae-Lae Village at 1.550 mg/kg [15]. In general, the heavy metal content of Hg in marine sediments in the coastal area of Makassar City has exceeded the threshold of 0.13 mg/kg which can pose a high risk to human health in the vicinity [15] .

A preliminary study in Rappokalling Village, Makassar City through interviews with one of the gold craftsmen workers in Rappokalling Village found that the gold craftsmen had been around for a long time without wearing the masks. Based on these data, this study aimed to determine the knowledge about health risks related to Hg contamination of the gold craft industry workers in Rappokalling Village, Makassar City. Due to the fact that the knowledge of gold industry workers related to the occupational security and health is a factor that can reduce work accidents, it can be interpreted that the knowledge of gold industry workers is an important part for workers as a form of response to occupational security and health which is influenced by the knowledge of each worker.

2. Methodology

This study uses an observational survey design that aims to measure the level of knowledge, and practice of gold industry workers on health risks related to mercury exposure, especially for household-scale gold craftsmen in Rappokalling village, Makassar City. This research was conducted in Rappokalling Village in 2019, Makassar City with the descriptive observational approach which aims to identify the knowledge and practice related Hg exposure of the gold craftsmen in household scale. Data was obtained from the direct interviews through questionnaires with the total samples of 52 respondents. The data is processed using the SPSS application and applies the stages of editing, coding, scoring, processing and cleaning [16].

3. Result

This research was conducted in one area as a gold processing center in Makassar City since 1924. Rappokalling District supplies gold handicrafts to Sulawesi and Java islands. This research is a descriptive observational research method with the aim of knowing the knowledge of workers, and the practice of preventing health risks related to mercury contamination in household scale gold craftsmen in Rappokalling Village, Makassar City.

Based on Table 1, data obtained in the highest age group is 26-35 years accounting for 29 respondents (55.8%). The number of respondents who had the elementary education as the highest education was 24 respondents (46.2%). The higher number of the respondents worked as the gold craftsmen making accessories which accounts for 34 respondents (65.4%). The data on the highest length of the workgroup is the group of 6-10 years of work as many as 17 respondents (32.7%).

Table 1. Characteristics of respondents in Rappokalling Village, Makassar City.

Characteristics of Respondents	n=52	%
Age (Years old)		
16-25	7	13.5
26-35	29	55.8
36-45	16	30.7
Education		
Elementary School	24	46.2
Junior High School	23	44.2
Senior High School	5	9.60
Occupation		
Accessories maker	34	65.4
Grinder	18	34.6
Working duration (year)		
1-5	5	9.60
6-10	17	32.7
11-15	12	23.1
16-20	9	17.3
21-25	9	17.3

Table 2. Frequency distribution of respondents based on worker knowledge regarding Hg contamination in Rappokalling Village, Makassar City in 2019.

Workers' Knowledge	n	%
Enough	34	65.4
Less	18	34.6
Total	52	100

Table 2 indicates that 34 of 52 respondents (65.4%) have sufficient knowledge and the 18 respondents (34.6%) have less knowledge.

Table 3. Overview of the Frequency Distribution of Worker Risk Prevention Practices in Rappokalling Village, Makassar City, 2019.

Workers' Practice	n	%
Secure	0	0
Unsecure	52	100
Total	52	100

Based on Table 3, the data obtained from 52 respondents indicated that risk prevention practices employed workers Rappokaling goldsmiths at all insecure as many as 52 respondents (100%), since it is not using PPE.

Table 4. Overview of the knowledge of the workers with risk prevention practices in Rappokalling Village, Makassar City, 2019.

Workers' knowledge	Workers' Risk Prevention Practices		Total
	Secure	Unsecure	

	n	%	n	%	n	%
Enough	0	0	34	65.4	52	100
Less	0	0	18	34.6	52	100
Total	0	0	52	100	52	100

Table 4 shows that unsecure work practices were performed by 34 respondents (65.4%) who are knowledgeable enough and they had the practice of preventing worker risks. Another 18 respondents (34.6%) had less knowledge and also engaged in unsecure worker risk prevention practices.

4. Discussion

4.1. Knowledge Level

The results showed that from 52 respondents there were 34 (65.4%) respondents who belonged to the category of sufficient knowledge and 18 (34.6%) of respondents who belonged to the category of insufficient knowledge. This shows that a person's knowledge not only be seen from the level of education but can also be obtained or is influenced by several factors among which are the external factors including the environmental factor or factors of habit [17].

According to the researcher, although many of the respondents have low education, they are used to working as gold craftsmen. It can be seen from the results of interviews with respondents as many as 17 (32.7%) of respondents who had been working for 6-10 years even 9 (17.3%) of respondents who worked for 21-25 years. Although many are quite knowledgeable, due to the habits or environmental factors, they do not care about the dangers of mercury or they think everything is normal [18].

In addition, respondents with the highest recent education were Elementary School as many as 24 (46.2%) respondents who were in the early phase before adolescence where they began to get to know the environment around them. So, this can trigger their curiosity about the work that is being done around them that will benefit them. Without thinking about the risks that will occur so that they do not use PPE at work [19].

4.2. Risk Preventing Practices

The results showed that from 52 respondents there were 52 (100%) respondents who were not secure. This is because many gold craftsmen do not use PPE. Therefore, according to researchers with no use of PPE, respondents or goldsmith workers in the village Rappokkaling Makkasar city can easily be directly contaminated with Hg.

The results also showed that of the 52 respondents, 34 respondents (100%) worked as accessories makers and the lowest was respondents who worked as grinders as many as 18 (34.6%) respondents. According to researchers, the types of work that are at risk of unsecure with Hg are those who work as grinders or those that separate gold from soil because they use Hg to separate gold from soil. Meanwhile, those who only work as accessories makers do not use Hg in their work because they only process gold into accessories such as rings, etc., although there is a possibility that they will also be contaminated with mercury if the environment around them is at risk, especially the craftsmen who do not use PPE [20].

Most miners who have work activities in the form of mixing Hg have the highest percentage of receiving mercury poisoning [21]. Workers are at risk for direct exposure to Hg. Such exposure can occur at the stage of mixing Hg and squeezing [22]. Furthermore, exposure can also occur in the combustion process where the Hg vapor from the combustion can be inhaled directly by the workers [23].

5. Conclusion

This study assessed the knowledge level of goldsmith workers as well as the practices of PPE used in work place to prevent the Hg contamination. According to the results of workers' knowledge and worker practices, all workers who have sufficient knowledge are lack of occupational security in their daily works. However, unsecure work practices can cause the gold craftsmen to be exposed to or contaminated with Hg, especially most of the gold craftsmen who do the milling can be directly contaminated with mercury during gold separation. This is caused by habits or environmental factors. Respondents with sufficient knowledge do not guarantee that workers are safe from mercury craftsmen but must also be seen from the understanding of work practices or the use of personal protective equipment while working.

Recommendation

The results of this study recommend to the government to help goldsmiths to create a household gold craft industry model that is environmentally friendly and does not endanger public health. Goldsmiths should pay more attention to occupational health and safety aspects in the gold processing process, discipline to use personal protective equipment to minimize Hg exposure. In addition, people living around the household-scale gold processing industry participate in reducing the impact of mercury exposure by adopting a clean, healthy and safe lifestyle.

References

[1] Charisman F G, Nurul H and Helmawati H 2021 ANALISIS PENGEMBANGAN SEKTOR PARIWISATA KEC.SIBERUT BARAT DAYA KABUPATEN KEPULAUAN MENTAWAI SEBAGAI DESTINASI UNGGULAN DI SUMBAR

[2] Ninomiya T, Ohmori H, Hashimoto K, Tsuruta K and Ekino S 1995 Expansion of Methylmercury Poisoning Outside of Minamata: An Epidemiological Study on Chronic Methylmercury Poisoning outside of Minamata *Environmental Research* **70** 1 pp 47-50 doi: 10.1006/ENRS.1045

[3] Ekino S, Susa M, Ninomiya T, Imamura K and Kitamura T 2007 Minamata disease revisited: An update on the acute and chronic manifestations of methyl mercury poisoning *Journal of the Neurological Sciences* **262** 1–2 pp 131-144 doi: 10.1016/J.JNS.2007.06.036

[4] Wingfield J C 2013 Ecological processes and the ecology of stress: the impacts of abiotic environmental factors *Functional Ecology* **27** 1 pp 37-44 doi: 10.1111/1365-2435.12039

[5] Crespo-López M E *et al.*, 2009 Mercury and human genotoxicity: Critical considerations and possible molecular mechanisms *Pharmacological Research* **60** 4 pp 212-220 doi: 10.1016/J.PHRS.2009.02.011

[6] JDuruibe J O, Ogwuegbu M O C, and Egwurugwu 2007 Heavy metal pollution and human biotoxic effects *International Journal of Physical Sciences* **2** 5 pp 112–118

[7] Armah F A, Quansah R, and Luginaah I 2014 A Systematic Review of Heavy Metals of Anthropogenic Origin in Environmental Media and Biota in the Context of Gold Mining in Ghana *International Scholarly Research Notices* **2014** pp 1-37 doi: 10.1155/2014/252148

[8] Geosciences | Free Full-Text | Mercury Contamination of Cattle in Artisanal and Small-Scale Gold Mining in Bombana, Southeast Sulawesi, Indonesia. https://www.mdpi.com/2076-3263/7/4/133 (Accessed on Dec. 23, 2021)

[9] Pencemaran G Merkuri terhadap Masalah Kesehatan Penambang dan Masyarakat di Sekitar Aliran Sungai Behe Bulan Juli -Agustus, Venny Grishela V and E. Tamba 2016 Gambaran Pencemaran Merkuri terhadap Masalah Kesehatan Penambang dan Masyarakat di Sekitar Aliran Sungai Behe Bulan Juli *Jurnal Kedokteran Meditek* **23** 61 doi: 10.36452/JKDOKTMEDITEK.V23I61.1459

[10] Mona R K, Pontoh J and Yamlean P V Y 2018 Analisis Kandungan Merkuri (Hg) pada Beberapa Krim Pemutih Wajah Tanpa Ijin BPOM yang Beredar di Pasar 45 Manado *PHARMACON*, vol. 7 no. 3 doi: 10.35799/PHA.7.2018.20454

[11] Reza, Karimuna S R and Fachlevy A F 2016 Analisis Perbedaan Potensi Risiko Keterpaparan Merkuri pada Masyarakat di Desa Tahi Ite Kecamatan Rarowatu Kabupaten Bombana *JURNAL ILMIAH MAHASISWA KESEHATAN MASYARAKAT* **1** 4 pp 1-13 (Accessed onDec. 23, 2021) [Online]

[12] Basri, Sakakibara M and Sera K 2020 Mercury in Soil and Forage Plants from Artisanal and Small-Scale Gold Mining in the Bombana Area, Indonesia *Toxics* **8** 1 p 15 doi: 10.3390/TOXICS8010015.
[13] Basri, Sakakibara M and Sera K 2017 Current Mercury Exposure from Artisanal and Small-Scale Gold Mining in Bombana, Southeast Sulawesi, Indonesia—Future Significant Health Risks *Toxics* **5** 1 p 7 doi: 10.3390/TOXICS5010007.
[14] Basri, Sakakibara M and Ratnawati 2017 Economic features of the artisanal and small-scale gold mining industry in Bombana, Southeast Sulawesi, Indonesia *IOP Conference Series: Earth and Environmental Science* **71** p 012016 doi: 10.1088/1755-1315/71/1/012016
[15] Aenia N 2017 Biokumulasi logam berat merkuri (Hg) pada ikan ekor kuning (caesio cuning) di perairan pulau Lae-lae Makassar
[16] PASANDA A G 2016 PERBEDAAN PENGETAHUAN, SIKAP DAN PERILAKU PENJAMAH MAKANAN SESUDAH DIBERIKAN PENYULUHAN PERSONAL HYGIENE DI HOTEL PATRA JASA SEMARANG
[17] Pratiwiuniversitas N K 2017 PENGARUH TINGKAT PENDIDIKAN, PERHATIAN ORANG TUA, DAN MINAT BELAJAR SISWA TERHADAP PRESTASI BELAJAR BAHASA INDONESIA SISWA SMK KESEHATAN DI KOTA TANGERANG *Pujangga* **12** p 31 doi: 10.47313/PUJANGGA.V1I2.320
[18] Mastuti A 2017 PROBLEMATIKA SAMPAH DAN UPAYA DAI UNTUK MENJAGA KEBERSIHAN LINGKUNGAN DESA WAY HUWI KECAMATAN JATI AGUNG KABUPATEN LAMPUNG SELATAN
[19] Muhamid R, Tambunan W and Fatimahhayati L D 2018 Analisis Risiko Keselamatan dan Kesehatan Kerja Kegiatan Bongkar Muat Pupuk," *Jurnal INTECH Teknik Industri Universitas Serang Raya* **4** 2 pp 45-52 doi: 10.30656/INTECH.V4I2.924
[20] Junita N R 2014 Risiko keracunan Merkuri (Hg) pada pekerja Penambangan Emas Tanpa Izin (PETI) di desa Cisarua Kecamatan Nanggung Kabupaten Bogor Tahun 2013 (Accessed on Dec. 23, 2021). [Online]. Available: https://repository.uinjkt.ac.id/dspace/handle/123456789/24270
[21] Gafur N A, Sakakibara M, Sano S and Sera K 2018 A Case Study of Heavy Metal Pollution in Water of Bone River by Artisanal Small-Scale Gold Mine Activities in Eastern Part of Gorontalo, Indonesia *Water 8* **10** 11 p 1507 doi: 10.3390/W10111507
[22] Abbas H H, Sakakibara M, Sera K and Arma L H 2017 Mercury Exposure and Health Problems in Urban Artisanal Gold Mining (UAGM) in Makassar, South Sulawesi, Indonesia *Geosciences* **7** 3 p 44 doi: 10.3390/GEOSCIENCES7030044
[23] Ekawanti A and Krisnayanti B D 2015 Effect of Mercury Exposure on Renal Function and Hematological Parameters among Artisanal and Small-scale Gold Miners at Sekotong, West Lombok, Indonesia *Journal of Health and Pollution* **5** 9 pp 25-32 doi: 10.5696/2156-9614-5-9.25

Study of Rhodamine B Dyes Content in Snacks of Karuwisi Traditional Market Makassar, South Sulawesi, Indonesia

Andi Tilka Muftiah Ridjal[1], Andi Yulia Kasma[1], Aminullah[1] and Basri[1]

[1] Sekolah Tinggi Ilmu Kesehatan Makassar, Jl. Maccini Raya No. 197 Makassar 90231, South Sulawesi Indonesia

Corresponding Author: tilka.muftiah@gmail.com

Abstract. Snack is food that is purchased in a ready-to-consume form and is widely preferred because of its relatively cheap price, easy to obtain, attractive and varied appearance. Rhodamine B dye is one of the colouring substances prohibited for food and is declared as hazardous substance according to Regulation of Health Minister Indonesia No. 722/Menkes/Per/IX/1988 regarding dyes that are declared dangerous and prohibited in Indonesia as well as in the Decree of the Directorate General of POM No. 239/Menkes/Per/V/1985 regarding certain dye stuffs declared as hazardous materials. This study aimed to recognize the levels of Rhodamine B dye in snacks traded. This was descriptive survey research with laboratory examination by analysing Rhodamine B dye on snacks by the results of UV-Vis spectrophotometric method. Population in this study were all snacks marketed in Karuwisi Traditional Market, while the sample was nine snacks consisting of: chili sauce, layer cake, red chocolate cream, red curly crackers, cassava chips, red jelly, red popcorn, cassava crackers, and macaroni crackers because the nine samples were physically identical to Rhodamine B colouring agent, namely the use of a striking red or pink colour. The sampling technique was total sampling. The results revealed that of the nine samples, only one sample was positive for Rhodamine B, namely cassava cracker containing 7.960µg/g. It concluded that red cassava cracker was a wholesale snack at Karuwisi Traditional Market which is positive for Rhodamine B.

1. Introduction

Snack foods play an important role in providing energy and other nutrients for consumers. Globally, food-borne diseases are reported to increase significantly every year [1]. WHO data reveals that more than 50% of cases of diarrhea occurred are caused by contaminated food and lead to 550 million illnesses and 230,000 deaths [1][2]. In addition, Africa and Southeast Asia are regions that perform a high incidence and mortality rate due to food poisoning. Developing countries have a 4 times higher risk of experiencing food poisoning than developed countries. This risk is caused by low community hygiene such as inadequate water preparation, low levels of community knowledge and unsupported food safety regulations [2].

Snacks are one of the ready-to-eat foods found in traditional market and are routinely consumed by most people. According to the National Poisoning Information Center (SikerNas) of the Food and Drug Monitoring Agency of Indonesia (BPOM-RI), food poisoning in Indonesia based on case reports since 2014 tends to fluctuate. In 2014 the incidence of food poisoning was 974 cases and tended to decrease to 697 cases in the following year. Meanwhile, in 2016 food poisoning increased to 791 cases. Throughout July to September 2017, SIKerNas collected incident cases from online mass media registered with the press council. Incidents of poisoning caused by snacks of street vendors were 6 with

88 victims, packaged processed foods were 2 with 37 victims, fresh food was 1 with 7 victims and 1 of them died, and unknown cause was 1 with 7 victims and 1 of them died [3]. Data from the Food and Drug Supervisory Agency (BPOM) in 2012 in Paratmanitya and Aprillia (2016) depicted that the incidence of poisoning due to consuming food was highest position by 66.7%. One of the causes of food poisoning is the presence of chemical contaminants in food, such as borax, formalin and rhodamine. In these data, there were 2.93% of snack food samples in school children containing borax, 1.34% of formalin, and 1.02% of rhodamine-B [4].

This study aimed to recognize the levels of Rhodamine B dye in snacks at Karuwisi Traditional Market Makassar. While the specific purpose was determining the contains of Rhodamin B dye in nine samples of snacks consisted of: chili sauce, layer cake, red chocolate, red curly crackers, cassava chips, red jelly, red popcorn, cassava crackers, and macaroni crackers.

2. Materials and Methods

The type of research is a descriptive survey with laboratory tests to determine the content of Rhodamin B dye in snacks. Sample examination will be carried out at Health Laboratory Center Makassar (BBLK).

The population is all snacks sold in Karuwisi Traditional Market. While the sample is nine types of snacks such as: chili sauce, layer cake, red chocolate, red curly crackers, cassava chips, red jelly, red popcorn, cassava crackers, and macaroni crackers, because all of these samples are physically identical to the dye Rhodamine B, with bright red or pink color, selected by total sampling.

The procedure for bringing samples of snacks from the market to the laboratory is as follows:

a. Initial recording of all types of sampled snacks.
b. Samples are packed in sterile and dry plastic containers (Zip Lock).
c. The container is labeled including the name of samples and codes.
d. The container is packed safely during the transportation process.

This research procedure includes:

a. Observation
   In this study, researchers observed using a planned procedure, namely seeing and recording the number and level of certain activities related to the problem.
b. Documentation
   This aimed to find data concerning variables in the form of notes, transcripts, books, newspapers, magazines, agendas and so on. In this study, researchers took data through written and electronic documents using a camera.
c. Analysis
   Coloring Rhodamine B on snacks was analyzed in UV-Vis spectrophotometry method.

To determine the content of dye, Rhodamine B color standard series was used.

a. Tools:
   1) Erlenmeyer 250 ml, 500 ml and 1000 ml
   2) 100 ml and 1000 ml beakers
   3) Glass funnel
   4) Pipette
   5) Stirring rod
   6) Stamper mortar
   7) Measuring cup
   8) Analytical balance
   9) Hot plate
   10) Fat-free wool yarn
   11) Whatman filter paper
   12) UV-Vis Spectrophotometry
b. Materials/reagents used:
   1) Acetic acid

2) Ammonia
3) Ethanol 70%
4) Food dye standard solution (Rhodamine B)
5) Aquadest water
6) Elution solution (n butane : ethyl acetate : ammonia = 10 : 4: 5)

Furthermore, the samples that have gone through the extraction step with standard procedures are measured by UV-Vis Spectrophotometry to assess the content of Rhodamine B. Calibration curve is applied with the regression equation $y = bx \pm a$ in order to calculate the levels of Rhodamine B.

3. Result and Discussion

There are nine types of snack samples suspected of containing Rhodamine B because of their striking reddish or pink color, namely: chili sauce, layer cake, red chocolate, red curly crackers, cassava chips, red gelatin/jelly, red popcorn, cassava crackers and macaroni crackers. The process of laboratory examination on samples took a long time at Health Laboratory Center Makassar. This is due to the large number of samples examined at Health Chemical Installation during the COVID-19 pandemic. The results of the laboratory examination were issued on November 10, 2020 with the following results:

Table 1. Results of Examination of Rhodamin B in Snacks at Karuwisi Traditional Market

Samples	Rhodamine B (μg/g)
Cassava crackers	7,960
Red curly crackers	Negative
Red popcorn	Negative
Red jelly	Negative
Layer cake	Negative
Cassava chips	Negative
Macaroni crackers	Negative
Chilli sauce	Negative
Red chocolate	Negative

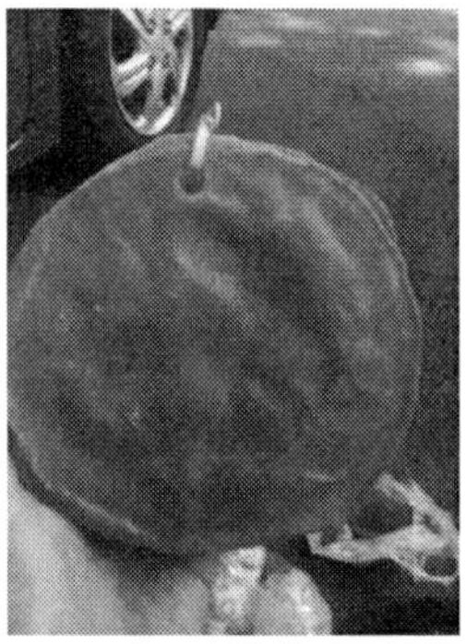

Figure 1. Cassava crackers

The results of the laboratory examination showed that from the nine samples, there was only one positive sample containing Rhodamine B, namely cassava cracker by 7.960μg/g.

Snack according to FAO (Food and Agriculture Organization) is food prepared and sold by street vendors on the streets and in other public crowded places which are directly eaten or consumed without further processing or preparation [5][6]. Food poisoning is the emergence of clinical symptoms of a disease or other health problems due to contaminated food. Foods that cause poisoning are usually contaminated with physical, microbial or chemical elements in dangerous doses. This condition is due to food management that does not meet health requirements and ignore the rules of food hygiene and sanitation. There are several causes of food poisoning, namely [7]:

a. Natural food ingredients, namely foods that naturally contain toxins, such as poison mushrooms, puffer fish, green cassava, poison yam.
b. Microbial infection, which is caused by bacteria in the digestive tract of food that enter the body or ingestion of microbes in large numbers, which then live and multiply, such as *Salmonellosis* and *Streptococcus*.
c. Microbial toxins, namely toxins produced by microbes in food that enter the body in dangerous amounts.
d. Chemicals, namely hazardous materials in food that enter the body in dangerous amounts, such as arsenic, antimony, cadmium, pesticides with symptoms of respiratory depression to coma or die.
e. Allergies, namely allergens in food that cause sensitive reactions to susceptible people, such as histamine in shrimp, tuna, cooking spices and so on.

The use of synthetic dyes in food, although it has a positive impact on producers and consumers, including making a food more attractive, evens out the color of food, and restores the color of the basic ingredients that are lost or changed during processing, but it may also damage human health.

Rhodamine B is a synthetic dye in the form of a crystalline powder, green or reddish purple in color, odorless, and a bluish red/strong fluorescence solution in water and comes from the cationic dyes group. Rhodamin B is widely used as a dye for paper, textiles, wool, silk and as a reagent for the analysis of antimony, cobalt, bismuth and others [8].

a. Symptoms when exposed to Rhodamine B:
    1) If swallowed, it may cause irritation to the digestive tract and symptoms of poisoning and red or pink urine.
    2) If in contact with skin, it may cause irritation to the skin.
    3) If in contact with eyes, it may cause eye irritation, redness, or edema of the eyelids.
    4) If inhaled, it may cause irritation to the respiratory tract.
b. The characteristics of foods containing Rhodamine B:
    1) The color looks bright (colorful).
    2) There is a slight bitter taste (especially in syrup or lemonade).
    3) There is itching in the throat after consuming it.
    4) The smell is not natural as food.
c. Actions when exposed to Rhodamine B:
    1) If exposed to skin, remove clothing, jewelry, or shoes contaminated with Rhodamine B. Wash skin with soap and running water until clean for approximately 15 to 20 minutes, if necessary, contact a doctor.
    2) In case of contact with eyes, rinse with running water or physiological saline solution, blinking the eyes until it is confirmed that the remaining Rhodamine B is no longer present/clean, if necessary, consult a doctor.
    3) If swallowed and vomiting occurs, place the head lower than the hips to prevent vomiting entering the respiratory tract.
    4) If the victim is unconscious, tilt the head to the side or to one side.

The government has issued a regulation regarding the violation of the use of Rhodamine B in medicine, food and cosmetics through the Minister of Health Regulation No. 239/MenKes/Per/V/Year 1985 concerning Certain Dyes Declared as Hazardous Substances. In addition, specifically for food, Ministerial Regulation No. 722/MenKes/Per/IX/Year 1988 concerning Food Additives stipulates that these compounds may not be used as food coloring. For food coloring, it is recommended to use natural or synthetic dyes based on Ministerial Regulation No. 722/MenKes/Per/IX/Year 1988 concerning Food Additives [8].

The Food and Drug Supervisory Agency (BPOM) has socialized the public to be smart in choosing food. By knowing the characteristics of food with Rhodamine B, it is hoped that people should aware in choosing foods that are safe for their health. Foods containing Rhodamine B will have a striking red/pink color, uneven coloration, and slightly bitter taste if consumed. Foods that have certification from BPOM will be guaranteed not to contain harmful ingredients, while products containing Rhodamine B will not include codes, labels, brands, or distribution permits from BPOM [9].

Rhodamine B compounds show both acute and chronic effects if ingested. In acute conditions, it can be irritating to the digestive tract and may cause toxic effects. In addition, exposure from vegetable dyes containing Rhodamine B in excess may lead to red or pink urine. While the chronic impact, if Rhodamine B is widely absorbed by the digestive tract, it will show strong protein binding. High concentrations in food if given to rats can cause liver damage [8].

Crackers are one of the favorite foods by many people, while producers often add various kinds of colors to crackers, especially red in order to attract attention. The dyes applied may come from natural and synthetic dyes, but there are still manufacturers which use prohibited synthetic dyes.

Identification of Rhodamine B in crackers in North Jakarta Market also showed that of the 24 samples of crackers identified, there were two samples that were positively identified using Rhodamine B dye. Onion cracker Cap Kambing Kembar from Sunter Podomoro Traditional Market and onion cracker (large curly round) from Warakas Traditional Market because visually they show a pink stain and fluoresces under UV light 254nm and its Rf value is same as the comparison standard for Rhodamine B, each of which is 0.8 for eluent I and 0.89 for eluent II. Therefore, the sample was declared positive for containing synthetic dye Rhodamine B [10].

Hedie and jusnita (2017) stated that the use of these prohibited compounds by producers is due to the price being more affordable, attractive and durable compared to natural dyes. In addition, the color appearance produced by natural dyes is very different from synthetic dyes which is very prominent and attracts consumers, while natural dyes produce a pale red so that consumers are less interested, and difficult to obtain. In contrast to synthetic dyes, many are found on the market at affordable prices and their use is quite practical [10].

Widaryanto (2018) also declared that the lack of knowledge about the dangers of synthetic dye additives in food by producers is a factor that is often found in the field. Of the 10 samples of colored crackers studied, there were six samples with negative results (-) which were indicated by a clear color after passing the chromatographic test, while the other four samples showed positive results (+) which were indicated by pink color. This is because most of the cracker producers in Tanjung Anyar Traditional Market Mojokerto, have switched to using permitted food colorings such as anthocyanin, curcuma, carmine, caramel, erythrosine, and tartrazine [11].

The content of Rhodamin B in cassava crackers has also been found in Malang City Traditional Markets. There are 6 samples of cassava crackers, 2 samples from each of 3 markets, namely Besar Market, Sukun Market and Kedungkandang Market. The results revealed that 2 samples were positive contained Rhodamine B [12]. Similarly, red crackers at Antasari Traditional Market Banjarmasin, found 1 out of 5 samples (korek api cassava crackers) [13]. Otherwise, different results were shown in samples of prawn crackers at Masomba Traditional Market in Palu, which were safely to consume because all the crackers sold did not contain Rhodamin B [14].

Therefore, consumers who have the right to food safety should know the dangers of food products contaminated with toxic hazardous materials. Zazili and Hartono (2016) have implemented a consumer empowerment model against the dangers of food products contaminated with chemical additives through direct counseling to the public using lectures, discussions, and audio-visual presentations [15].

The information submitted is in the form of ways to identify food products that are indicated to be contaminated or use food additives that are prohibited such as Formalin, Borax, Rhodamin B, and Metanil Yellow. In order to increase legal knowledge, knowledge of consumer rights is regulated and guaranteed by Law Number 8 of 1999 concerning Consumer Protection. It is essential to convey information regarding consumer rights so that consumers are able to know and understand that the act of using prohibited food additives is an act that violates the law, and then consumers may sue business actors who violate consumer rights. Awareness, legal compliance and consumer behavior also determine the practice of law protection and enforcement [15].

4. Conclusions

It concluded that red cassava crackers are snacks sold at Karuwisi Traditional Market Makassar which are positive for Rhodamin B dye. It is encouraged to consumers to aware of the dangers of red snacks which may contain Rhodamine B. Moreover, traders should not use Rhodamine B as a food additive because of the numerous acute and chronic impacts that can damage health. In addition, BPOM should periodically carry out inspections to snacks or foods suspected containing hazardous materials in traditional markets.

Acknowledgements

This project was funded by Indonesian Ministry of Research, Technology and Higher Education (Kemenristekdikti) as well as technical support from Health Laboratory Center Makassar and School of Health Science (STIK) Makassar.

References

[1] N. A. Abdul-Mutalib, A. N. Syafinaz, K. Sakai, and Y. Shirai 2015 An overview of foodborne illness and food safety in Malaysia *Int. Food Res. J.*, **22** 3 pp 896–901
[2] WHO 2015 WHO's first ever global estimates of foodborne diseases find children under 5 account for almost one third of deaths, Geneva.
[3] BPOM 2016 Sentra Informasi Keracunan (SIKER) Nasional
[4] Y. Paratmanitya and A. Veriani 2016 Kandungan bahan tambahan pangan berbahaya pada makanan jajanan anak sekolah dasar di Kabupaten Bantul," *J. Gizi dan Diet. Indones. (Indonesian J. Nutr. Diet.* **4** 1pp 49
[5] H. S. P. Manalu and A. Suudi 2017 Kajian Implementasi Pembinaan Pangan Jajanan Anak Sekolah (PJAS) untuk Meningkatkan Keamanan Pangan: Peran Dinas Pendidikan dan Dinas Kesehatan Kota," *Media Penelit. dan Pengemb. Kesehat.* **26** 4 pp 249–256
[6] A. Pakhri, C. Chaerunnimah, and R. R 2018 Edukasi Gizi terhadap Pengetahuan dan Kebiasaan Jajan pada Siswa SMP Negeri 35 Makassar *Media Gizi Pangan* **25** 1pp 77
[7] S. Rahmi 2018 Cara Memilih Makanan Jajanan Sehat dan Efek Negatif yang Ditimbulkan Aapabila Mengkonsumsi Makanan Jajanan yang Tidak Sehat Bagi Anak-anak Sekolah Dasar
[8] Badan POM RI 2008 Rodamin B (Rhodamine B)
[9] Y. S. Putri 2016 Uji Kualitatif Penggunaan Rhodamin B Pada Saus Cilok Di Pedagang Kaki Lima Se-Kota Mataram
[10] Hedie and N. Jusnita 2017 Identifikasi Rhodamin B pada Kerupuk yang Beredar di Pasar Jakarta Utara dengan Metode Kromatografi Lapis Tipis, Jakarta
[11] B. E. Widaryanto 2018 Identifikasi Pewarna Rhodamin B pada Kerupuk Berwarna yang Dijual di Pasar Tanjung Anyar Kota Mojokerto, Sekolah Tinggi Ilmu Kesehatan Insan Cendekia Medika Jombang
[12] U. Wuryani 2013 Identifikasi Rhodamin B pada Kerupuk Singkong dengan Metode Kromatografi Lapis Tipis dan Densitometri (di Pasar Kota Malang), Malang
[13] A. Khufah, R. E. D.Y, and N. Aisyah 2017 Identifikasi dan Penetapan Kadar Rhodamin B dalam Kerupuk Berwarna Merah yang Beredar di Pasar Antasari Kota Banjarmasin
[14] M. Farid, A. Bungawati, and A. R. A. C. Noor 2019 Analisis Kandungan Rhodamin B Pada Kerupuk Udang Di Pasar Masomba Palu," *J. Kolaborasi Sains* **1** 1 pp 597–604
[15] A. Zazili and Hartono 2016 Model Pemberdayaan Konsumen terhadap Ancaman Bahaya Produk Pangan Tercemar Bahan Berbahaya Beracun di Provinsi Lampung," *J. Huk. IUS QUIA IUSTUM* **23** 3 pp 391–414

The influence of decision making of farmers choosing rice varieties on rice productivity levels in Central Lampung Regency

K K Rangga[1], H Yanfika[1], R A Mangesti[1], Rudy[2], R Bursan[3], A Rahmat[4], and D A Adipathy[5]

[1] Study Program of Agriculture Extension, Faculty of Agriculture, University of Lampung, Sumantri Brojonegoro Street 1, Bandar Lampung, Lampung, Indonesia, 35145
[2] Department of Constitutional Law, Faculty of Law, University of Lampung, Sumantri Brojonegoro Street 1, Bandar Lampung, Lampung, Indonesia, 35145
[3] Department of Management Economics, Faculty of Economics and Business, University of Lampung, Sumantri Brojonegoro Street 1, Bandar Lampung, Lampung, Indonesia, 35145
[4] Research Center for Limnology, The Indonesian Institute of Science, Bogor Highway Jakarta, Cibinong, Bogor, West Java 16911
[5]Department of Magister Economics, Faculty of Economics and Business, University of Lampung, Sumantri Brojonegoro Street 1, Bandar Lampung, Lampung, Indonesia, 35145

Corresponding Author: korrangga@gmail.com

Abstract. This study aims to determine the influence of decision making of farmers choosing rice varieties on rice productivity levels. The data analysis method used in this study is descriptive and quantitative analysis. This research was conducted in Bangun Rejo Subdistrict, Central Lampung Regency. The number of respondents in this study was 73 samples. The data used is primary and secondary data. The data analysis in this study was conducted with multiple linear regression tests. The results showed that the decision-making process of choosing rice varieties by farmers is done by considering several things, namely the suitability of varieties to the needs of farmers, the suitability of varieties with the experience of farmers, the level of complexity in the application of varieties, as well as the relative advantages of varieties. The decision-making of farmers choosing rice varieties positively affects 90.4 percent of rice productivity. The more appropriate farmers choose rice varieties that suit the needs, conditions of the land, type and intensity of pest attacks in the area, it will support the acquisition of high productivity.

1. Introduction

Rice is a very important food crop in the world after wheat and corn. Indonesian people make rice as one of the main food sources in the form of rice consumed daily. Various efforts have been made to increase rice production, but still not able to meet the national rice needs. One of the causes of unmet national rice needs is due to the decrease in the amount of production in October to December 2018.

Rice production produced in October to December 2018 was 3,900,000 tons while consumption in October to December 2018 amounted to 7,450,000 tons, resulting in a shortage of 3,510,000 tons. This decline continues until 2019. There was a decrease in rice production by 7.75 percent or by 2,630,000 tons. The decrease in rice production is due to a long drought and not optimal cultivation efficiency conducted by farmers [1].

Bangun Rejo subdistrict is one of the sub-districts in Central Lampung Regency and is the sub-district with the highest rice productivity rate with rice productivity of 53.59 tons per hectare [2]. The high productivity rate is due to farmers who have started to apply technology to increase production. One of the technologies used is the use of superior rice seed varieties as a production component. There are three types of rice varieties grown by farmers, namely hybrid superior rice varieties (HSV), superior rice varieties inhibrida, and local rice varieties. Based on data by the Department of Agriculture Food Crops and Horticulture (2019) for the last five years farmers in Bangun Rejo Subdistrict dominant to plant superior rice varieties inhibrida. inhibrida superior rice varieties have a potential yield of 8–10 tons per hectare, but based on data by the Center for Agricultural Extension Bangun Rejo District (2017) it is known that the rice harvest obtained by farmers in Bangun Rejo Subdistrict is 4–5 tons per hectare [3]. Based on this, it is known that there is a gap between the potential that should be produced and the results obtained by farmers, but this does not change the decision of farmers to continue to use rice seed varieties inhibrida as inputs in their farming [4].

The act of choosing and using rice varieties begins with the decision-making process in choosing rice varieties has a very important role in farming activities. This is because varieties contribute the most in increasing rice production by 56.1 percent [5]. Decision making is influenced by social, educational, and economic environmental factors. Based on this, research needs to be done on "Decision Making Farmers Choose Rice Varieties in Bangun Rejo Subdistrict, Central Lampung Regency" to know the influence of decision-making of farmers choosing rice varieties on rice productivity levels.

2. Research Methods

This research was conducted in Bangun Rejo Subdistrict, Central Lampung Regency. The determination of the location of the research was done deliberately (purposive) with the consideration that Bangun Rejo Subdistrict is the sub-district with the highest rice productivity level in Central Lampung Regency. The research was conducted using survey methods and involved as many as 73 respondents. The types of data used in research are primary data and secondary data. Primary data is obtained through direct interviews with respondents using questionnaires (list of questions) that have been prepared. Secondary data was obtained from agencies related to research such as the Central Bureau of Statistics, the Department of Agriculture for Food Crops and Horticulture, and the Agricultural Extension Center of Bangun Rejo Subdistrict. Respondents in this study were rice farmers spread in Sidomulyo Village, Sinar Luas Village, Tanjung Jaya Village, and Tanjung Pandan Village. The population is 1,592. The sample determination refers to Sugiarto (2003) as follows:

$$n = \frac{NZ^2S^2}{Nd^2 + Z^2S^2}$$

$$\frac{1.592(1,96)^2(0,05)}{1.592(0,05)^2 + (1,96)^2(0.05)}$$

$$\frac{305,79}{3,98 + 0,19}$$

= 73

Information:

n = number of samples

N = number of rice farmers population

d = sample variation (5%= 0.05)

Z = confidence level (95%= 1.95)

S = degree of deviation (5%=0.05)

Based on the calculation above obtained the number of samples obtained as many as 73 respondents, then taken the number of respondents proportionally refers to Nazir (1988) with the following formula:

$$ni = \frac{Ni}{N} x\ 73$$

$$n1 = \frac{423}{1592} x73 = 19$$

$$n2 = \frac{302}{1592} x73 = 14$$

$$n3 = \frac{520}{1592} x73 = 24$$

$$n4 = \frac{347}{1592} x73 = 16$$

Information:
n1= Number of samples in Sidomulyo Village
n2= Number of samples in Sinar Luas Village
n3= Number of samples in Tanjung Jaya Village
n4= Number of samples in Tanjung Pandan Village

Sampling is done using simple random sampling using raffle. The number of samples for Sidomulyo Village is as many as 19 farmers and the number of samples for Sinar Luas Village is as many as 14 farmers. The number of samples for Tanjung Jaya Village is as many as 24 farmers and the number of samples for Tanjung Jaya village is as many as 16 farmers [6]. The data analysis method used in this study is descriptive and quantitative analysis. Descriptive analysis is used to look at the stages of decision-making of farmers in choosing rice varieties. Quantitative analysis is used to determine the factors that influence farmers' decisions in choosing rice varieties. Quantitative analysis is conducted using multiple linear regression analysis tests [7]. The multiple linear regression equations defined are as follows:

$$Y = a+b1X1+b2X2+b3X3+b4X4+\varepsilon$$

Information:
Y = Decision of farmers to choose varieties
a = Coefficient of constants
b1= Regression coefficient
X1 = individual factors
X2 = environmental factor
X3 = social factors
X4 = economic factors
ε = Error, variable interference

3. Result and Discussion

3.1. Respondents' General Circumstances

This study involved respondents as many as 73 rice farmers in Bangun Rejo Subdistrict, Central Lampung Regency. Respondents were rice farmers spread in Sidomulyo Village, Sinar Luas Village, Tanjung Pandan Village, and Tanjung Jaya Village with proportional sampling for Sidomulyo Village as many as 19 farmers, Sinar Luas Village as many as 14 farmers, Tanjung Jaya Village as many as 24 farmers, and Tanjung Pandan Village as many as 16 farmers. The identity of respondents in this study included age, level of formal education, and area of arable land. Here are the identities of respondents in this study.

Table 1. Division Identity of the respondent.

	Classification	Total	Percentage (%)
Age	0-14 age	0	0
	15-64 age	63	86.31
	>65 age	10	13.69
Education	Primary School	23	31.50
	Junior High School	33	45.20
	Senior High School	17	23.30
Area of arable land	0,50-1,00 Ha	44	60.27
	1,01-1,50 Ha	15	20.55
	1,51-2,00 Ha	14	19.18

Based on Table 1 it is known that most respondents are in the productive age range with an age range of 15-64 years. Of the 86.31per cent in the productive age range and 13.69 per cent in the non-productive age range in addition, the majority of respondents were in junior high school education with a percentage of 45.20 per cent, and as many as 31.50 per cent of respondents were in primary school education level, 23.30 per cent were in high school level. The level of education will affect the ability of farmers to capture the information provided. Education serves to develop abilities, improve the quality of life and human dignity both individually and socially [8].

Based on the table above, it is also known that most of the respondents have a land area with a narrow classification of 60.27 percent and the remaining 20.55 percent have land with a medium classification, 19.18 percent have land with a wide classification. This is due to the area of land that is divided through generations, causing the area of land owned will be narrower for each generation.

Table 2. Classification of varieties planted by respondents.

Varieties	Number of respondents (Soul)	Average length of use (Years)
Inhibrida	64	12.09
Hybrid	9	10

Based on Table 2 it is known that the majority of respondents prefer to plant varieties of inhibrida. As many as 80 percent of respondents planted rice varieties of inhibrida. Some types of rice varieties of inhibrida planted by respondents are: Inpari 30, Ciherang, Inpari 31, and Mekongga. This variety of inhibrida has been used on average for 10-15 years. It appears that farmers are more likely to choose the inhibrida variety because this variety has long been used by farmers. The duration of this use indirectly gives farmers a sense of trust in the variety, so the majority of farmers prefer to use rice varieties inhibrida.

3.2. Analysis of the effect of decision-making farmers choose rice varieties on rice productivity level

Decision making is an alternative selection process of several alternatives that exist to be followed up as a solution to the problem. Decision making is an action that is done over and over again. Decision making to choose rice varieties is very important in farming activities. According to Noviyanti, et al (2020) varieties contributed the most in increasing rice production by 56.1 percent [7].

The effect of decision making on rice productivity levels in this study was analyzed through multiple linear regressions with SPSS 22 aids. The following is the result of an analysis of the influence of decision making on rice productivity levels. The effect of Decision-Making Farmers Choosing Rice Varieties on Rice Productivity Levels is known through simple linear regression analysis with SPSS 22 tools. The free variables in this study are the decision-making of farmers choosing rice varieties (X) and the bound variables in this study are Rice Productivity Level (Y). The regression results obtained can be seen in Table 3.

Table 3. The results of the regression analysis of the influence of individual decision-making farmers choose rice varieties on the level of rice productivity.

Variable	Coefficient	Significant	VIF
Constant (C)	1.016	0.014	
decision to choose rice varieties (X)	0.088	0.000	1.000
F count	666.129	0.000^b	
R square	0.904		
Adjusted R-Square	0.951		

According to Table 3, it is also known that the R Square value is 0.904. This means that the influence of independent variables (the decision to choose rice varieties) on dependent variables (rice productivity rate) is 90.4 percent. Farmers' decision making in choosing varieties positively affects the level of rice productivity, meaning that the more appropriate farmers in deciding which varieties to plant, the higher the level of productivity obtained. It is adapted to the condition of arable land, the quality and quantity of grain, the resistance of varieties to pest and disease attacks, the ease of accessing seeds, as well as the adaptability of varieties to environmental changes.

4. Conclusion

Farmers' decision making in choosing rice varieties has a positive effect on rice productivity. The more appropriate farmers choose rice varieties that suit the needs, conditions of the land, types and intensity of pests and diseases that attack in the area, it will support the acquisition of high rice productivity.

References

[1] Central Bureau of Statistics 2019 Comparison of Rice Production and Consumption in Indonesia
[2] Central Bureau of Statistics 2018 Comparison of Rice Production and Consumption in Indonesia
[3] Agricultural Extension Center Bangun Rejo Subdistrict 2017 Activities to Increase Rice Production of Inhibrida Rice Seed Subsidy Program BPP Office Bangun Rejo
[4] Department of Agriculture Food Crops and Horticulture Central Lampung Regency 2019 SP Padi
[5] Nazir M 1988 Research Method Ghalia Indonesia Jakarta
[6] Sugiarto D 2003 Sampling Technique PT Gramedia Pustaka Utama Jakarta
[7] Noviyanti S, Kusmiyati and Dwiwanti S 2020 Adoption of Innovation in The Use of New Superior Varieties of Rice Fields (Oryza Sativa L.) in Cilaku Subdistrict Cianjur Regency West Java Province. *Journal of Research Innovation.* **1** pp 771-782
[8] Listiana I *et al* 2019 *J. Phys.: Conf. Series* **1467** 012025

Authentication of pure and adulterated Sumbawa monofloral honey using ultraviolet-visible spectroscopy

D Suhandy[1], Kusumiyati[2], S Kuncoro[3], W Rahmawati[3] and M Yulia[4]

[1] Spectroscopy Research Group (SRG), Laboratory of Bioprocess and Postharvest Engineering, Department of Agricultural Engineering, Faculty of Agriculture, The University of Lampung, Bandar Lampung, Lampung, Indonesia
[2] Crop Science Department, Faculty of Agriculture, Padjadjaran University, Bandung, Indonesia
[3] Department of Agricultural Engineering, The University of Lampung, Jl. Prof. Dr. Soemantri Brojonegoro No.1, Bandar Lampung, 35145, Indonesia
[4] Department of Agricultural Technology, Lampung State Polytechnic, Jl. Soekarno Hatta No. 10, Rajabasa Bandar Lampung, 35141, Indonesia

Corresponding Author: diding.sughandy@fp.unila.ac.id

Abstract. Sumbawa monofloral honey is one of the most expensive honeys in Indonesia. In this present study, ultraviolet-visible (UV-vis) spectroscopy was utilized to discriminate between authentic pure Sumbawa monofloral honey and its adulterated one. The adulterated honey samples were created by intentionally mixing pure authentic Sumbawa monofloral honey with high fructose corn syrup (HFCS) at a level of 10% (v/v). A total of 50 samples of pure and adulterated samples were prepared. The UV-vis spectral data of each sample were obtained by using a UV-vis spectrometer in the range of 190-1100 nm. The result of PCA (principal component analysis) using the first two PCs with the CEV (cumulative explained variance) of 93% shows that the pure and adulterated Sumbawa honey samples could be separated. All pure samples lied in the negative of PC1 (PC1<0) and all adulterated samples were in the positive of PC1 (PC1>0). In conclusion, this result suggested us the possible application of UV-vis spectroscopy for authentication of Sumbawa honey adulterated with HFCS in a simple and low-cost method.

1. Introduction

Honey is one of the most popular natural sweetening materials which is rich in sugar, organic acids, enzymes, vitamins, and flavonoids [1]. It also provides various human health benefits and medicinal values and is widely used as anti-microbial, anti-inflammatory, anticarcinogenic, antiatherogenic, anti-thrombotic, and anti-oxidant agents [2]. Indonesia with huge areas of protected and limited production forest has been an important source of several monofloral *Apis dorsata* honey such as *Acacia mangium*, *Acacia crassicarpa*, *Calliandra callothyrsus*, *Hevea brasiliensis*, *Eucalyptus sp*, and *Ceiba pentandra*.

Honey authentication is an important issue in food analysis. The high nutritional value and distinct flavors of honey is the main reason of frequent honey adulteration [3]. The difficulty in detection is also another factor making adulteration widespread. The most common type of adulteration is the addition of sugar to honey with syrup from two types of plants: C4 plants (corn syrup, sucrose syrup, fructose syrup, maltose syrup, etc.) and C3 plants (rice syrup, beet syrup, etc.) [4]. High fructose corn syrup

(HFCS) is a cheap artificial sweetening and a well-known syrup for honey adulteration. At the present, the standard analytical method for authentication of syrup-adulterated honey is stable carbon isotope ratio analysis [4]. However, this method is quite expensive, laborious, and not suitable for routine analysis.

Several previous works on honey authentication using different spectral-based analytical methods have been well reported, from Raman spectroscopy to UV-vis spectroscopy [5-7]. Li *et al.* [5] utilized Raman spectroscopy and partial least squares-linear discriminant analysis (PLS-LDA) to detect honey adulteration with a total accuracy of 75.6%~97.8% depending on different types of adulterants. NIR and discriminant partial least squares (DPLS) method had been used for the determination of Chinese honey adulterated with HFCS with 90%-95% of the correct classification rate [6]. Recently, Suhandy and Yulia used UV-vis and fluorescence spectroscopy to authenticate several Indonesian agricultural products such as honey [7-8], coffee [9-10], and tea [11]. The present study aimed to investigate the application of UV-vis spectroscopy and PCA to authenticate between the pure and adulterated Sumbawa honey.

2. Materials and methods

2.1. Samples

Pure Sumbawa monofloral honey samples were obtained directly from two bee-keepers in Batulanteh regency, Sumbawa. The samples are *Apis dorsata* monofloral honey of *Ceiba pentandra* which was harvested on 12 September 2020 from a protected forest in Batulanteh regency, Sumbawa, Indonesia. About 1 kg of pure Sumbawa honey samples was transported to the laboratory at Lampung University and were stored at about 9-10°C before further analysis. The adulterated honey samples were created by intentionally mixing pure authentic Sumbawa monofloral honey with high fructose corn syrup (HFCS) at a level of 10% (v/v). A total of 50 samples were prepared (pure and adulterated). Figure 1 shows the pure and adulterated Sumbawa honey samples. The color of both samples was difficult to be differentiated by direct human visual inspection.

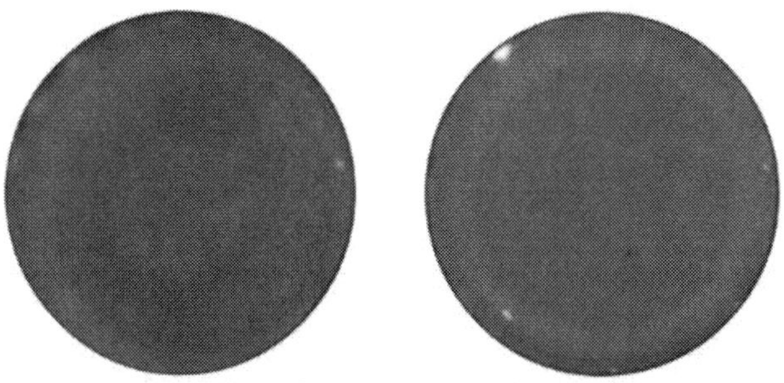

Figure 1. The appearance of pure and adulterated Sumbawa honey samples.

2.2. Spectral data acquisition and preprocessing

A benchtop UV-visible spectrometer from 190-1100 nm equipped with a quartz cuvette was used to obtain spectral data of pure and adulterated honey samples. For spectral measurement, all samples (pure and adulterated) were diluted with distilled water at a proportion of 1:30 (v/v). The samples were measured in the following parameters: fast scanning and transmittance mode with a 1 nm interval. A reference of water was measured before sample measurement. To reduce undesirable systematic noise, original spectra obtained from spectral measurement were preprocessed simultaneously using three different algorithms namely moving averaging smoothing (MAS), standard normal variate (SNV), and Savitzky-Golay 1st derivation.

2.3. *Data analysis*

Data spectra of pure and adulterated Sumbawa honey samples were used to perform a principal component analysis (PCA) with a full cross-validation method. Two plots of PCA were used: Hotelling's T^2 ellipse plot and scores plot of PC1 and PC2 (PC1xPC2). Hotelling's T^2 ellipse is widely used for detecting shifts in the mean when two or more interrelated variables are being analyzed simultaneously. This plot utilizes the combination values from the mean and the dispersion of the analyzed variables. It also is useful to identify and remove potential outlier samples [12-13]. The percentages of cumulative explained variance (CEV) of the PC were used to determine the optimal number of PC included in the calculation of PCA. The result of PCA was acceptable when the CEV is 90% or more [14].

2.4. *Software*

All calculations of spectral preprocessing, outlier detection and principal component analysis (PCA) were performed in The Unscrambler 9.8 (CAMO AS, Oslo, Norway) under Windows 10.

3. Results and discussion

3.1. *Spectral analysis of original and preprocessed spectra*

Figure 2 shows the original data spectra of fifty samples (pure and adulterated) in the wavelength of 190-1100 nm. It is clear that in general, spectral data of pure Sumbawa honey is higher than that of adulterated one. The difference was observed in the wavelength window of 250-300 nm. The presence of high fructose corn syrup in Sumbawa honey decreased the absorbance intensity of pure Sumbawa honey. However, we noticed one sample of pure Sumbawa honey with lower absorbance than that of adulterated Sumbawa honey. The preprocessed spectral data of pure and adulterated Sumbawa honey was shown in Figure 3. A sharp spectral difference was observed at 280 nm. This wavelength was related to the absorbance of the aromatic amino acids in honey [8]. The previous work on UV-visible spectral measurement on corn syrup reported an intense peak at 285 nm [15].

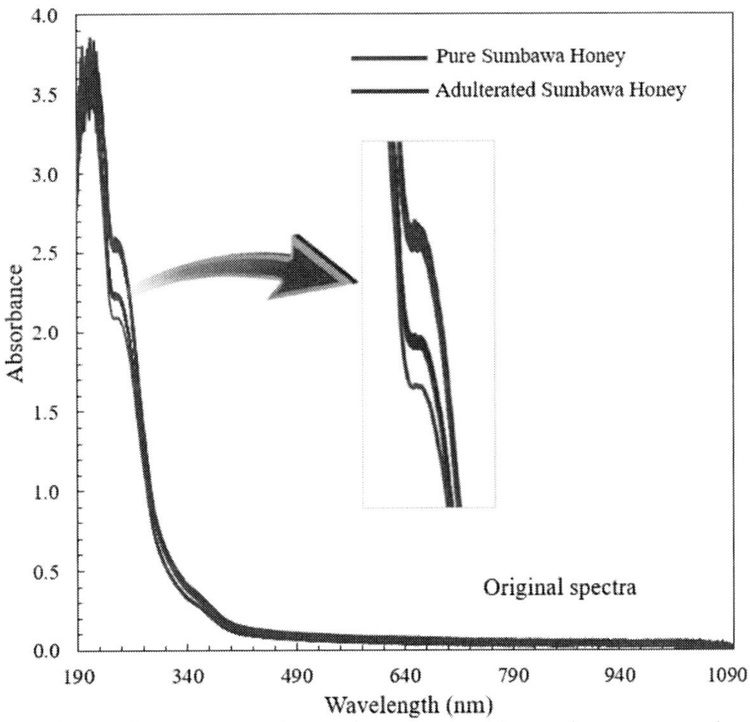

Figure 2. Original spectral data of pure and adulterated Sumbawa honey samples in the window of 190-1100 nm.

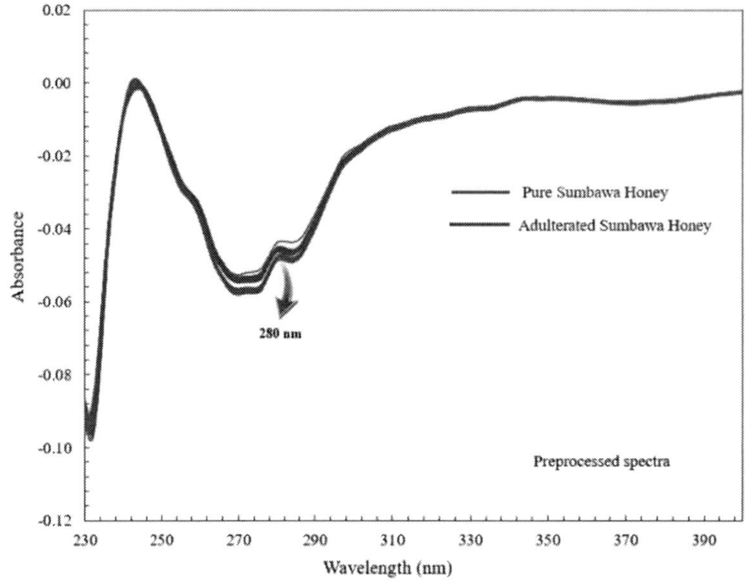

Figure 3. Preprocessed data spectra of pure and adulterated Sumbawa honey samples in the window of 230-400 nm.

3.2. *Outlier detection using Hotelling's T^2 ellipse*

Figure 4 shows the result of PCA with Hotelling's T^2 ellipse using preprocessed data spectra of fifty samples (pure and adulterated) in the wavelength of 230-400 nm. One pure Sumbawa sample was located outside the confidence ellipse and considered an outlier. This sample was associated with the pure Sumbawa sample with the lowest absorbance intensity as depicted in Figure 2. This sample was removed from further analysis.

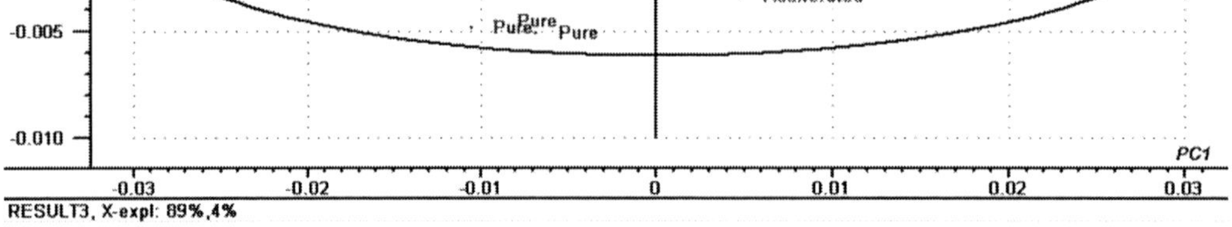

Figure 4. PCA scores plot of all samples using preprocessed spectral data in the window of 230-400 nm with 95% confidence interval of the Hotelling's T^2 ellipse. Samples outside the confidence ellipse are considered outliers.

3.3. Results of PCA

Figure 5 shows the result of PCA analysis using 49 samples (after removing 1 outlier sample). The result of PCA using the first two PCs with the CEV (cumulative explained variance) of 93% shows that the pure and adulterated honey samples could be separated along the *x*-axis (PC1). All pure samples lied in the negative of PC1 (PC1<0) and all adulterated samples were in the positive of PC1 (PC1>0). It showed that the differentiation between the pure and adulterated Sumbawa honey samples could be established with 100% of accuracy (calculated at 5% of significance level). The previous works reported similar results. Suhandy and Yulia [8] showed a PC1xPC2 scores plot to discriminate several authentic Indonesian monofloral and multifloral honey with different botanical, entomological, and geographical origins. Laser-induced breakdown spectroscopy (LIBS) coupled with PCA could be used to discriminate several multifloral honeys from different origins [16].

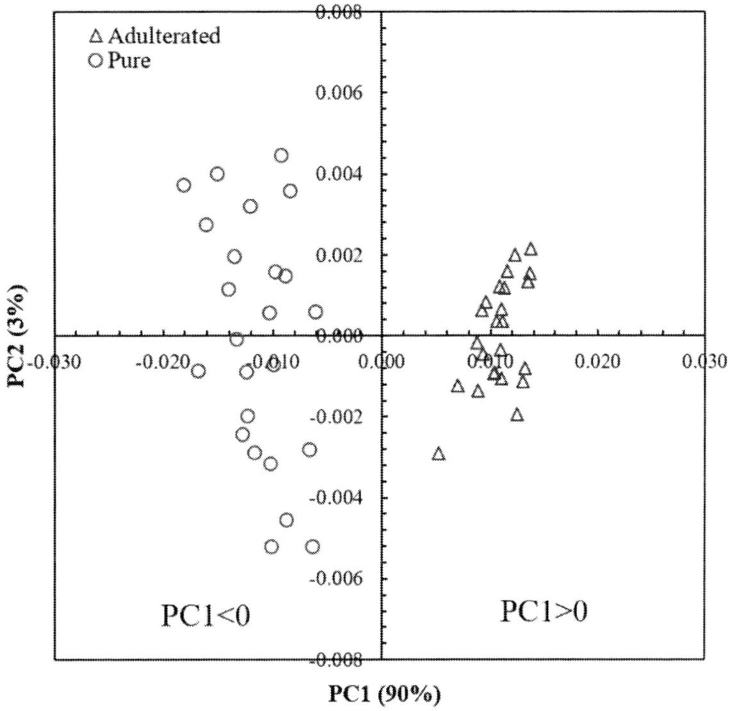

Figure 5. The score plots of PCA results (PC1xPC2) generated from preprocessed data in the window of 230-400 nm.

4. Conclusion

The discrimination of pure and adulterated Sumbawa honey was well demonstrated with this UV-vis spectroscopy method. It was also noted that Hotelling's T^2 ellipse is a useful plot to detect a potential outlier sample before further analysis. Using preprocessed spectra, a sharp spectral difference between the pure and adulterated Sumbawa honey samples could be identified at a wavelength of 280 nm. However, the samples used in the present study were limited. More investigation with more samples (with various levels of adulteration) was necessary to confirm this finding.

Acknowledgements

This work was fully supported by the Ministry of Education, Culture, Research and Technology /BRIN, Republic of Indonesia (Applied Research Grant 2020-2022).

References

[1] Meda A, Lamien C E, Romito M, Millogo J and Nacoulma O G 2005 *Food Chem.* **91**: 571–7
[2] Siddiqui A J, Musharraf S G, Choudhary M I and Rahman A 2017 *Food Chem.* **217**: 687–98
[3] Maione C, Barbos Jr F and Barbosa R M 2019 *Comput. Electron. Agric.* **157**: 436–46
[4] Huang F, Song H, Guo L, Guang P, Yang X, Li L, Zhao H and Yang M 2020 *Spectrochim. Acta A Mol. Biomol. Spectrosc.* **235** 118297
[5] Li S, Shan Y, Zhu X, Zhang X and Ling G 2012 *J. Food Compos. Anal.* **28**: 69–74
[6] Chen L, Xue X, ye Z, Zhou J, Chen F and Zhao J 2011 *Food Chem.* **128**: 1110–4
[7] Suhandy D and Yulia M 2021 *AIP Conf. Proc.* **2342** 100004
[8] Suhandy D and Yulia M 2021 *Molecules* **26** 915
[9] Yulia M, Asnaning A R and Suhandy D 2018 *IOP Conf. Ser.: Earth Environ. Sci.* **147** 012010
[10] Suhandy D and Yulia M 2018 *IOP Conf. Ser.: Mater. Sci. Eng.* **334** 012059
[11] Suhandy D and Yulia M 2019 *IOP Conf. Ser.: Earth Environ. Sci.* **258** 012042
[12] Karunathilaka S R, Yakes B J, He K, Chung J K and Mossoba M 2018 *Heliyon* **4** e00806
[13] Basri K N, Hussain M N, Bakar J, Sharif Z, Khir M F A and Zoolfakar A S 2017 *Spectrochim. Acta A Mol. Biomol. Spectrosc.* **173** 335
[14] Yulia M and Suhandy D 2018 *MATEC Web Conf.* **197** 09003
[15] Guellis C, Valério D C, Bessegato G G, Boroski M, Dragunski J C and Lindino C A 2020 *J. Food Compos. Anal.* **89** 103466
[16] Zhao Z, Chen L, Liu F, Zhou F, Peng J and Sun M 2020 *Sensors* **20** 1878

UV spectroscopy for discrimination of two arabica coffee cultivars in West Java Indonesia: a feasibility study

M Yulia[1], K R Ningtyas[1], S Suharyatun[2], W Rahmawati[2] and D Suhandy[3]

[1] Department of Agricultural Technology, Lampung State Polytechnic, Jl. Soekarno Hatta No. 10, Rajabasa Bandar Lampung, 35141, Indonesia
[2] Department of Agricultural Engineering, The University of Lampung, Jl. Prof. Dr. Soemantri Brojonegoro No.1, Bandar Lampung, 35145, Indonesia
[3] Spectroscopy Research Group (SRG), Laboratory of Bioprocess and Postharvest Engineering, Department of Agricultural Engineering, Faculty of Agriculture, The University of Lampung, Bandar Lampung, Lampung, Indonesia

Corresponding Author: meinilwitayulia@polinela.ac.id

Abstract. Arabica coffee variety is one of the key parameters that highly influence the cup quality of arabica coffee. The objective of this current research is to evaluate the possible application of UV spectroscopy for the discrimination of arabica coffee with different cultivars. Green beans from two arabica coffee cultivars (*Coffea arabica*) of Typica and Sigarar Utang were collected from the same origin in Papandayan mountain West Java, Indonesia. The samples were subjected to the same postharvest treatments (wet cherry processing method). All samples were roasted in medium roasting with 200°C for 16 minutes using a portable roasting machine. A total of 40 samples of Typica and Sigarar Utang were provided by weighing 1 gram of coffee powder (mesh 40) for each sample. The extraction of coffee samples was performed based on previously reported work. The UV-visible spectral data of aqueous coffee samples were acquired by using a benchtop of UV-vis spectrometer in the range of 190-1100 nm. The first two PCs (principal components) with 94% of CEV (cumulative explained variance) could be used to separate between the Typica and Sigarar Utang samples. The Typica samples were situated on the left of PC1 (PC1<0) and Sigarar Utang samples were on the right of PC1 (PC1>0). In the future, it is promising to apply UV spectroscopy for simple and reliable discrimination of arabica coffee cultivar.

1. Introduction

Coffee is one of the important agricultural products in Indonesia with total area production of 1.3 million hectares (ha) in 2020. The development of better farming practices and proper coffee genotype selection has promoted an improvement in the final coffee cup quality and thus increase the popularity of specialty coffee products, especially for Arabica coffee [1-2]. In Indonesia, several arabica cultivars are available and planted widely such as arabica Typica, Yellow Bourbon, Catimor, and Sigarar Utang. In general, there are two types of analytical methods for coffee cultivar identification: traditional laboratory-based chemical methods and spectroscopy-based analytical methods. The first method including instrumental neutron activation analysis (INAA) [3], liquid chromatography coupled with UV spectrophotometry [4], and sensory data [5]. The spectroscopy-based analytical methods have been reported for coffee variety and cultivar identification from near-infrared spectroscopy to fluorescence spectroscopy [1, 6-14].

However, there is no reported work on the Indonesian cultivar coffee identification using UV spectroscopy.

Coffee authentication both for arabica and robusta coffee through UV spectroscopy has been well reported. Yulia *et al.* [15] reported the promising application of UV spectroscopy and the SIMCA (soft independent modeling of class analogy) method for the classification of ground roasted decaffeinated coffee. UV spectroscopy and PLS-DA (partial least-squares discriminant analysis) method were demonstrated to be suitable for authentication of organic Lampung robusta ground roasted coffee [16]. Suhandy and Yulia [17] utilized support vector machine regression (SVMR) coupled with UV spectroscopy for Luwak coffee content measurement in coffee blends. A similar approach had been used to identify and discriminate between fresh and expired ground roasted robusta coffee [18].

The objective of the research was to develop a simple and relatively fast analytical method using UV spectroscopy and PCA (principal component analysis) for the discrimination of different cultivars of ground-roasted arabica coffee from Indonesia.

2. Materials and methods

2.1. Samples

Two cultivars of arabica coffee of Sigarar Utang and Typica were prepared in this research. Both cultivars were Java Preanger arabica coffee which was harvested in 2020 from the same origin in Papandayan mountain West Java, Indonesia. Roasting of coffee samples was conducted using a portable roaster at a medium level with 200°C for 16 minutes. A total of 40 samples of Typica and Sigarar Utang were provided by weighing 1 gram of coffee powder (mesh 40) for each sample. The extraction of coffee samples was performed based on previously reported work [19]. Figure 1 shows arabica Sigarar Utang and arabica Typica coffee samples before extraction. It is hard to directly discriminate between the two samples by using human naked eyes.

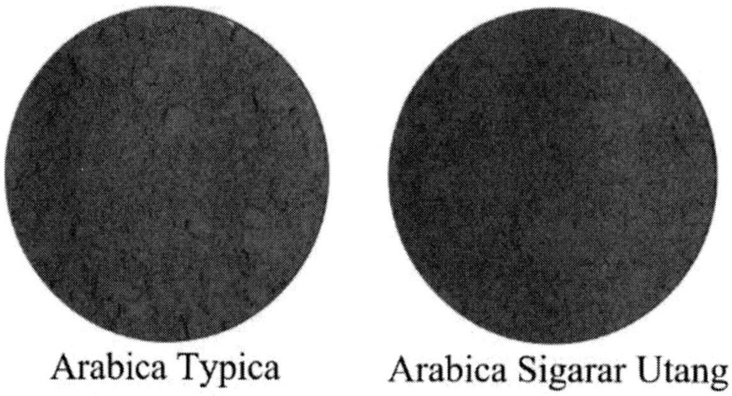

Figure 1. The visual appearance of arabica Sigarar Utang and arabica Typica samples before extraction.

2.2. Spectral data acquisition and preprocessing

The UV-visible spectra of aqueous coffee samples were measured by using a benchtop of a UV-visible spectrometer in the window of 190-1100 nm. Distilled water was used as a reference and it was measured before sample measurement. Three spectral preprocessing were used: moving averaging smoothing (MAS) to improve the signal-noise ratio (SNR), standard normal variate (SNV) to minimize the effects of light scattering, and Savitzky-Golay 1st derivation to enhance a small difference between spectral data of arabica Sigarar Utang and arabica Typica.

2.3. Data analysis

Discrimination of the two cultivars of arabica Sigarar Utang and arabica Typica was conducted using PCA (principal component analysis). First, the PCA was calculated using original spectral data in the window of 190-1100 nm and an influence plot was plotted. This plot is leverage in the x-axis and residual x-variance in the y-axis. Residual x-variance is used to measure the variation that is not taken into account by the PCA model. Leverage is used to measure how extreme a data point is compared to the majority. We assigned a potential outlier for any samples with both high residual x-variance and high leverage value. The second plot is Hotelling's T^2 ellipse. This plot is also useful to identify and remove potential outlier samples [20-21]. Using those two plots, detection of outlier samples was possible. Further PCA calculation was performed using all samples of arabica Sigarar Utang and arabica Typica without outlier samples. Scores plot of the first two PCs (principal components) was created. To select the appropriate number of PC included in the calculation of PCA, the percentages of cumulative explained variance (CEV) of the PC were used (CEV should be 90% or more [18]). Software of The Unscrambler 9.8 (CAMO AS, Oslo, Norway) was used to calculate spectral preprocessing, outlier detection, and PCA analysis.

3. Results and discussion

3.1. Spectra of arabica Sigarar Utang and Typica

Spectral data of all samples of arabica Sigarar Utang and arabica Typica in the window of 190-1100 nm was depicted in Figure 2. The absorbance intensity of arabica Sigarar Utang is higher than that of arabica Typica especially in the window of 250-400 nm which is in line with the result of the previously reported study [19]. Figure 3 shows the preprocessed spectra of the samples in the window of 250-400 nm. In this window, several wavelengths with high peaks were observed. Those peaks were in line with the findings of the previously reported works. For instance, the positive peak at a wavelength of 270 nm was in correspondence with the absorbance of caffeine [22]. The positive peak at 315 nm and the negative peak at 345 nm is related to the absorbance of caffeic acid or trigonelline [23].

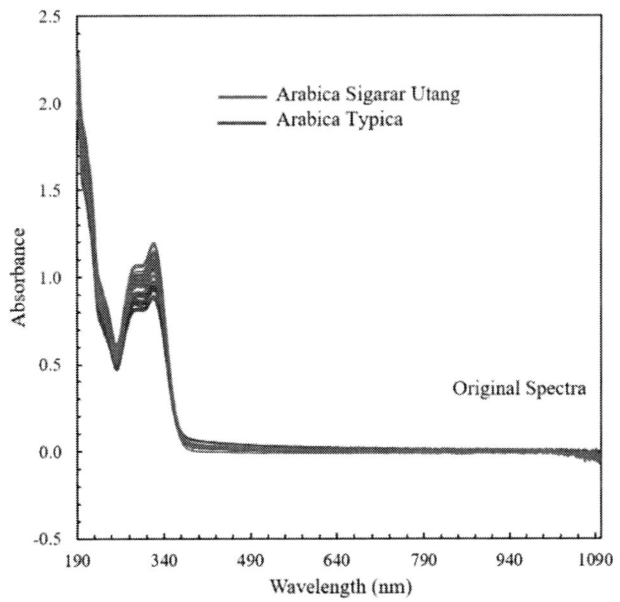

Figure 2. Spectra of arabica Sigarar Utang and Typica coffee samples obtained directly from UV-visible spectral measurement (without preprocessing) in the window of 190-1100 nm.

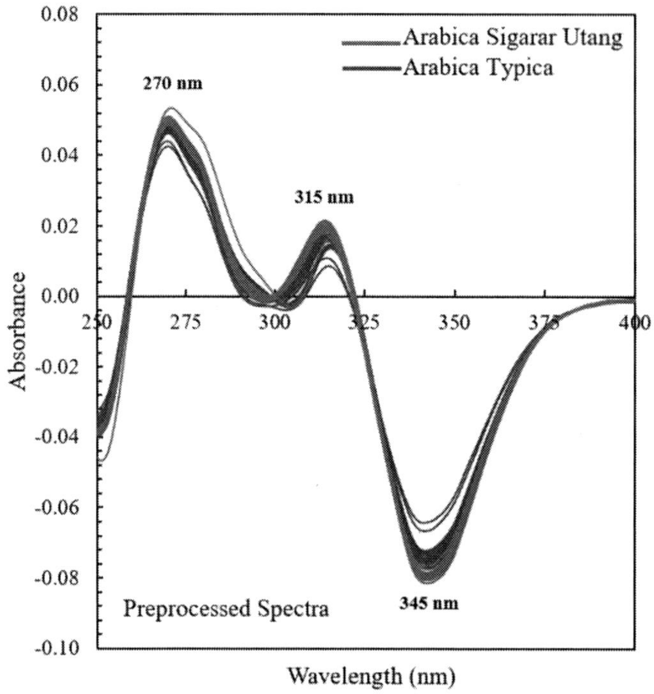

Figure 3. Preprocessed spectra of arabica Sigarar Utang and Typica coffee samples in the window of 250-400 nm.

3.2. Detection of outlier sample

Figure 4 shows the plot of influence showing leverage in the x-axis and residual x-variance in the y-axis coming from PCA calculation using original spectra in the window of 190-1100 nm. This plot can be used to identify an occurrence of potential outlier samples. Any samples with high values both in leverage and residual x-variance can be associated with a potential of outlier samples. One sample of arabica Sigarar Utang sample (SU1a) has a high value both in leverage and residual x-variance. Two samples of arabica Typica (AT1a and AT1b) have a high value of leverage but an acceptable value for residual x-variance. However, the result of Hotelling's T^2 ellipse as depicted in Figure 5 showed that the two samples of AT1a and AT1b were lied outside the confidence ellipse and hence confirmed as potential outlier samples. All potential outlier samples were excluded from further PCA calculation.

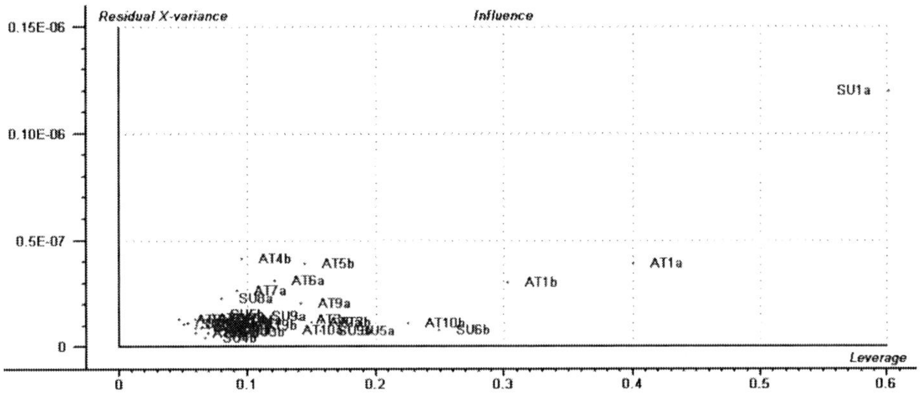

Figure 4. A plot of leverage versus residual x-variance calculated using all samples of original spectra in the window of 190-1100 nm.

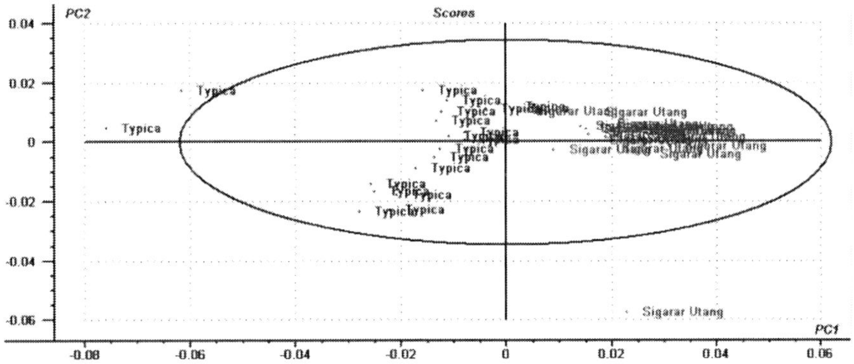

Figure 5. PCA scores plot of all samples (40 samples) of arabica Sigarar Utang and Typica using original spectral data in the window of 190-1100 nm with 95% confidence interval of the Hotelling's T^2 ellipse. Samples outside the confidence ellipse are considered outliers.

3.3. PCA calculation

The potential outlier samples (one sample of arabica Sigarar Utang and two samples of arabica Typica) were removed from data analysis. The result of PCA calculation without outlier samples was plotted in Figure 6. The calculation was performed using preprocessed spectral data of 37 samples of arabica Sigarar Utang and arabica Typica in the window of 250-400 nm. The first two PCs (principal components) with 94% of CEV (cumulative explained variance) could be used to separate between the Sigarar Utang and Typica samples. The Sigarar Utang samples were on the right of PC1 (PC1>0) and Typica samples were situated on the left of PC1 (PC1<0) resulted in 100% of accuracy (calculated at 5% of significance level). PCA and UV spectroscopy was effective to do the unsupervised classification of Indonesian specialty arabica coffee concerning the origin of the cultivar. Our finding was in line with several previously reported works. Edelmann *et al.* [24] utilized UV-visible spectroscopy and chemometrics to discriminate Austrian red wines with different cultivars with an acceptable result. Zhang *et al.* [9] used mid-infrared spectroscopy and PCA to distinguish four coffee varieties including Typica and Catimor from China with a promising result.

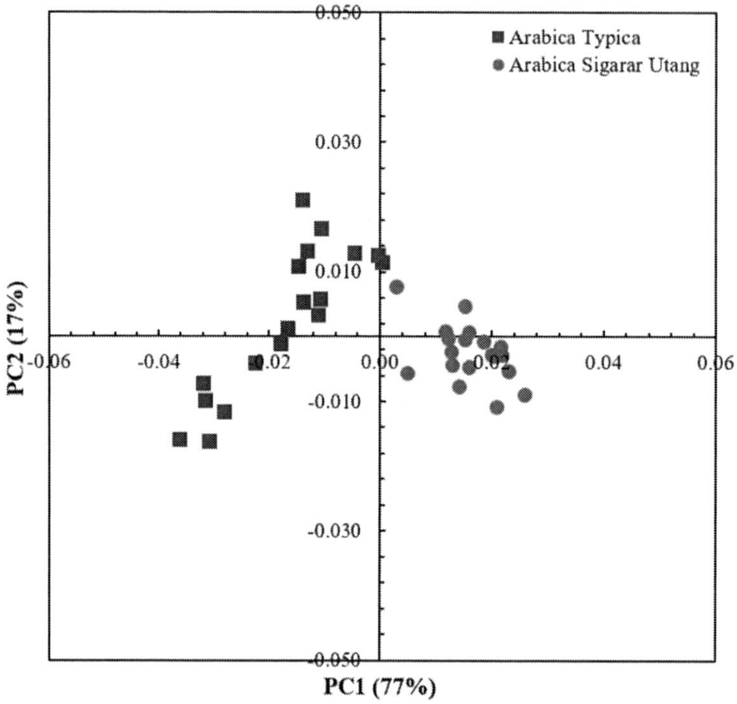

Figure 6. The score plots of PCA analysis without outlier samples (PC1xPC2) calculated using preprocessed spectral data in the window of 250-400 nm.

4. Conclusion

Feasible discrimination of two arabica cultivar coffees from West Java, Indonesia has been well demonstrated using UV spectroscopy. One important step before PCA calculation is outlier sample detection. Here, we showed that plot of residual *x*-variance versus leverage and the plot of Hotelling's T^2 ellipse is quite appropriate to be used for detecting a potential outlier sample. In the future, it is promising to apply UV spectroscopy for simple and reliable discrimination of arabica coffee cultivar. However, this present research used a small number of samples with the same geographical origin. Therefore, to clarify this finding, the next research should be conducted with more samples with more cultivars from different geographical origins.

Acknowledgements

The authors acknowledge the Ministry of Education, Culture, Research and Technology/BRIN, the Republic of Indonesia for financial support (PKPT Grant 2020-2022).

References

[1] Marquetti I, Link J V, Lemes A L G, Scholz M B S, Valderrama P and Bona E 2016 *Comput. Electron. Agric.* **121**: 313–9
[2] Moreira I and Scarminio I S 2013 *Talanta* **107**: 416–22
[3] Fernandes E A D N, Tagliaferro F S, Azevedo-Filho A and Bode P 2002 *Accred. Qual. Assur.* **7**: 378–87
[4] Alonso-Salces R M, Serra F, Reniero F and Heberger K 2009 *J. Agric. Food Chem.* **57**: 4224–35
[5] Wada K, Ohgama S, Sasaki H, Shimoda M and Osajima Y 1987 *Agric. Biol. Chem.* **51**: 1745–52
[6] Bona E, Marquetti I, Link J V, Makimori G Y F, da Costa Arca V, Guimarães Lemes A L, Ferreira J M G, dos Santos Scholz M B, Valderrama P and Poppi R J 2017 *LWT-Food Sci. Technol.*

76: 330–6
[7] dos Santos Scholz M B, Kitzberger C S G, Pereira L F P, Davrieux F, Pot D, Charmetant P and Leroy T 2014 *J. Near Infrared Spectrosc.* **22**: 411–21
[8] Esteban-Díez I, González-Sáiz J M, Sáenz-González C and Pizarro C 2007 *Talanta* **71**: 221–9
[9] Zhang C, Wang C, Liu F and He Y 2016 *J. Spectrosc.* **2016** 7927286
[10] Monakhova Y B, Ruge W, Kuballa T, Ilse M, Winkelmann O, Diehl B, Thomas F and Lachenmeier D W 2015 *Food Chem.* **182**: 178–84
[11] Luna A S, da Silva A P, da Silva C S, Lima I C A and de Gois J S 2019 *J. Food Compos. Anal.* **76**: 44–50
[12] Figueiredo L F, Borém F M, Almeida M R, de Oliveira L F C, Alves A P C, dos Santos C M and Rios P A 2019 *Food Chem.* **288**: 262–7
[13] Zhang C, Shen T, Liu F and He Y 2018 *Sensors* **18** 95
[14] Suhandy D and Yulia M 2018 *IOP Conf. Ser.: Mater. Sci. Eng.* **334** 012059
[15] Yulia M, Asnaning A R and Suhandy D 2018 *IOP Conf. Ser.: Earth Environ. Sci.* **147** 012010
[16] Yulia M and Suhandy D 2019 *J. Phys.: Conf. Ser.* **1341** 022006
[17] Suhandy D and Yulia M 2019 *J. Phys.: Conf. Ser.* **1341** 022006
[18] Yulia M and Suhandy D 2018 *MATEC Web Conf.* **197** 09003
[19] Suhandy D and Yulia M 2018 *MATEC Web Conf.* **197** 09002
[20] Karunathilaka S R, Yakes B J, He K, Chung J K and Mossoba M 2018 *Heliyon* **4** e00806
[21] Basri K N, Hussain M N, Bakar J, Sharif Z, Khir M F A and Zoolfakar A S 2017 *Spectrochim. Acta A Mol. Biomol. Spectrosc.* **173** 335
[22] López-Martínez L, López-de-Alba P L, García-Campos R and León-Rodríguez L M D 2003 *Anal. Chim. Acta* **493**: 83–94
[23] Suhandy D, Yulia M and Kusumiyati 2018 *AIP Conf. Proc.* **2021** 060010
[24] Edelmann A, Diewok J, Schuster K C and Lendl B 2001 *J. Agric. Food Chem.* **49**:1139–45

Income and welfare level of cassava farmers during covid-19 pandemic era in Lampung Province

W A Zakaria[1], L S M Indah[1*], T Endaryanto[1], and D Hermawan[2]

[1] Department of Agribusiness, Faculty of Agriculture, University of Lampung, Bandar Lampung.
[2] Public Administration Departement, University of Lampung, Bandar Lampung.

* Corresponding Author: lidya_masindah@yahoo.com

Abstract. Indonesia is currently experiencing the Covid-19 pandemic which first started in March 2020. This outbreak continuously spreads throughout the regions including Lampung Province and has adversely affected economic activity of the society which included household income and welfare of cassava farmers. This study aims at analyzing the level of household income and welfare of cassava farmers during Covid-19 Pandemic Era in Lampung Province. This study applied survey method and conducted in the Regency of Central Lampung and East Lampung as cassava production center area in Lampung Province. A total of 120 people consisted of 60 people from each regency were selected through the method of simple random sampling. Data collection was conducted in July 2020. Data analysis was done using quantitative descriptive method. The level of household welfare of cassava farmers was analyzed according to the indicator of BPS, and Sajogyo. Result of the study showed that farmers experienced problems in accessing capital, production facility, and market of product during the Covid-19 era. Income gained from cassava farming dominated household income. However, there were only 31,67 percent of cassava farmers in Central Lampung and 5 percent in East Lampung who were classified as non-poor, while the rest were found to be near-poor and suffered from poverty. Food expenditure still dominated the total household expenditure and expenditure for personal hygiene was non-food expenditure that was considered quite high.

1. Introduction

Food security is one of the main priorities in national development. The subsector of food crops plays essential role to achieve national food security by supplying both raw materials for industry and food needs for the society. Moreover, employment absorption in this subsector is quite high. The program of National Food Security as one solution to overcome food insecurity in Indonesia. In fact, reduction of food insecurity will only occur if food security is achieved at household level [1]. One of important commodities in food crop subsector is cassava. Moreover, cassava farming that is optimally performed might become one of production increase sources to fulfil food needs of the society besides creating income source for farmers [2].

Lampung Province is the largest cassava production area in Indonesia. At the level of province, Central Lampung Regency is the main cassava production area in Lampung Province. In 2016, total cassava harvested area in Central Lampung Regency reached 68.720 ha with cassava production of 1.730.156 ton, placing Central Lampung Regency not only as regency with the largest cassava production in Lampung, but also in Indonesia. In 2018, actual average cassava production ranged of 23,5 ton per ha which was still considered low if compared with the potential production of 30-35 ton per ha [3]. Low productivity was caused by the use of less superior variety, delayed fertilizer application, inappropriate dose, and limited irrigation facility [4].

The characteristics of cassava farmers in farming business reflect moral economy which is characterized by kinship, group, or the tradition of mutual assistance-based existence without maximum use of modern technology due to limited capital [5]. The value of RC on total cost obtained by cassava farmers in Lampung Province was 2,02, indicating that cassava farming was feasible [3]. However, the value could not yet interpret the welfare level of cassava farmer in Lampung Province. Factors that affected cassava farmer welfare included education, total household income, land ownership, years of farming experience, and production cost [6].

Non-technical problems faced by farmers in performing farming business will inhibit farmers from effectively and efficiently conducting farming, thus affecting their level of welfare. Cassava farming in Lampung Province is considered technically efficient, yet improvement is still necessary [7]. Problems faced by farmers included the limited use of production technology due to lack of knowledge, limited farm capital, farm management inefficiency, the absence of market assurance which resulted in low price received by farmers due to low bargaining power of farmers to processing factory, declining of soil fertility, thus lowering cassava productivity, and increase in input requirement and cost spent on cassava farming [3]. Based on the description, analysis of cassava farmer welfare in Lampung Province is necessarily done.

2. Research Methods

This study was conducted in Lampung Province. Research location was purposively determined by consideration that Lampung Province is the largest cassava production area in Indonesia. Research was carried out from April-September 2020. This study applied survey method in the location of cassava production center in Lampung Province. A total of 120 cassava farmers were selected as research sample through simple random sampling. Data analysis was done using the method of quantitative descriptive, namely by performing analysis of welfare level according to the indicator of BPS and Sajogyo.

Analysis tools applied to measure the level of cassava farmer welfare included five indicators introduced by Badan Pusat Statistik (Statistics Indonesia) 2015 which consisted of indicators of population, health and nutrient, education, housing and environment, and other social aspects. There are two classifications of welfare used in this study, namely non-poor household and poor household. Variables of observation along with classification and score which represented the weight of indicator classification were also presented.

Level of household welfare according to Sajogyo is seen from the annual household expenditure per capita, that is total expenditure of farmers, both food and non-food expenditure, divided by the number of dependents. Total household expenditure was further converted to unit equaled to rice per kilogram on the basis of applicable price in research location to determine welfare level of farm household [8].

3. Results and discussion

3.1 The role of household economy of cassava farmers

Household economic resilience of cassava farmers is an important factor which ensures the sustainability of farmer welfare. Poor socio-economic condition has insisted farmers to rationally make decision to increase their income by working in non-cassava farming sector. This way, their needs are expected to be met. Fluctuated cassava price along with uncertainty of production input availability and limited capital have made farmers to make such effort to increase their welfare. Welfare of farmer highly affected the source of household income. Contribution of household income of cassava farmer is listed in Table 1.

Table 1. Contribution of household income of cassava farmers in Lampung Province.

Source of Household Income	Central Lampung		East Lampung	
	Income	Percentage	Income	Percentage
Income from cassava farming (on-farm)	21.318.907	46,43	21.355.564	51,51
Income from non-cassava farming (on-farm)	12.112.609	26,38	884.333	2,13
Farm income from outside of farming activity (off farm)	6.043.091	13,16	8.578.400	20,69
Income from non-farming business (non Farm)	6.438.667	14,02	10.637.333	25,66
Total	45.913.274	100	41.455.630	100

Source: Primary Data, processed (2020)

Based on Table 1, income gained from cassava farming was still found to be dominant, both by cassava farmers in Central Lampung (46,43 %) and East Lampung (51,51%). This finding was in line with study that report the highest income contribution of farm household was from cassava farming [3]. Similarly, farmer income in East Lampung was earned from cassava farming. It shows that agriculture still plays important role in the economy of farm household [9]. Conversely, income from outside of farming activity is an important factor which influences the welfare of farmers [10]. This type of income is obtained from labor, poultry, sheep, and cattle farmers, drivers, traders, and other business. Result of study showed that income from outside of cassava farming only had a small contribution to the economy of farm household. For cassava farmers in Central Lampung, income from off-farm activity contributed the lowest, while the lowest income contribution in cassava farmers in East Lampung was earned from non-cassava farming of 2,13 %. This result was due to covid-19 pandemic which has been going for more than 6 months, causing farmers unwillingly leave their house besides the PSBB (social distancing) regulation.

This pandemic situation will affect welfare level of cassava farmers. Furthermore, the regulation of PSBB or social distancing has created several impacts, such as hindering the distribution of production facility and transportation, also causing limited availability of production facility, thus lowering cassava production and resulting in low welfare of farmer. For the last 20 years, or since 2018, there was declining welfare level of farm household in Lampung Province [11].

Analysis tools used to measure welfare level of cassava farmers in Lampung Province was in accordance with the five indicators of Badan Pusat Statistik (2014) which included population, health and nutrient, education, housing and environment, and other social aspects. Several welfare indicators of cassava are listed as follows:

Table 2. Achievement of BPS indicators 2015.

	Central Lampung			**East Lampung**		
Indicator	**Class**			**Class**		
	B	**C**	**K**	**B**	**C**	**K**
Population	✓			✓		
Health and nutrient		✓			✓	
Education	✓			✓		
Housing and environment		✓			✓	
Other social concerns		✓			✓	

Source: Primary data, processed (2020)

Remarks:

B: Good

C: Moderate

K: Poor

Based on Table 2, it is known that the value of indicator of population and education both in Central Lampung and East Lampung was classified as good. However, this result was not in line with the study who reported that education level of farmer was still low, yet this study obtained high value for the indicator of education [11]. Furthermore, cassava farmers both in Central Lampung and East Lampung obtained a moderate score in term of health and nutrient, housing and environment, and other social aspects. Therefore, welfare level of cassava farmers in both regions were considered similar.

The indicator of health of farm household also belonged to moderate category since most farm households both in Central Lampung and East Lampung still found difficulties to access health facility along with high cost of health service, thus sick members of farm household were not immediately taken to the public health center (puskesmas) or hospital. Knowledge and skill of farmers to meet nutrient requirement for family was still low as similarly found by study who report that the majority of cassava farm household was in the condition of food insecurity [12].

3.2 Analysis of household welfare according to the Theory of Sajogyo

The level of household level according to Sajogyo is seen from the amount of household expenditure per capita per year [8]. The annual household expenditure per capita is total expenditure of farmers, both food and non-food expenditure, divided by the number of dependents. In line with the study who report that measurement of welfare level is possibly done based on household expenditure and income of farm household [13] [14] [15]. Total household expenditure was further converted to unit equaled to rice per kilogram on the basis of applicable price in research location of Rp9.450,00 to determine welfare level of farm household. Welfare distribution of cassava farm household in Lampung Province is showed below.

Table 3. Welfare distribution of cassava farm household in Lampung Province in 2020.

Group	Interval of score (equal to rice/year)	Percentage (%)	
		Central Lampung	East Lampung
Poorest	<180 kg	0,00	
Very poor	181-240 kg	0,00	13,33
Poor	241-320 kg	3,33	15,00
Near-poor	321-480 kg	11,67	30,00
Adequate	481-960 kg	53,33	30,00
Non-poor	>960 kg	31,67	5,00
Total		**100**	**100**

Source: Primary data, processed (2020)

Based on Table 3, most (53,33%) cassava farmers in Central Lampung belonged to the adequate category. Moreover, cassava farmers in East Lampung were dominated by those included in near-poor (30%) and adequate (30%) category. This is because the household income of cassava farmers in Central Lampung is higher than in East Lampung. In addition, Central Lampung is the parent district with a much more complete and better regional infrastructure than East Lampung Regency, which is an expansion area of Central Lampung. Group distribution of welfare level of respondent farmers is presented in Table 4.

Table 4. Group distribution of welfare level of farm household.

Category of Welfare	Central Lampung		East Lampung	
	Total (household)	%	Total (household)	
Non-poor	49	81,67	57	95,00
Poor	11	18,33	3	5,00
Total	**60**	**100**	**60**	**100**

Source: Primary data, processed (2020)

Based on Table 4, cassava farmers were mostly not poor, both those in Central Lampung and East Lampung. Percentage of cassava farmers included in non-poor category in East Lampung were greater than those in Central Lampung. The reason is that 95 percent of cassava farmers in East Lampung belonged to non-poor category, while in Central Lampung, it was only 81,67 percent.

More cassava farmers in East Lampung were found to have better life than those in Central Lampung by reason of higher household expenditure of cassava farmers in Central Lampung that reached Rp14.542.884,81 for food expenditure and Rp11.647.070,00 for non-food expenditure. This result of study was in line with a research which showed that most cassava farmers in Sukadana Subdistrict of East Lampung Regency were included in non-poor category [9]. This situation was due to the fact that 100 percent of cassava farmers in East Lampung already used the Thailand variety, while 50% of cassava farmers in Central Lampung still used the Cassesart variety. Adoption of technology by applying superior variety will increase farmer welfare. Table of household expenditure of cassava farmers in Lampung Province is presented below [14] [15].

Table 5. Household expenditure of cassava farmers in Lampung Province in 2020.

Type of Expenditure	Central Lampung	East Lampung
	Value (Rp)	Value (Rp)
Food Expenditure	14.542.884,81	14.208.647,92
Non-food expenditure	11.647.070,00	14.212.766,67
Total Expenditure	26.189.954,81	28.421.414,58

Source: Primary data, processed (2020)

Based on Table 5, household expenditure of cassava farmers in East Lampung was higher than those in Central Lampung, respectively of Rp28.421.414,58 and Rp26.189.954,81. Moreover, food household expenditure of cassava farmer in Central Lampung was found to be higher than that of cassava farmers in East Lampung, namely Rp14.542.884,81. However, non-food expenditure of cassava farmers in East Lampung was Rp14.212.766,67. The highest non-food expenditure was spent by farm households both in Central Lampung and East Lampung on fuels and goods and services [16].

4. Conclusion

Based on the result of study, it is concluded that income from cassava farming still played important role in the household economy of cassava farmers. During the Covid-19 era, farmers experienced difficulties to access capital, production facility, and market of product. Income gained from cassava farming was found to dominate household income. However, there were only 31,67 percent of cassava farmers in Central Lampung and 5 percent in East Lampung who were classified as non-poor, while the rest were near-poor and lived in poverty. Food expenditure was still dominant and expenditure for personal hygiene was non-food expenditure that was considered quite high. The level of welfare of cassava farmers in Lampung Province according to the measurement of BPS and Sajogyo, and subjective well-being indicators was included in non-poor category.

References

[1] A. Azwardi, H. F. Widyasthika, R. C. Saleh, and N. Adnan 2019 Household Food Security: Evidence From South Sumatera *JEJAK* **12** 2 doi: 10.15294/jejak.v12i2.20264.

[2] L. Geo 2020 ANALYSIS OF PRODUCTION AND INCOME OF CASSAVA FARMING *Int. J. Res. -GRANTHAALAYAH* **8** 4 doi: 10.29121/granthaalayah.v8.i4.2020.21.

[3] W. A. Zakaria, T. Endaryanto, L. S. Mas Indah, I. R. Mellya Sari, and A. Mutolib 2020 PENDAPATAN DAN KESEJAHTERAAN RUMAH TANGGA PETANI UBIKAYU DI PROVINSI LAMPUNG *J. Agribisnis Indones* **8** 1 doi: 10.29244/jai.2020.8.1.83-93.

[4] G. Suja, K. Susan John, and J. Sreekumar 2010 Soil test based nutrient management for short-duration cassava (Manihot esculenta): Analysis of growth, biomass, yield, quality, nutrient uptake and soil nutrient status *Indian J. Agron* **55** 4

[5] A. Murdyastuti 2016 Aksesibilitas Petani Singkong terhadap Kebijakan Pemerintah dalam Mengentaskan Kemiskinan di Kabupaten Bondowoso Jawa Timur *J. Strateg. dan Bisnis* **4** 2

[6] O. Igweoscar 2014 Effect of Contract Farming on Productity and Welfare of Cassava-Based Farmers in South Eastern Nigeria *Eur. J. Bus. Manag* **6** 7

[7] K. Murniati, W. A. Zakaria, T. Endaryanto, and L. S. M. Indah 2021 Analysis of production efficiency and income to support sustainability of cassava farming in Lampung Tengah District, Lampung Province *IOP Conference Series: Earth and Environmental Science*, 2021 **739** 1 doi: 10.1088/1755-1315/739/1/012048

[8] A. Mahmud 2019 Rekonstruksi Pemikiran Sajogyo Tentang Kemiskinan dalam Perspektif Agraria Kritis *BHUMI J. Agrar. dan Pertanah* **5** 1 doi: 10.31292/jb.v5i1.322

[9] M. Iqbal, Agum, H. Lestari, Dyah, Aring, and A. Soelaiman 2014 Pendapatan dan Kesejahteraan Rumah Tangga Petani Ubi Kayu di Kecamatan Sukadana Kabupaten Lampung Timur *J. Ilmu Ilmu Agribisnis* **2** 3

[10] Abdullah *et al* 2019 Determinants of commercialization and its impact on the welfare of smallholder rice farmers by using Heckman's two-stage approach *J. Saudi Soc. Agric. Sci* **18** 2 doi: 10.1016/j.jssas.2017.06.001

[11] W. A. Zakaria, T. Endaryanto, L. S. M. Indah, and A. Mutolib 2020 The Economic role of cassava in farmers' households in Central Lampung Regency, Lampung Province *E3S Web of Conferences*, 2020 **153** doi: 10.1051/e3sconf/202015303008

[12] K. Indira Putri, K. Murniati, R. Adawiyah Jurusan Agribisnis, F. Pertanian, U. Lampung, and J. Soemantri Brojonegoro No 2019 Pola Konsumsi Dan Tingkat Ketahanan Pangan Rumah Tangga Petani Ubi Kayu Di Kecamatan Terusan Nunyai Kabupaten Lampung Tengah *Jiia*, **7** 3

[13] S. Asfaw 2010 Estimating Welfare Effect of Modern Agricultural Technologies: A Micro-Perspective from Tanzania and Ethiopia *Chronicpoverty.Org*

[14] C. A. Afolami, A. E. Obayelu, and I. I. Vaughan 2015 Welfare impact of adoption of improved cassava varieties by rural households in South Western Nigeria *Agric. Food Econ* **3** 1 doi: 10.1186/s40100-015-0037-2.

[15] A. Kolapo and A. J. Kolapo 2021 Welfare and productivity impact of adoption of biofortified cassava by smallholder farmers in Nigeria *Cogent Food Agric* **7** 1doi: 10.1080/23311932.2021.1886662.

[16] Wan Abbas Zakaria, Teguh Endaryanto, Muhammad Ibnu, and Lina Marlina 2019 Kesediaan Petani Melakukan Kemitraan Dimasa Datang: Analisis Heckprobit Pada Petani Ubi Kayu Di Provinsi Lampung *J. Trop. Upl. Resour. (J. Trop. Upl. Res.)* **1** 1 doi: 10.23960/jtur.vol1no1.2019.7

Business models on peatlands to prevent land and forest fires

D Yuniati[1], H Khotimah[1], Y Lisnawati[1] and H Meinas[2]

[1.] Forest Research Development Centre, Ministry of Environment and Forestry, Jalan Gunung Batu No.5, Bogor, 16610, Jawa Barat, Indonesia.
[2.] Indonesian Conservation Community Warsi, Jalan Inu Kertapati No.12, Pematang Sulur Telanaipura, Kota Jambi Provinsi Jambi, 36124, Indonesia.

Corresponding Author: dhanyyuniati@yahoo.co.id

Abstract. Peatlands nowadays, most of it has been degraded due to land and forest fires. The Indonesian government through The Peatland Restoration Agency has attempted to carry out peat restoration through three pillars. Revitalization is one of the restoration pillars besides rewetting and replanting, it seeks to improve the community livelihoods. The objectives of this study are to identify business models that have developed in the community and to analyze the mechanisms that occur in those business models in preventing forest and land fires. The research was conducted at the Peat Hydrological Unit of Mendahara-Batanghari, Jambi Province. In-depth interviews were conducted with respondents who were obtained by the purposive sampling method. Data and information then were analyzed descriptively qualitatively. Based on the field research this study found that the business models which can prevent forest and land fires are the rice field and horticulture business model, *areca nut* cultivation and utilization, *liberica* coffee agroforestry, pineapple cultivation, honey/beekeeping, *jelutong* agroforestry, ecotourism, and carbon trading. This study concludes that these business models have similar characteristics in preventing forest and land fires, namely land use adjusting the peat depth zoning with native peat species, non-timber forest products, and agroforestry patterns; intensive land management; utilization of biomass so that it does not become fuel during the dry season; maintaining peatlands in humid conditions through water management, and maintaining forest cover.

1. Introduction

Peatlands are an essential resource [1] because they have a role in storing carbon, water management, endemic and unique sources of biodiversity, and a source of community livelihoods [2]. Indonesia is a tropical country that has large peat reserves [1]. The area of peatland in Indonesia reaches 13.34 million hectares [3]. Currently, the condition of peatlands in Indonesia has experienced a lot of degradation. According to [4], peatlands in Indonesia are in a degraded state covering an area of 6.66 million hectares. Peatland degradations are caused by, among others, land fires, errors in water management, and mining activities [5]. In 2015, fires broke out in Indonesia burned about 2.6 million forests and land, which is about 33 percent of this area is a peatland. These fires are estimated to produce 800 mega to 1.6 Giga metric tons of carbon dioxide equivalent [6]. These fires caused degraded peatlands area increased significantly.

The Government of Indonesia, through the Peat Restoration Agency (BRG), is undertaking peatland restoration efforts to improve the degraded peat land ecosystem. There are 3 (three) pillars in restoration, rewetting, replanting, and revitalization [6]. Rewetting peatlands is carried out so that the peatlands return to ideal moisture [7]. Revegetation is an action to restore land cover through replanting activities with native plants (endemic) and adaptive plants on open peatlands, enrichment planting in degraded peat forest areas, and improving and applying techniques seed-dispersing agents. dispersal techniques) to encourage the regeneration of peat vegetation [8]. Revitalization is carried out to empower the local community's economy [7]. In the restoration process, the three pillars are carried out in an integrated process [9]. The implementations of peat restoration were carried out in landscape unit called the Peat Hydrological Unit (PHU), which includes protection and cultivation functions [9]. The PHU took place various activities, both aimed at protecting and utilizing peatlands.

According to their depth the typologies of peatlands are grouped as follows: shallow peat (50 - 100 cm), medium peat (100 - 200 cm), deep peat (200 - 300 cm), and very deep peat (> 300 cm) [10]. Based on Government Regulation of the Republic of Indonesia Number 57 Year 2016 concerning Amendments to Government Regulation Number 71 of 2014 concerning Protection and Management of Peat Ecosystems. Peat ecosystem functions include for protection function at peat depth > 3 meter and for cultivation function in the peat < 3 meters. Peatlands developments in Indonesia are carried out by dividing the area into two, namely conservation areas which focused on rehabilitation and saving peatlands, and cultivation areas which are used for agricultural utilization and revitalization [11].

Peatlands have been widely used for settlement, infrastructure, plantations, and agriculture [12]. Plantation and agricultural cultivation activities are the main businesses of the community on peatlands [5]. Some peatlands for agriculture and plantations which were cleared in the 1970-1990s, are deep and very deep peatlands (400-700 cm thick) [13]. Cultivation activities are usually carried out by draining peatlands [14]. The leading cause of forest and land fires is the miss management of peatlands that have been trained to become flammable materials [15]. In addition, there are positive local customs and wisdoms that have long developed in the community in utilizing peatlands. These things foster community knowledge to manage and place business activities on peatlands. As a result, several business models based on land use, commodities, ecosystems, and environmental services have been developed. Some of them were identified as preventing peatland damage, preventing fires, and becoming a source of livelihood for the community.

The purpose of this research is to identify business models that developed on peatlands by the community, which are alleged to prevent forest and land fires and at the same time could improve the economic condition of the community. The best practices of these business models are expected to be lesson learnt to revitalization efforts in other peat land areas.

2. Methodology

2.1. Time and location of research

This study was conducted on 20-23 October 2020. The research location is in the Mendahara-Batanghari Peat Hydrological Unit, Tanjung Jabung Timur Regency, Jambi Province.

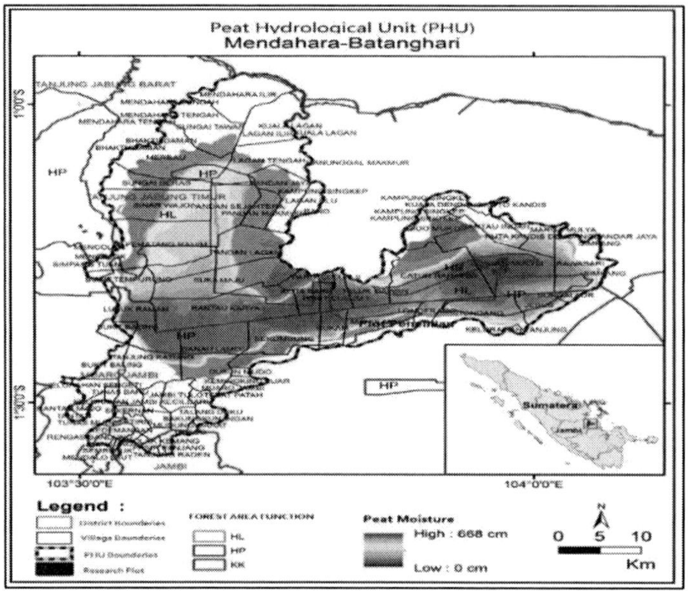

Figure 1. Map of research locations located in KGH Mendahara Batanghari, Tanjung Jabung Timur Regency, Jambi Province, Indonesia.

2.2. Data Collection Methods
The types of data used are primary and secondary data mostly in the form of qualitative data. This study conducted in-depth interviews with two people as representatives from each type of business, so there are 18 respondent of business actors from nine business models. As well as added respondents from KHG Manager from FMU Unit XIV Tanjung Jabung Timur. The forms of business are an individual business, Forest Farmers Group, and Forest Women Farmers Group. Determination of respondents purposively based on the role and type of business related to resource management in peat. Meanwhile, secondary data were obtained from books, reports, internet sites, and data and information from affiliated institutions. Data collection methods used were in-depth interviews, field observations, and exploration of secondary data.

2.3. Data analysis method
The business models identified are based on people's preferences. Based on the results of interviews and literature studies, a qualitative descriptive analysis was carried out to describe the mechanisms that run on these business models to prevent forest and land fires, and at the same time to improve community welfare.

3. Result and discussion

3.1. Rice paddy and horticulture cultivation business model
Lowland rice cultivation and horticulture business models are well developed on peatlands. Specifically the land availability for rice cultivation business models are very limited compared to the whole area of peatland. The small amount of lands means that only tiny part of the community lands are cultivated for lowland rice. Lowland rice cultivation is only carried out on thin peat, which is usually close to rivers. Lowland rice cultivation activities produce rice which has a vital role in maintaining food security and independence.

Lowland rice cultivation activities are carried out alternately with horticultural crops. In the rainy season the farmers will cultivate rice plants, while in the dry season they will cultivate horticulture. The types of horticulture are spinach, kale, corn, soybeans, eggplant, chilies, long beans, and mustard greens.

The activities of drying and burning of land are not required in lowland rice cultivation and horticulture. In addition, the remain decomposed leaves and stumps from rice cultivation can be used as input for horticultural cultivation so that it does not need to be burned. As we know, fires are usually caused by land preparation activities by burning on drained peatlands. Regarding forest and land fires, [16] stated that forest and peatland fires in South Sumatra often originate from burning activity in the land preparation process for rice cultivation on peatlands. [17] stated that the primary source of fire came from shifting cultivation carried out by the communities. The burning activity can spread to other areas if there are no management and supervision of the landowners. The rice and horticultural cultivation activities are carried out throughout the year so that there is no time lag for the land to be empty without management from the landowner. Based on research conducted by [18], one of the triggering factors for peat fires in Musi Banyuasin Regency is abandoned open access area. Thus, developing a business model for lowland rice farming and horticulture can prevent forest and land fires.

3.2. Areca nut cultivation business model

Areca nut is one of the primary plantation commodities for farmers in Jambi Province, apart from coconut and rubber [19]. The areca-producing centers areas are in Tanjung Jabung Barat and Tanjung Jabung Timur Regencies. The areas of areca nut in Jambi Province reached 19,968 ha. It covered about 9,882 ha in Tanjung Jabung Barat and about 8,894 ha in Tanjung Jabung Timur [20]. Initially, the *areca* nut trees were used to be a land divider or intercropping plant which were cultivated between oil palms. However, nowadays, people are starting to grow areca nuts in monoculture pattern. Some farmers have begun replacing unproductive oil palm trees with areca nut trees.

Areca nut cultivation requires the construction of primary and secondary trenches. The secondary canals are equipped with watergates to regulate the water discharge in the channels so that the surrounding peatland conditions are constantly moist. Humid conditions are needed so that *areca* plants can grow well. Thus, *areca* nut cultivation can keep the peatlands moist. Besides that, farmers usually plant vegetables such as corn, cassava, chilies, and other vegetable crops among *areca* plants. The *areca* nut harvest period is every 20 days. But farmers will visit their gardens more often because they have to maintain the vegetables among the areca plants, which have shorter harvest period. The frequent visits of farmers to the farm will intensify the maintenance of lands. So, when there is a threat of fire it will be quicker to be identified and controlled. Therefore the effort to keep peatlands in moist conditions and intensive management will protect peatlands from fires.

3.3. Business model of making plates from betel nut midrib

A business model utilizing by-products is starting to develop; one of them is a business model that uses betel nut midrib. Betel nut midribs are usually just thrown away but are now used as an innovative product in form of plates. This innovation was initiated by the Faculty of Agricultural Technology, Jambi University with the Rumah Jambe-e production. This invention was to assemble a betel nut midrib pressed by machine into a plate that could be used as a container for foods or fruits. Unlike the press machine originating from India, which required a large power, this press machines from Rumah Jambe-e are made with natural gas which are easy to be used by people in a home industrial scale. However, this business model is still running on a home industry scale by operating five hotpress machines with a production capacity of 5,000 pcs/smonth/machine.

Appropriate technology and the potential for abundant raw materials in the location areas are good opportunities for increasing the added value of betel nut midrib and as an alternative of income livelihood. In addition, the use of dry betel nut midrib can also reduce the organic fuels that can trigger forest and land fires. As stated by [16], abundant biomass in the dry season can trigger forest fires.

3.4. The agroforestry business model of liberika coffee

Liberika coffee has begun to be developed by many people in Tanjung Jabung Barat and Tanjung Jabung Timur, Jambi Province due to its high potential economic values. Unlike *Arabica* and *Robusta* coffee, Liberica coffee can grow well on peatlands. Agronomically, *Liberika* coffee has some advantages such as it can grow well on marginal lands especially on peatlands and it is resistant to leaf rust disease and coffee fruit borer attacks (BPTP, 2014). There are only a limited high economic species that can grow well on peatlands [21]. So, *Liberika* coffee can be used as an alternative commodity to increase people's income. *Liberika* coffee trees are usually cultivated with other plants that serve as the shade.

Liberika coffee trees are grown together with areca nut as the shade [22] or agroforestry with coconut trees [23]. With several canopy strata in the agroforestry cropping pattern, the land unit will function as a good receiver, storage, distributor, and release agent for water irrigation [24]. Thus the agroforestry pattern in this *Liberika* coffee cultivation can conserve water by controlling water flows during the rainy season and maintaining the condition of peatlands to be moist during the dry season. Furthermore, the moist condition of peatlands during the dry season can prevent fires.

3.5. Pineapple cultivation business model

Peat pineapple or pineapple tangkit (*Ananas comosus* (L) merr) is a type of pineapple that can adapt well and thrive in burned peatlands. Pineapple is an adaptive plant in acid swamps (pH 2-3), poor drainage, and thick peat [25]. Pineapple has excellent economic values because the pineapple business covers activities from upstream to downstream. Pineapple can be consumed without being processed first. The processed products are also varied, such as pineapple lunkhead, pineapple jam, pineapple chips, pineapple syrup, etc. There are also many side products and services from pineapple cultivation, such as pineapple fiber, bromeline for medicine, and ecotourism.

The pineapple cultivation process starts from land preparation. After the land conditions become clean, then soil compaction is carried out. Compaction made the soil pores become tiny so that it can reduce evaporation. The trenching activities are carried out around the planting area with a width of 2 meters and a depth of 2 meters and are not connected to other trenches. In the planting area, a canal is also made with a width of 50 cm and a depth of 50 cm. Soils which are always maintained in moist conditions as well as the trenches are always filled with water can prevent forest fires.

Pineapple harvests are carried out a year after planting. While waiting for the harvest time, the farmers maintain intensively the pineapple farm. After the first harvest, pineapple fruits can be harvested every week for a year ahead. Maintenances are also carried out so that pineapples can produce fruits optimally. Intensive care, monitoring, and utilization throughout the year on land for pineapple cultivation will prevent fires.

3.6. Honey beekeeping business model

The species is *Apis mellifera* L. This species is not a native Indonesian honey bees but originates from Europe, first imported to Indonesia in 1972 by Boyscouts from Australia [26]. This introduction aims to increase domestic honey production in Indonesia because this honey bees have a much higher productivity level than local *Apis cerana* F [27]. Community groups carry out honey beekeepings around the Wira Karya Sakti Industrial Plantation Forest areas.

The honey beekeeping activity is a program of the Peat Restoration Agency (BRG) to empower farmer groups around forest areas. the honey beekeeping provided alternative livelihoods for people without cutting trees in the forests or burning the peatlands. the honey beekeeping required foods derived from plant or tree flowers. The flowers of the acacia trees can be used as food for bees. However, the fire will destroy the acacia plant and its flowers. Fires also affect the migration of honey bees [28] and cause bees to leave the bee hives. Thus, farmers must protect the environment around their cultivation from fires to maintain food supplies and the presence of honey bees.

3.7. Jelutong agroforestry development business model

The swamp *jelutong* (*Dyera polyphylla* Miq.) is a native species of peatland forest. It can be an alternative for people besides oil palm [29]. This type of tree can be used for wood and sap. The woods are used as raw material for making pencils, carving, plywood, partition boards, ceilings, and clogs (a type of sandal made of wood). The sap can be used as raw material for chewing gum, paint, electric cable insulators, and export commodities.

Before, *jelutong* sap could be used as raw material for chewing gum, the demand for *jelutong* sap in 2014 are still good. But after that, the economic values of *jelutong* sap are no longer exists because the emergence of synthetic materials which can substitute its function. Therefore, *jelutong* cultivation is carried out in an agroforestry pattern with high economic value commodities under its shade to increase economic value, such as pepper, pepper, rattan, rhizome, etc.

As one of the native species of peatland, the cultivation of swamp *jelutong* does not require environmental engineering in the form of land draining. The agroforestry pattern encourages the community to carry out an intensive land management. Therefore, *jelutong* cultivation using agroforestry can improve environmental conditions and the economic conditions of the community. The integrated management system can prevent forest and peatland fires. Agroforestry *jelutong* business model is suitable for peat with a peatland depth of >3 meters.

3.8. Ecotourism business model based on peat ecosystem conservation

The business model for utilizing environmental services in the peat ecosystem has begun to develop by taking advantage of protected peat forests, which still have a good ecosystems. In 2017 Pematang Rahim Village obtained a Social Forestry permit with a Village Forest scheme covering 1,185 ha. The permit grants are expected to strengthen the community groups in protecting peat areas from fires and peatland damage. Besides that, it is also to support government programs to reduce greenhouse gas emissions and provide positive benefits for the community. Furthermore, village forest management is planned to be directed towards the ecotourism development. The peculiarities from the village forest in peatland areas can be appointed as ecotourism landmarks. This ecotourism business model will rely on preserving the ecosystem to support this business. Realizing the sustainability of the ecosystem as much as possible will protect the area from the threat of fire. Therefore, the ecotourism business model will be in line with fire control program.

3.9. Carbon trading business model

The carbon trading business model for the REDD scheme was started in 2013 by the community accompanied by the Indonesian Conservation Community Warsi. The Bujang Raba community applies the Village Forest and Community-based Forest Management scheme to have a prominent role in mitigation by maintaining forest cover and preventing logging from storing carbon in the form of carbon stock and preventing it from releasing into the air. This activity is a form of community role in mitigating and adapting to climate change. This is motivated by the threat of land conversion and deforestation into Industrial Plantation Forests and oil palm plantations against the impact of the ecological disaster that will be caused. Communities with awareness want area management that prioritizes sustainability and sustainable values. The carbon trading business model is one of the reward schemes to increase and maintain carbon stocks in stands and land management units. Fire is one of the events that can cause the release of carbon stocks. Thus, the carbon trading business goes hand in hand with preventing forest and land fires.

4. Conclusion

This study concluded that the business models that develops in the community which can prevent forest and land fires have particular characteristics such as (1) land preparation mechanisms without drying and burning, (2) land protecting to prevent abandoned open access areas, (3) agroforestry patterns that can intensify land management and add economic values, (4) Utilization of biomass which was previously an organic fuel for fires, (5) maintaining soil moisture through compaction for soil pore reduction and water management, (6) planting native species combined with economic value commodities with agroforestry patterns, (7) utilizing of the uniqueness of ecosystems, and (8) increasing public awareness of the environment through preventing the release of carbon stocks. These characteristics and the best practices of business models in peatland areas in PHU Mendahara-Batanghari can be lesson learnt and good examples to develop business models and to implement revitalization efforts in other peatland areas in Indonesia.

5. Acknowledgment
The authors would like to thank The Peatland Restoration Agency which has funded this research, The Head of KPHP Tanjung Jabung Timur and his staff for their cooperation so that this research can be carried out well, The Indonesian Conservation Community Warsi for their assistance and cooperation, The Village Head and his apparatus and the people of Sungai Beras Village, Sidodadi Village, Koto Kandis Dendang Village and Rantau Kariya Village Honey Farmers for the data and information provided.

6. References
[1] Purnomo H 2017 *Mengapa lahan gambut penting* (Bogor: Center for International Forestry Research) (in Bahasa Indonesia)

[2] Hapsari K A, Biagioni S, Jennerjahn T C, Reimer P, Saad A, Sabiham S, Behling H 2018 Resilience of a peatland in Central Sumatra, Indonesia to past anthropogenic disturbance: Improving conservation and restoration designs using palaeoecology *J. of Ecology* **106** (6) 2473-90

[3] BBSDLP 2019 *Balitbangtan Luncurkan Peta Gambut, Luas Turun 1.5 juta ha* (Jakarta: Gatra.com) (in Bahasa Indonesia)

[4] Wahyunto, Kuntjoro D, Sugiarti T, Jianto 2013 *Lahan Gambut di Kabupaten Kapuas Hulu, dan Potensinya untuk Pertanian* (Jakarta: Badan Litbang Pertanian Kementerian Pertanian) (in Bahasa Indonesia)

[5] Masganti M, Wahyunto W, Dariah A, Nurhayati N, Yusuf R 2014 Karakteristik dan potensi pemanfaatan lahan gambut terdegradasi di Provinsi Riau *J. Sumberdaya Lahan* **8** (1) (in Bahasa Indonesia)

[6] BRG 2016 *Mengawali Restorasi Gambut Indonesia* (Jakarta: Badan Restorasi Gambut) (in Bahasa Indonesia)

[7] Wicaksono A 2019 Kolaborasi multi aktor dalam program restorasi gambut di Provinsi Riau *J. Administrasi dan Kebijakan Publik* **4** (2) pp 99-113 (in Bahasa Indonesia)

[8] Dohong A, Sitepu D 2019 *Modul Pelaksanaan Kegiatan Revegetasi di Lahan Gambut* (Jakarta: Badan Restorasi Gambut) (in Bahasa Indonesia)

[9] Gunawan H, Afriyanti D 2019 Potensi perhutanan sosial dalam meningkatkan partisipasi masyarakat dalam restorasi gambut *J.Ilmu Kehutanan* **13** (2) pp 227-36 (in Bahasa Indonesia)

[10] Agus F, Subiksa I M 2008 *Lahan Gambut: Potensi untuk Pertanian dan Aspek Lingkungan* (Bogor: World Agroforestry Centre) p 36 (in Bahasa Indonesia)

[11] Suriadikarta D A 2012 Teknologi pengelolaan lahan gambut berkelanjutan *J. Sumberdaya Lahan Pertanian* **6 (2)** pp 197-211 (in Bahasa Indonesia)

[12] Maulana M R P, Herawati H, Kartini K 2021 Pengelolaan lahan gambut secara partisipatif studi kasus Desa Wajok Hilir *JeLAST J. PWK, Laut, Sipil, Tambang* **8** (1) (in Bahasa Indonesia)

[13] GoI (Government of Indonesia - The Netherland) 2008 Master plan for the rehabilitation and revitalization of the ex mega rice project in Central Kalimantan report first draft for consultation July (unpublished)

[14] Nurkholis A, Rahma A D, Widyaningsih Y, Maretya D A, Wangge G A, Widiastuti A S, Suci A, Abdillah A 2018 Analisis temporal kebakaran hutan dan lahan di indonesia tahun 1997 dan 2015 (studi kasus Provinsi Riau) *INA-Rxiv* doi:10.31227/osf.io/cmzuf. (in Bahasa Indonesia)

[15] Arifudin A, Syahza A, Kozan O, Mizuno K, Mizuno K, Isnaini ZL, Iskandar W, Hadi S, Asnawi A , Natasya A. Dinamika Penggunaan, Kebakaran, dan Upaya Restorasi Lahan Gambut: Studi Kasus di Desa Tanjung Leban, Bengkalis *Unri Conf. Series: Agriculture and Food Security* (Pekanbaru, Indonesia 26 September 2018) (in Bahasa Indonesia)

[16] Tampubolon J, Aluyah C, Heptiana E 2019 Persepsi masyarakat Desa Riding Kabupaten Ogan Komering Ilir terhadap upaya pencegahan kebakaran di lahan gambut *J. Sylva* **7** (2) pp 49-57 (in Bahasa Indonesia)

[17] Akbar A, Sumardi S, Hadi R, Purwanto P, Sabarudin M S 2011 Studi sumber penyebab terjadinya kebakaran dan respon masyarakat dalam rangka pengendalian kebakaran hutan gambut di areal mawas Kalimantan Tengah *J. Penelitian Hutan Tanaman* **8** (5) pp 287-300 (in Bahasa Indonesia)

[18] Tata H L, Pradjadinata S 2017 Native species for degraded peat swamp forest rehabilitation *J. Silvikultur Tropika* **7** (3) pp S80-S82

[19] Nur M 2016 Hubungan produksi buah dengan beberapa karakter vegetatif dan generatif populasi pinang betara melalui analisa koefisien lintas *Bul. Palma* **14** (1) pp 34-40 (in Bahasa Indonesia)

[20] Nasamsir N, Harianto H 2018 Pertumbuhan dan produktivitas lahan tumpang sari tanaman pinang (areca catechu l.) dan kopi (coffea sp.) *J. Media Pertanian* **3** (2) pp 61-71 (in Bahasa Indonesia)

[21] Tata H L, Bastoni, Sofiyuddin M, Mulyoutami E, Perdana A , Janudianto 2015 *Jelutong Rawa, Teknik Budidaya dan Prospek Ekonominya* (Bogor: World Agroforestry Centre) (in Bahasa Indonesia)

[22] Gusfarina DS 2014 *Mengenal Kopi Liberika Tungkal Komposit (Libtukom)* (Jambi: Balai Pengkajian Teknologi Pertanian Jambi) (in Bahasa Indonesia)

[23] Waluyo E A, Nurlia A Yeareditor^editors. Potensi pengembangan kopi liberika (coffea libericca) pola agroforestry dan prospek pemasarannya untuk mendukung restorasi lahan gambut di sumatera selatan. Pengembangan ilmu dan teknologi pertanian bersama petani lokal untuk optimasi lahan sub optimal. *Proc. Seminar Nasional Lahan Suboptimal* 2017 (in Bahasa Indonesia)

[24] Widiyanto A 2013 *Agroforestry dan Peranannya dalam Mempertahankan Fungsi Hidrologi dan Konservasi* (Tokyo: National Graduate Institute for Policy Studies) (in Bahasa Indonesia)

[25] Noor M 2004 *Lahan Rawa Sifat dan Pengelolaan Tanah Bermasalah Sulfat Masam: wetlands Characteristics and Management of the Sulphate Acid Soils* (Jakarta: PT RajaGrafindo Persada) (in Bahasa Indonesia)

[26] Hadisoesilo S 1991 Jenis-jenis lebah madu (species of honey bees). *J. Komunikasi* **5** (4) pp 5-6 (in Bahasa Indonesia)

[27] Verma L 1991 *Beekeeping in Integrated Mountain Development: Economic and Scientific Perspectives (Icimod Senor Fellowship Series, No. 4* (Newcastle: Aspect Publications Ltd)

[28] Rahmad B, Damiri N, Mulawarman M 2021 Jenis lebah madu dan tanaman sumber pakan pada budi daya lebah madu di hutan produksi subanjeriji, Kabupaten Muara Enim, Sumatera Selatan *J.l Penelitian Kehutanan FALOAK* **5** (1) pp 47-61 (in Bahasa Indonesia)

[29] Giesen W 2015 Utilising non-timber forest products to conserve Indonesia's peat swamp forests and reduce carbon emissions *J. of Indonesian Natural History* **3** (2) pp 10-19

Synergy between Disaster Preparedness Area program with local institutions for community disaster preparedness

Suradi[1] , S Sumarno[1] , Sugiyanto[1] , T Nainggolan[1] , R Murni[1] , RG Erwinsyah[1] and L Andriyani[1]

[1] Research and Development Center for Social Welfare, Ministry of Social Affairs, Cawang Kencana Building, East Jakarta 13630, Indonesia

Corresponding Author: erwinsyah@kemsos.go.id

Abstract. The Government of Indonesia through Ministry of Social Affairs launched the Disaster Preparedness Area Program (*Kawasan Siaga Bencana* or KwSB) in 2019 to build community preparedness for natural disasters. This program is a forum for early community-based disaster management synergy with an integrated approach that involves various stakeholders in areas with high risk of natural disasters. Meanwhile social reality exposes that historically, people living in areas prone to natural disasters have had their own local institutions and knowledge on dealing with natural disasters. Several research results indicated there is a gap between the government programs and the social realities in society. Accordingly, this research will identify the consequences of the KwSB Program on existing local institutions in the community before this program was launched. That very problem is answered through qualitative approach with data collected from literature studies, observations, in-depth interviews, and Focus Group Discussions (FGDs). Furthermore, the collected texts and narratives are analyzed by categorization and then comparing the existence and actions of local institutions before and after the KwSB Program. The results show that there are consequences due to the presence of the program on the social reality of local institutions, especially in the reconfiguration of social networks in the community.

1. Introduction

Natural disasters are global issue since they occur in most parts of the world. The disaster resulted in enormous impacts and risks for various dimensions of life, and threatened the survival of humans and other living things [1–4]. For Indonesia, as of February 2021, 237 natural disasters have occurred. These natural disasters include floods with 227 incidents, cyclones 66 incidents, landslides 60 incidents, earthquakes 7 incidents, tidal waves or abrasion 7 incident, and forest and land fires 4 incidents [5].

Handling these natural disasters requires an integrated and holistic approach that involves various parties [6–12]. The government through the Indonesian Ministry of Social Affairs has developed a social protection program for victims of natural disasters, which is technically organized by the Directorate of Social Protection for Natural Disaster Victims (*Direktorat Perlindungan Sosial Korban Bencana Alam* or in short; Dit-PSKBA) [13–15]. One of them is community-based disaster management in an area known as the Disaster Preparedness Area Program (*Program Kawasan Siaga Bencana* or in short KwSB), with the aim of increasing the culture of disaster preparedness in the community [16].

Several research results indicated there is a gap between the government programs and the social realities occurring in society [17–19]. The targeted communities are not a blank canvas that can be filled with anything, because they already have their own dynamics and uniqueness before the presence of government programs [17, 20]. Accordingly, this research will identify the consequences of the KwSB on existing local institutions in the community before this program was launched. In line with the explanation, the problem of this research is how KwSB builds synergy with related parties to build community preparedness to face disasters. Thus, the purpose of this study is to describe the activities of KwSB to build synergy with related parties to build community preparedness to face disasters.

2. Material and methods

Conceptually, the KwSB is an institution or organization that has a clear structure with special management, with all activities driven by a number of personnel in collaboration with a number of stakeholders who are partners. Thus, preparedness is the result, consequence, or even product of the KwSB as an institution with all the factors that influence it. This explanation is in line with the notion of preparedness as a series of activities carried out to anticipate disasters through organizing bwith appropriate and efficient steps.

Based on these considerations, this study uses a qualitative approach with descriptive research methods. This approach provides a detailed description of a specific situation, social setting or social relations [21]. The qualitative approach itself is a way to ascertain objects in which researchers study things or various phenomena in this world in their natural environment, and tries to get an understanding of these phenomena based on the meanings of people who live or experience them first-hand [22].

Primary data was collected by means of observation, in-depth interviews, and Focus Group Discussions. Secondary data was collected by means of documentation studies, by reading and reviewing books, papers, news, and regulations. The data collected were analyzed descriptively qualitatively according to the research objectives. Furthermore, the collected texts and narratives are analyzed by categorization and then comparing the existence and actions of local institutions before and after the Disaster Preparedness Area Program.

The research location was chosen purposively based on the consideration of the ongoing program implementation. The selected research locations are natural disaster-prone areas, such as Sunda Strait [23–25] and southern coast of Java Island with megathrust threat [26–28], that have already implemented the program [16]. As shown in **Table 1** below.

Table 1. Fieldwork locations.

Province	Regency	District	Village/Sub-District
S.R. Yogyakarta	Kulon Progo	Galur	Banaran
Banten	Pandeglang	Carita	Carita
West Java	Pangandaran	Parigi	Kalipucang
East Java	Malang	Tirtoyudo	Purwodadi
Lampung	South Lampung	Ketapang	Sumur
Gorontalo	Bone Bolango	Bulawa	Mopuya

3. Result and discussion

3.1. *KwSB overview*

The central government through the Ministry of Social Affairs conducts KwSB Program with the aim of: (a) improving public understanding of disasters; (b) improving the culture of disaster preparedness in the community; (c) improving community preparedness indicators, including the public, students, educators, and educational institutions located in disaster-prone locations; and (d) improving the readiness of disaster management officers and administrators.

The establishing process began with outreach activities, training and simulations involving relevant stakeholders in disaster-prone locations, who are also the administrators of KwSB. Of the six research locations, only Kulon Progo KwSB already has a decree (*Surat Keputusan* or SK) on the formation of KWsB issued by the Head of the Social Service for Women's Empowerment and Child Protection in

Kulon Progo Regency (*Dinas Sosial Pemberdayaan Perempuan dan Perlindungan Anak Kabupaten Kulon Progo*) dated September 14, 2020, while five KwSB in other locations do not yet have a decree.

In general, the establishment of the KwSB received positive responses. Informants from all research locations, especially stakeholders at the village level, positively welcomed the presence of KwSB. Informants stated that the presence of KwSB complements or strengthens the existing programs, by supplementing physical supplies or buffer stock. However, it is still limited to the community around the location of the establishment and the local Social Service (*Dinas Sosial*) who are indeed involved in the formation process as working partners of the Ministry of Social Affairs in the regions. The low insight of this formation is not only caused by the age of KwSB which has not yet reached one year, but also because of the vacuum state of KwSB itself, in the sense that it has not carried out institutional activities, so that it is not yet widely known, except for informal activities carried out by management administrators along with their daily activities.

This vacuum state is also related to the process of establishing KwSB through information dissemination activities, training and disaster management simulations involving the community. However, the activity lasted very short (three days), so the results obtained were less than optimal. The KwSB Administrators expressed his disappointment by saying "there were trainings, but that was it, the main thing is that it is carried out. We were even given tools, but we didn't know how to use them. Our people are not trained to use it."

After the establishment, no monitoring was carried out by involved parties in the form of further guidance activities to ensure the functioning of the KwSB. This means that the KwSB has not yet been fully formed. KwSB Kulon Progo, which has formal legality, is still experiencing difficulties due to limited authority, unavailability of budget, and no secretariat. The social buffer stock which is used as a place to store disaster mitigation equipment is still staying at the Karang Taruna secretariat. Meanwhile, in other locations, KwSB which does not yet have a SK is more difficult to obtain a budget from the regional government because it does not have a legal standing.

This basic problem makes it difficult for KwSB to coordinate with government agencies and other parties engaged in disaster activities. Several informants indicated that the administrators had doubts about managing the KwSB. The doubt is related to the existing government structure with all the bureaucratic provisions that are felt to be difficult to penetrate. The same complaint appeared in the FGD forum when participants said "the KwSB should have been supported by a decree from the Regency Government so that its position is stronger". While another said, "the unclear regulation and control of KwSB is felt to be a moral burden for the administrators."

But at the community level, in fact efforts to build community preparedness already exist through the relevant agencies with the presence of the Disaster Resilient Village (*Desa Tangguh Bencana* or Destana) formed by the Regional Disaster Management Agency (*Badan Penanggulangan Bencana Daerah* or BPBD), Disaster Preparedness Youth (*Taruna Siaga Bencana* or Tagana) formed by the Ministry of Social Affairs, and similar institutions formed by the community. The presence of KwSB is expected to further strengthen efforts to build preparedness. However, the disaster care group in the FGD complained by saying "...so the facts are like this, sir. No matter how many programs, each agency makes a program. When it comes to the community, they're the same people, sir. It would be better if the government made one program but comprehensive. So the development is also easy."

3.2. KwSB synergies

Synergy is a form of a process or interaction that produces a harmonious balance so that it can produce something optimal. This activity runs informally in the community along with daily community activities involving KwSB management administrators.

Conceptually, the KwSB is not yet known by the wider community in the research location. However, the individuals who become the administrators of the KwSB carry out their moral responsibilities in carrying out their duties even though it has not been carried out formally. Along with daily activities, administrators interact with disaster care groups in responding to social situations to create disaster preparedness.

In the Ketapang area of South Lampung, disaster management groups run in sectors under each institution so that the development of the concept of a Disaster Preparedness Village (*Kampung Siaga Bencana* or KSB) into a Disaster Preparedness Area (KwSB) has not been realized, such as the KSBwith the Social Service, the National Police with Resilient Village (*Kampung Tangguh*), BPPD with the Destana, and others.

In Kalipucang Village, Pangandaran Regency, a KSB has already been formed. The task of this KSB is to carry out mapping and data collection, provide socialization related to natural disasters, guide community preparedness to face natural disasters, and provide services for victims of natural disasters. When carrying out these activities, KSB has established a synergy with the Community Early Preparedness Forum (*Forum Kesiapsiagaan Dini Masyarakat* or FKDM), which is an institution formed by the BPBD. In addition to coordinating with FKDM, KSB was also assisted and received reinforcement from Tagana and the KwSB. The disaster mitigation equipments managed by Tagana and KwSB are used jointly with KSB and FKDM in natural disaster management at the village level. The synergy between KSB and FKDM supported by Tanaga and KwSB succeeded in reducing the impact of natural disasters in Pengandaran Regency.

In Pandeglang. At the district level, cooperation between regional apparatus organizations (*organisasi perangkat daerah* atau OPD) has not yet been seen, although each OPD has carried out activities related to natural disasters, both in pre-natural disasters, natural disaster emergencies and post-natural disasters. However, each OPD implements it according to the activities that have been planned by each OPD. Meanwhile, at the sub-district and village levels, KSB in carrying out natural disaster management coordinates with local local organizations such as Destana, Balawista, Kompak, SAR Potential, and Tagana.

Kulon Progo synergies occur in establishing community preparedness in dealing with disasters at the community level. At the district level, there is an agreement between the Social Service and BPBD that preparedness is the main task of the BPBD, while emergency and rehabilitation are the affairs of the Social Service. There are several local institutions at the community level that are also engaged in disaster management. In the Banaran Village, there is an environmental care group that is active in planting mangroves in order to prevent abrasion on the coast as well as to protect the land from the threat of big waves. This group synergizes closely with KwSB, among others, by collaborating in planting mangroves in the estuary of Kali Progo. The synergy between KwSB and this group occurred because several activists of this environmental care group also served as administrators of the Galur KwSB. The network of environmental activist actors in this region is actually closely networked with actors who care about disaster issues. With the existence of KwSB, the existence of environmental care groups is further strengthened since they have access to the Social Service and the Ministry of Social Affairs. They are able to marry environmental conservation issues with natural disaster prevention so that the discourse on disaster management becomes more holistic starting from the pre, during, and post-disaster stages.

Various components, both institutions and individuals, have collaborated and synergized in community-based disaster management in Malang Regency. BPBD as the agency tasked with coordinating disaster management activities initiated the Destana while the Ministry of Social Affairs, the East Java Provincial Social Service and the Malang Regency Social Service established the KSB (there were five KSBs formed by the Ministry of Social Affairs and two KSBs formed by Provincial Social Service) as a form of community-based disaster management established by the government. Meanwhile, Indonesia Red Cross formed a Community-Based Assistance Alert Team (CBAT) for disaster-prone villages in Malang Regency. The South Malang area is a disaster-prone area, the types of disasters that often occur are floods, landslides and earthquakes and the potential for a tsunami disaster to occur because it is directly facing the Indian Ocean.

However, it has not been implemented well in Bone Bolango Regency. Synergy between institutions has not been well developed because each institution is still carrying out its duties and roles according to the duties and functions of the main organizations.

4. Conclusion

Based on the findings from the six research areas, it can be concluded that the synergy between the KwSB Program and local institutions occurs in areas that have historically been accustomed to dealing with natural disasters. In this area, there have been many local institutions that have concern for natural disasters. Meanwhile, in areas that have only recently begun to experience threats or have just experienced natural disasters, this synergy is less visible because local institutions have not been exposed.

Unfortunately, this synergy occurs because of the similarity of personnel between the KwSB management and local institutions, such as environmental activists who are also the KwSB administrators in Kulon Progo. However, changes that occur in local institutions are that with the actors becoming KwSB administrators, they are able to access new social capital, namely relations with regional and central governments as larger policy makers.

Acknowledgement

All authors are main contributors to this work. This work is supported and funded by Research and Development Center for Social Welfare, Ministry of Social Affairs.

References

[1] E. Rodriguez-Oreggia, A. de la Fuente, R. de la Torre, H. Morenp, and C. Rodriguez 2010 The Impact of Natural Disasters on Human Development and Poverty at the Municipal Level in Mexico, CID Working Paper No. 43, Feb. 2010. doi: 10.1002/pam.4050040209.

[2] O. Petrucci 2012 The Impact of Natural Disasters: Simplified Procedures and Open Problems in *Approaches to Managing Disaster - Assessing Hazards, Emergencies and Disaster Impacts*, InTech, 2012.

[3] S. M. S. Huq 2016 Community Based Disaster Management Strategy in Bangladesh: Present Status, Future Prospects and Challenges *Eur. J. Res. Soc. Sci.*, **4** 2 pp 22–35

[4] H. A. Rahman 2018 Community Based Approach Towards Disaster Management in Malaysia *Asian J. Environ. Hist. Herit.* **2** 2 pp 55–66

[5] Fajar Indonesia Network 2021 Awal 2021, 327 Bencana Alam Landa Indonesia https://fin.co.id/2021/02/09/awal-2021-327-bencana-alam-landa-indonesia/ (accessed Feb. 11, 2021).

[6] Y. Wulandari, S. A. H. Sagala, and G. B. Sullivan 2018 The Role of Community-Based Organization in Disaster Response at Mt. Sinabung *IOP Conf. Ser. Earth Environ. Sci.* **158** 1 p 012035 May 2018 doi: 10.1088/1755-1315/158/1/012035.

[7] L. Hanifa, A. Sadat, D. Mahmuda, A. Nazar, R. Jasiyah, R. S. Wijaya, A. A. Unde, N. Ichsani, S. Anwar, and R. A. F. Nurfida 2019 Handling disaster risks with the community-based approach *IOP Conf. Ser. Earth Environ. Sci.* **235** 1 p 012034 doi: 10.1088/1755-1315/235/1/012034.

[8] O. Rosmaladewi, I. Danuwikarsa, R. Robana, and P. Pranadikusumah 2019 Forest Village Community Empowerment through Multi-stake holder Partnership Program to form Sustainable Green Forests and Disaster Preparedness in Garut *IOP Conf. Ser. Earth Environ. Sci.* **298** 1 p 012036 doi: 10.1088/1755-1315/298/1/012036.

[9] E. Yulianti and Hastuti 2019 The role of women in deadling with risk of flood *IOP Conf. Ser. Earth Environ. Sci.* **271** 1p 012026 doi: 10.1088/1755-1315/271/1/012026.

[10] R. Afrian, J. Hariadi, B. Akob, and Z. R. Islami 2020 Local Culture Inventory for Disaster Mitigation Learning *IOP Conf. Ser. Earth Environ. Sci.* **412** 1 012017 doi: 10.1088/1755-1315/412/1/012017.

[11] I. L. Nugraheni and A. Suyatna, "Community Participation in Flood Disaster Mitigation Oriented on the Preparedness: A Literature Review," *J. Phys. Conf. Ser.*, vol. 1467, no. 1, 2020, doi: 10.1088/1742-6596/1467/1/012028.

[12] E. Maryani 2021 The role of education and geography on disaster preparedness *IOP Conf. Ser. Earth Environ. Sci.* **683** 1 p. 012043 doi: 10.1088/1755-1315/683/1/012043.

[13] Ministry of Social Affairs of The Republic of Indonesia 2011 Peraturan Menteri Sosial Republik Indonesia Nomor 128 Tahun 2011 Tentang Kampung Siaga Bencana (Minister of Social Affairs of The Republic of Indonesia Regulation No. 128 year 2011 about Disaster Preparedness Villages) [Online] Available: https://jdih.kemensos.go.id/dokumen/view?id=158.
[14] Ministry of Social Affairs of The Republic of Indonesia 2012 Peraturan Menteri Sosial Republik Indonesia Nomor 29 Tahun 2012 Tentang Taruna Siaga Bencana (Minister of Social Affairs of The Republic of Indonesia Regulation No. 29 Year 2012 about Disaster Preparedness Cadets) [Online] Available: https://jdih.kemensos.go.id/dokumen/view?id=1028.
[15] H. Habibullah 2013 Kebijakan Penanggulangan Bencana Berbasis Komunitas: Kampung Siaga Bencana dan Desa/Kelurahan Tangguh Bencana *Sosio Inf.* **18** 2 pp 133-150, Aug. 2013, doi: 10.33007/inf.v18i2.69.
[16] A. Mufida 2019 Ministry of Social Affairs Develops Disaster Preparedness Areas https://kemensos.go.id/en/ministry-of-social-affairs-develops-disaster-preparedness-areas-in-potential-megathrust-areas (accessed Mar. 23, 2021).
[17] T. M. Li 2007 *The Will to Improve: Govermentality, Development, and the Practice of Politics*. Durham: Duke University Press.
[18] B. Kusumasari and Q. Alam 2012 Bridging the gaps: the role of local government capability and the management of a natural disaster in Bantul, Indonesia *Nat. Hazards* **60** 2 pp 761-779 doi: 10.1007/s11069-011-0016-1.
[19] A. H. Simanjuntak and R. G. Erwinsyah, 2020 Kesejahteraan Petani dan Ketahanan Pangan pada Masa Pandemi Covid-19: Telaah Kritis terhadap Rencana Megaproyek Lumbung Pangan Nasional Indonesia *Sosio Inf.* **6** 2 pp 184-204 doi: 10.33007/inf.v6i2.2332.
[20] T. M. Li, 2005 *Transforming the Indonesian Uplands: Marginality, Power and Production*. London: Routledge.
[21] W. L. Neuman 2006 *Basics of Social Research: Qualitative and Quantitative Approaches*, 2nd ed.
[22] N. K. Denzin and Y. S. Lincoln, Eds. 2005 *The SAGE Handbook of Qualitative Research*, 3rd ed., **1** 1 Los Angeles: SAGE Publication.
[23] Alhamidi, V. H. Pakpahan, and J. E. S. Simanjuntak 2018 Analysis of tsunami disaster resilience in Bandar Lampung Bay Coastal Zone *IOP Conf. Ser. Earth Environ. Sci.* **158** 1 p 012037 doi: 10.1088/1755-1315/158/1/012037.
[24] W. L. Farahdita and H. S. R. Siagian 2020 Analysis of the area affected by the tsunami in Pandeglang, Banten: a case study of the Sunda Strait Tsunami *IOP Conf. Ser. Earth Environ. Sci.* **429** 1 p 012052 doi: 10.1088/1755-1315/429/1/012052.
[25] D. A. P. Sari and T. E. B. Soesilo 2020 Measuring Community Resilience to the Tsunami Disaster (Study of Sukarame Village, Carita District, Pandeglang Regency) *IOP Conf. Ser. Earth Environ. Sci.* **448** 1 p 012092 doi: 10.1088/1755-1315/448/1/012092.
[26] J. D. Prasetya, Y. N. Maharani, and I. Rahatmawati 2018 Mangrove ecotourism management at local community in Jangkaran, Kulonprogo, using hierarchy analysis *IOP Conf. Ser. Earth Environ. Sci.* **212** 1 p 012006 doi: 10.1088/1755-1315/212/1/012006.
[27] S. G. Marlyono and N. Nandi 2018 The Preparedness Level of Community in Facing Disaster in West Java Province *IOP Conf. Ser. Earth Environ. Sci.* **145** 1 p 012103 doi: 10.1088/1755-1315/145/1/012103.
[28] R. Wargadalam, H. Nakanishi, Y. Vidyattama, J. Black, and Y. Suenaga 2021 Tsunami evacuation decisions and behaviour: A case study of Pangandaran, Indonesia *IOP Conf. Ser. Earth Environ. Sci.* **630** 1 p 012023 doi: 10.1088/1755-1315/630/1/012023.

IOP Conf. Series: Earth and Environmental Science 1027 (2022) 012019 doi:10.1088/1755-1315/1027/1/012019

The Role of Crystal Fraction to Relative Magma Viscosity: An Approach to Understand the Explosive Caldera-Forming Volcanism of Raung Volcano

Firman Sauqi[1], Mirzam Abdurrachman[2] and Idham Andri Kurniawan[2]

[1] Doctoral Program of Geological Engineering, Faculty of Earth Science and Technology, Institut Teknologi Bandung, Jl. Ganesha No. 10, Bandung, 40132, Jawa Barat, Indonesia
[2] Department of Geological Engineering, Institut Teknologi Bandung, Jl. Ganesha 10 Bandung, 40132, Jawa Barat, Indonesia

Corresponding Author: firman.sauqi001@gmail.com

Abstract. Explosive volcanism is one of the deadliest volcanic hazards. The characteristics of the magma chamber that cause explosive eruptions need to be studied more deeply. Raung Volcano in East Java is an example of volcano that has a history of explosive eruptions that marked by a 2 km wide caldera. Petrographic analysis was carried out on selected 4 basaltic lava samples to determine the composition and proportion of minerals. Then, an analysis of the viscosity of the magma was carried out in relative terms using the shape and size configuration of the crystals. The comparison between the percentage of crystals and the relative viscosity value shows a positive correlation which indicates that there is an effect of the number of crystals on the viscosity of the magma. The increase in the viscosity of magma that occurred in the pre-caldera lava eruption of Raung Volcano indicates an intensive crystallization. This is supported by petrographic data which shows a porphyritic texture with a phenocryst of more than 40%. Plagioclase crystals become quite dominant crystals (> 30%). Based on this, it can be interpreted that the crystallization process occurred intensively in the shallow magma pocket of Raung Volcano.

1. Introduction

Raung Volcano in East Java is one of the most active stratovolcanoes in Indonesia. Raung is part of the Ijen Volcanic Complex, located in Jember, Bondowoso and Banyuwangi Regencies (Figure 1.). According to the Center for Volcanology and Geological Disaster Mitigation [1], the eruption of Raung Volcano was recorded as having experienced violent eruptions in 1586, 1597, 1638, 1953, and 1956 which caused thousands of casualties. After 1956, eruptions occurred almost every 10-20 years and since 2008 eruptions have occurred several times in a period of less than 10 years. The last eruption occurred in June to July 2015 with a strombolian type with the product of lava flow that extends inside the caldera and volcanic ash scattered over an area with a maximum radius of up to 200 km. Based on the historical record of the eruption, Raung Volcano still has the potential to experience eruptions in the future.

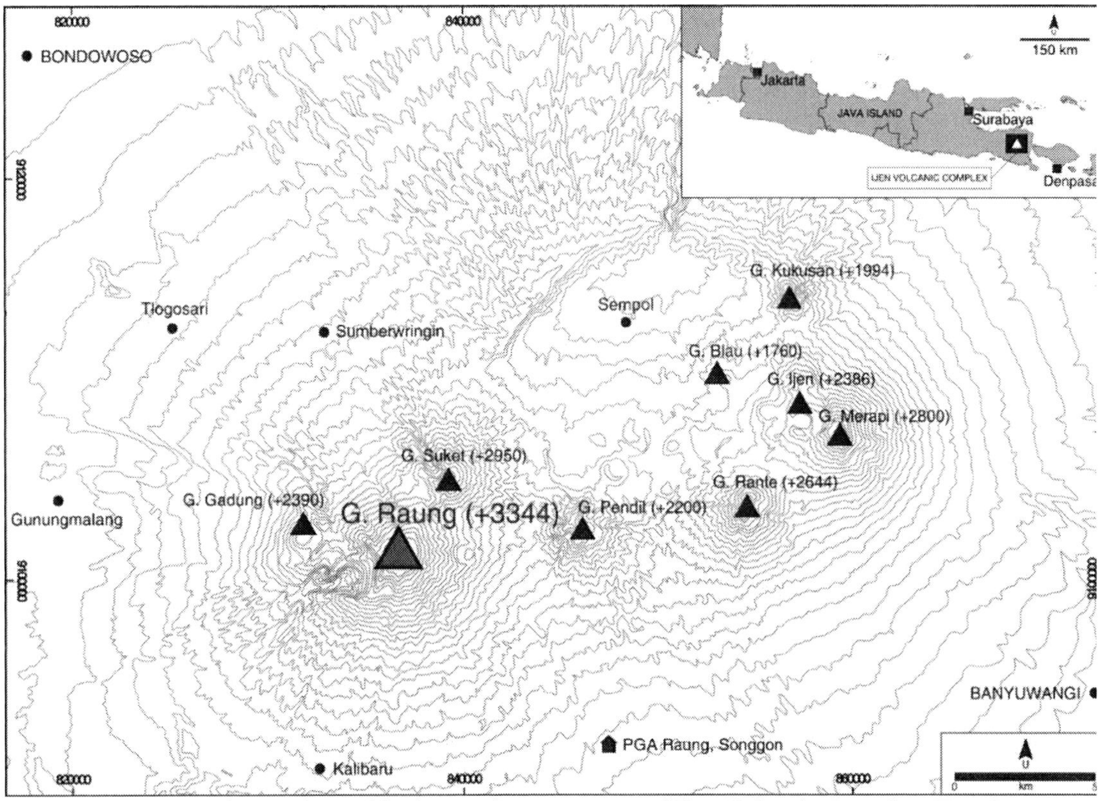

Figure 1. Raung Volcano situated in the southwestern part of Ijen Volcanic Complex, East Java.

1.1. Geological Setting

Raung Volcano is a stratovolcano with a height of 3344 meters above sea level that belongs to the Quartile volcano zone in Java [2]. Raung Volcano grows on granite rocks that form part of the southern mountain zone. The existence of granitic rocks from Gondwana as a constituent of continental crust under the mountains of southern Java and continuously to South Sulawesi. The bending to Sulawesi is right past the bottom of Raung Volcano, indicating that Raung Volcano is on top of a continental crust. The existence of Gondwana fragments is also estimated continuously to the southern mountains in western Java [3]. Ijen Volcanic Complex in the northeast of Raung Volcano, tectonically is an active continental type [4]. This indicates that Raung Volcano and Ijen Volcanic Complex grow on the same crust. Analysis of petrogenesis [4] shows that Raung Volcano is tectonically an active continental edge in the Subduction of the Indo-Australian Plate against fragments of the continental crust thought to be from Gondwana. This, geochemically, has confirmed the presence of a continental crust from Gondwana as the base of Raung Volcano.

2. Materials and Methods

The sampling location is the tip of the lava flow of the four pre-caldera lavas of Raung Volcano (Table 1). The samples used in this study were taken from the middle part of the lava body (Figure 2). The center of the lava is a less-than-refreshed part of external processes as the lava flows on the surface. Sampling is done with geological hammers and chisels. The sample measured approximately 15 cm x 15 cm with an average mass of 250 grams. The sample is the freshest part with a composition that describes the entire body of the lava.

Rocks sample obtained from the field already be prepared into thin section for petrographic analysis with a thickness of 0.03 mm. Through petrography analysis, already be done the analysis of the mineral presence which is then used for the distribution of crystal size.

Table 1. Lava sampling location according to previous research [4].

Rock Unit	Station	X (mE)	Y (mN)	Lithology	Distance from eruption centre (km)
RL4	SBM-020	826721	9108363	Basalt	10.7
RL3	LGN-014	833707	9110315	Basalt	11.2
RL2	RAU-016	832731	9113140	Basalt	14.7
RL1	KBG-006	826516	9114065	Basalt	18.4

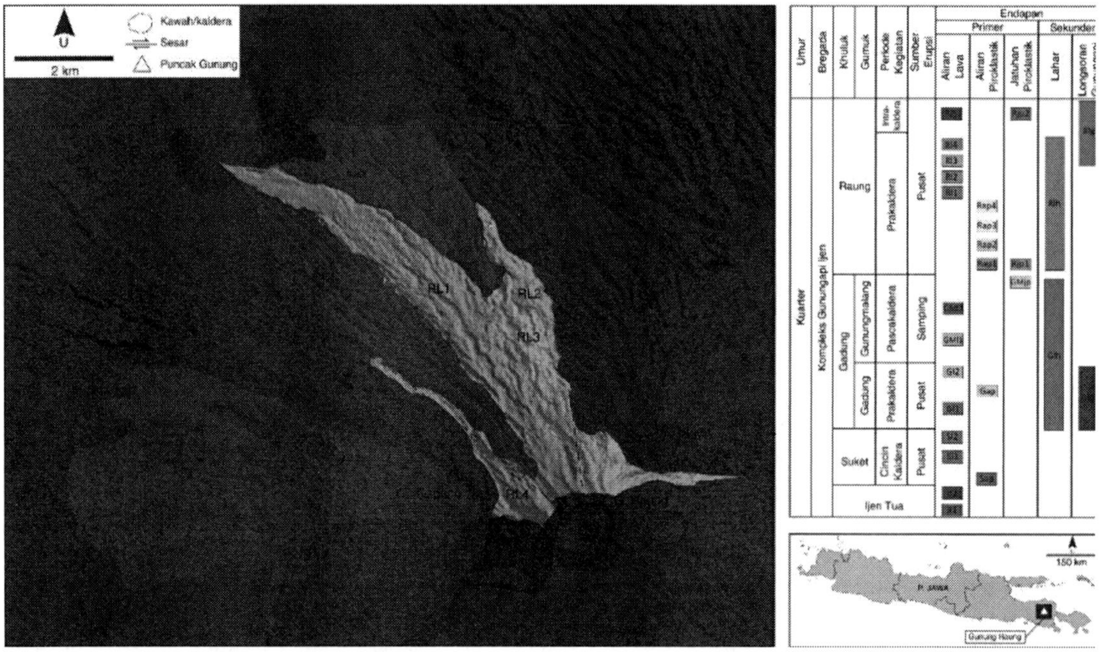

Figure 2. Geological map of Raung Volcano and the location of selected lava flow.

2.1. *Petrography Analysis*

Petrography observations were conducted in ITB Mineralogy and Petrography Laboratory with Ci-POL polarization microscope manufactured by Nikon, Tokyo, Japan. Calculation of the composition of mineral presence is carried out on observation with a magnification of 40 times. The magnification scale is based on average groundmass size that can be easily observed. The composition of this mineral, especially plagioclase, that has a continues magma cooling series, is used as a preliminary reference before performing an analysis of the distribution of crystal size. Plagioclase record the continues magmatic process. The percentage of abundance, size, and shape visually observed from petrographic observations will be able to provide an overview of the distribution of the size of the crystals. The proportion is based on area observed from crystal size distribution analysis. The dimension of observed area is based on scale of magnification of petrographic observation. The method used in quantifying the presence of minerals in this study is by visually calculating the fraction of mineral areas through petrographic observation. Based on the quantification, it is then determined the proportion of the crystal fraction and the non-crystalline fraction. The crystal fraction includes the percentage of each mineral observed. The non-crystalline fraction includes the percentage of volcanic glasses and the size of the vesicular.

2.2. Magma Viscosity Analysis

Magma viscosity is the physical properties of magma that describes the resistance caused by interparticle force within the magma [5-7]. The lower viscosity, the greater flow resistance in the magma will be smaller and make it easier for lava to flow. Viscosity is influenced by the abundance of crystals, the number of gas bubbles, and the temperature of magma. The influence of viscosity is thought to play an important role in the dynamics of magmatism systems and active volcano volcanic activity [6], including on Raung Volcano.

The distribution of the shape and size of crystals can draw the relative viscosity value of magma [5-7]. The viscosity of magma depends largely on the number of crystals and their crystal [8, 9]. The more crystals, the viscosity increases and the greater the aspect ratio or the flattened the crystal, the viscosity will decrease because the shape indicates high magma mobility [5, 7]. The shape of crystals is generally expressed in variations in aspect ratio and diameter equivalent, namely monodispere and polydispere [5, 5]. The aspect ratio of crystals states the comparison between the long axis and the short axis of the crystal at 2D projection [5, 7]. Meanwhile, the crystal equivalent diameter is the diameter obtained from the maximum projection of 2D crystals in the form of circles [8, 9]. The projection center point of the circle is the crystal symmetry rotary. The shape of monodispere crystals describes a low variation of crystals so that they have a more homogeneous crystal shapes. Polydispere crystal shapes describe variations in more heterogeneous crystal shapes. The shape of monodispere crystals makes the maximum packing fraction of crystals in magma will increase [5, 7]. The shape of the crystal can be expressed with the value of crystal polydispersity with the equation proposed by Klein [8, 9] as follows:

$$\gamma = \frac{s_p}{s_m} = \frac{\langle r\rangle\langle r^2\rangle}{\langle r^3\rangle} \quad \text{(I)}$$

with:
γ = crystal polydispersity
s_p = specific surface ratio of polydispere crystals
s_m = monodisper crystal specific surface ratio
r = crystal equivalent diameter

The polydispersity value is then combined with other parameters of the crystal size distribution analysis. The value of the crystal fraction is a percentage of the number of crystals in the thin incision photo. The aspect ratio of the average is the result of calculating the average aspect of the entire crystal ratio measured. Based on Klein [8, 9] experiments, the maximum pack fraction value on monodispere crystals is 0.637 and empirical constants α and b are used in equations. Based on the combination of all geometric parameters and polydispersity values of Equation I, the viscosity value of magma by Klein [8, 9] can be relatively determined using the following equation:

$$\eta_r = \left(1 - \phi\left[1 - \left(\left(1 - \left\{\phi_{m1} \exp\left[-\frac{(log_{10} r_p)^2}{2b^2}\right]\right\}\right)\gamma^{\alpha}\right)\right]^{-1}\right)^{-2} \quad \text{(II)}$$

With:
η_r = relative viscosity
φ = crystal fraction
γ = crystal polydispersity
r_p = aspect ratio of average crystals
φ_{m1} = 0.637 (maximum pack fraction of monodisper particles or current r_p=1)
α = 0.173

IOP Conf. Series: Earth and Environmental Science 1027 (2022) 012019 doi:10.1088/1755-1315/1027/1/012019

b = 1.08

3. Results

3.1. Petrography Overview

Based on petrographic observations that have been done (Table 2), in general in thin incisions lava samples can be divided into crystalline fractions and non-crystalline fractions. The crystalline fraction consists of minerals that are distinguished by their size into phenocryst, microphenocryst, and basic masses. The non-crystalline fraction consists of volcanic and vesicular glasses. The vesicular reflects the shape of the gas bubbles trapped during the magma freezing process. The proportion of the three fractions then calculated based on the fraction of the 2D area of the petrographic observation, and then presented as in Table 2. In general, thin incisions of lava samples (Figure 3.) porphyritic texture with phenocryst more than 40% consisting of plagioclase, clinopyroxene, olivine, and opaque minerals. The smaller crystals consist of microphenocryst and the base mass of plagioclases as well as clinopyroxene. Plagioclases are subhedral-anhedral with a size of 0.1-4 mm and some show special textures on plagioclase phenocryst. Clinopyroxene is subhedral-anhedral in shape and measures 1-2 mm. The anhedral olivine with its edges has been penetrate and filled with opaque minerals which are most likely ilmenite and magnetite. Volcanic glasses present as a base mass. The observed vesicular is circular in size 0.1-3 mm and there are also several elongated circular vesiculars that are thought to be the result of merging by several vesiculars that have smaller sizes.

Table 2. Proportion of crystalline and non-crystalline fractions of petrographic observations

Eruption Product	Crystalline Fraction (%)								Non-crystalline Fraction (%)	
	Plagioclase		Clinopyroxene		Olivine		Opaque		Vesicular	Glass
	F	M	F	M	F	M	F	M		
RL4	40	15	6	8	1	0	3	0	13	14
RL3	37	15	6	8	2	0	3	1	11	17
RL2	32	16	9	5	4	0	3	1	6	24
RL1	30	17	5	5	4	0	3	1	4	31

IOP Conf. Series: Earth and Environmental Science 1027 (2022) 012019 doi:10.1088/1755-1315/1027/1/012019

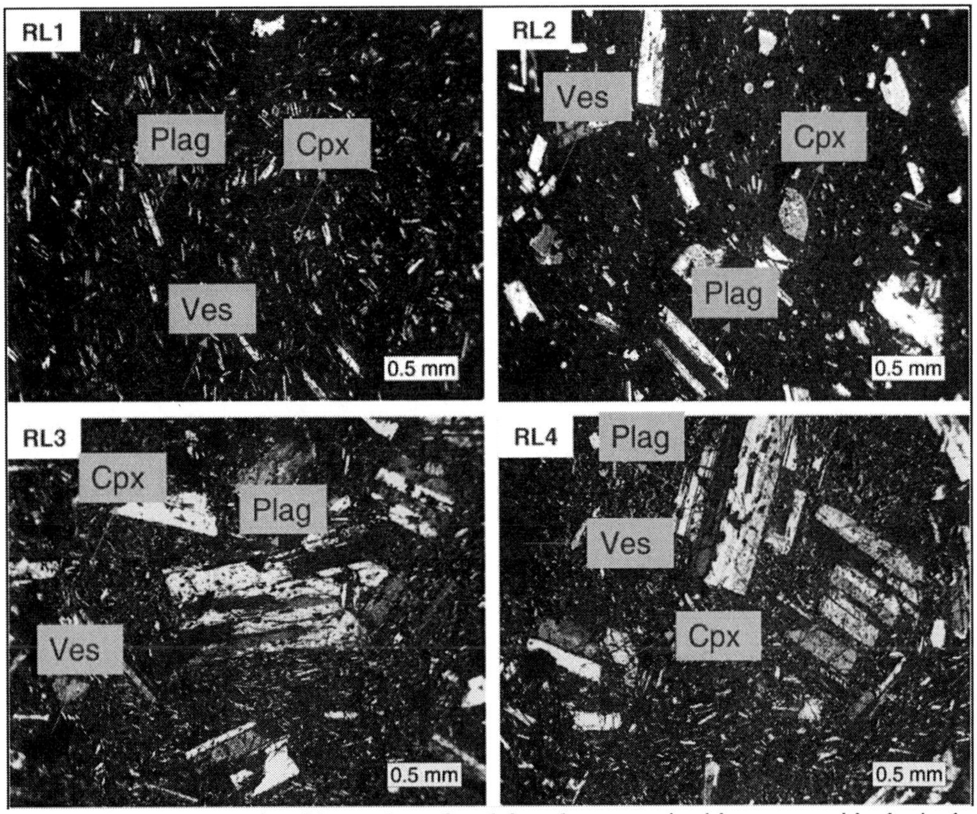

Figure 3. The representative thin section of each lava has a porphyritic texture with plagioclase phenocryst, clinopyroxene, olivine, and opaque minerals.

3.2. Magma Viscosity

The data of crystal geometric parameters that determine the relative viscosity of magma are crystalline fraction, polydispersity, and maximum packing fraction (Table 3). The pattern from old to young, the crystal fraction increases and is positively correlated to magma viscosity (Figure 4.). Meanwhile, polydispersity, which also affects the maximum pack fraction, shows no pattern based on the order of lava age. The relationship of polydispersity and the maximum pack fraction describing the shape of the crystals shows a negative correlation (Figure 4.) Relative viscosity is derived from the calculation using Equation II. The resulting relative viscosity value indicates an increasing pattern and has a positive correlation to the crystal fraction value.

Table 3. The geometric parameters of the crystals used in determining the relative viscosity value of magma in the entire lava.

Lava	Crystalline Fraction	Aspect Ratio	Polydispersity	Maximum Packing Fraction	Relative Viscosity
RL4	0.395	2.440	0.082	0.743	4.6
RL3	0.351	2.698	0.141	0.712	3.9
RL2	0.222	2.100	0.117	0.733	2.1
RL1	0.112	2.616	0.109	0.726	1.4

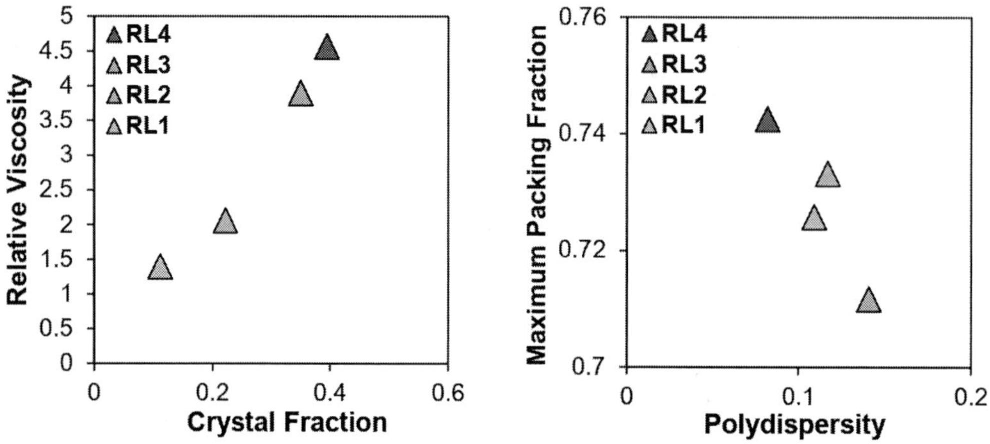

Figure 4. (a) Positive correlation between crystal fraction and relative viscosity and (b) negative correlation between maximum packing fraction against crystal polydispersity.

Magma viscosity increased from RL1 to RL4 by a ratio of 1.4:2.1:3.9:4.6 (Figure 5). This viscosity value is the relative value expressed in the ratio. The increase in viscosity from RL1 to RL2 and RL3 to RL4 showed relatively similar levels of improvement. However, in RL3 to RL4 the viscosity increase is quite significant reaching almost double.

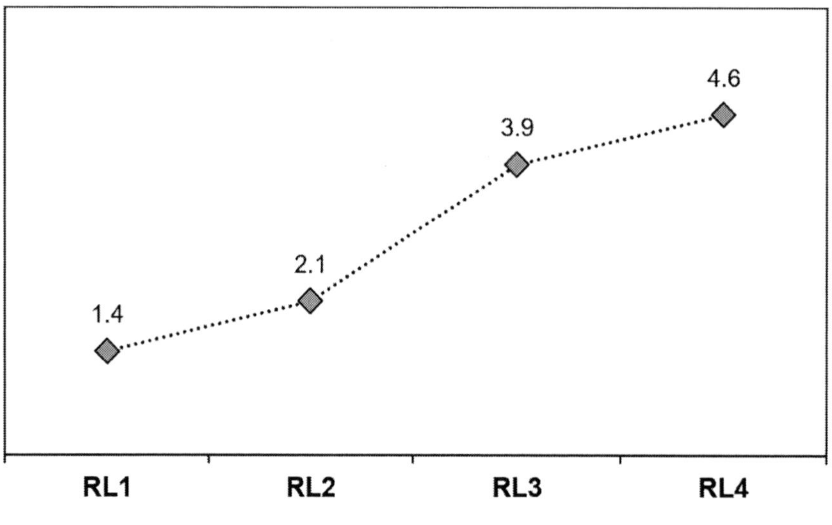

Figure 5. The change in the relative viscosity ratio of magma indicated by an increase pattern. The relative viscosity of magma increased most significantly in RL2 to RL3.

4. Discussion

Based on magma viscosity analysis, the increase in the relative viscosity value of magma has a positive correlation with the amount of crystal fraction (Figure 6.). The same is also illustrated in the rise of the crystal fraction. This indicates that the increase in magma viscosity is controlled by the crystalline fraction in the magma. The more crystalline fractions in magma, the higher the solid fraction will increase and will increase the value of particle friction [5,10]. The friction of these particles can make it difficult for magma to physically change its shape [5, 7]. This makes the viscosity of magma increasing. In addition to the proportion and size of crystals, the shape of crystals also affects the relative viscosity value of [7,8,10]. High relative viscosity in the youngest lava (RL4) is indicated by crystals with low polydispersity (Table 3.), which means the crystal has a relatively more homogeneous shape. The homogeneity of the crystal form, or so-called monodisperse, makes the maximum pack fraction of particles in magma increase and will make enough particles contained in magma [8, 9]. This makes the friction value of particles in magma increase [5,7,10] and magma viscosity increased.

The rapid increase in crystalline fractions in RL1 to RL3 may be influenced by the length of time magma is stored or the addition of crystals from more primitive magma. The injection of primitive magma from below will carry crystals that have formed in deeper magma sacs that usually have a fairly complex texture [5,6]. The addition of these foreign crystals is expected to begin to decrease in RL3 to RL4 and begin to show a stable phase. This is consistent with the observation of petrography in Figure 3 which indicates that foreign crystals with complex plagioclase textures are very much found in RL3, and the amounts are not much different in RL4.

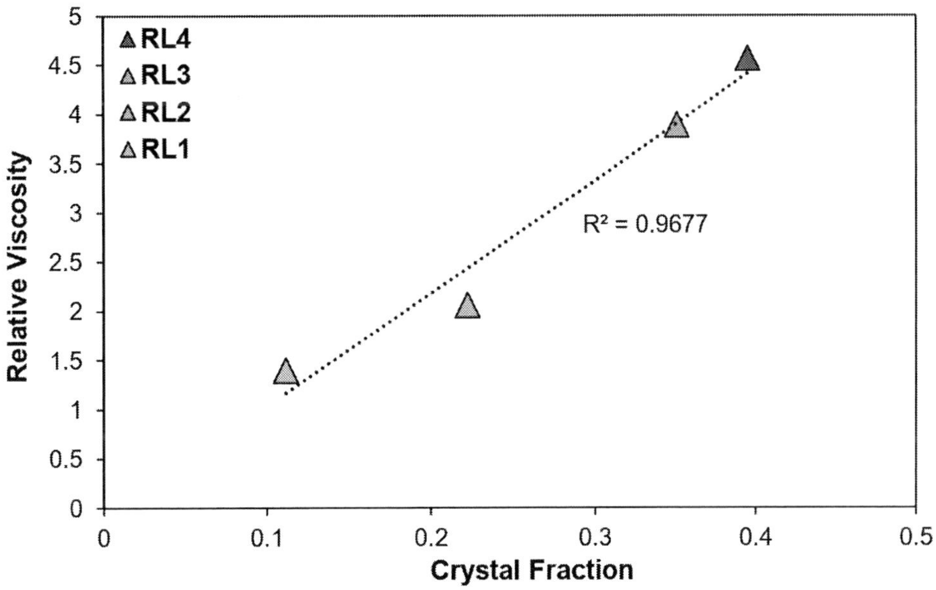

Figure 6. Graphic plot between crystalline fraction and relative viscosity indicating positive correlation.

The increase in magma viscosity that occurred in the eruption of Raung Volcano's pre-caldera lava indicates an intensive crystallization. It is supported by petrography data (Table 2.) which indicates porphyritic texture with phenocryst more than 40%. Plagioclase crystals become quite dominant crystals (>30%). Based on this, it can be interpreted that the crystallization process occurs intensively in the shallow magma chamber of Raung Volcano prior to the of caldera-forming eruption.

5. Conclusion

Based on the results of the analysis and the crystal fraction in the sample, a trend or pattern is found that is quite interesting. The pattern is that the greater the value of the crystal fraction, the value of the relative viscosity of the magma increases as well. More primitive magma injections are thought to play a role in intensifying side rock eroding so that the volume of shallow magma chamber increases, and magma storage time increases. In addition, magma injection also plays a role in the addition of larger foreign crystals to shallow magma chamber. The longer it is stored, making the magma have more time to undergo the crystallization process and form many crystals. The addition of crystals through magma injection and crystallization then increases the solid fraction inside the magma. It is this increase in solids fraction that makes the viscosity of magma increase. The high viscosity of magma tends to cause explosive volcanic eruptions.

6. References

[1] PVMBG 2005 *Data Dasar Gunungapi Raung,* Bandung

[2] Van Bemmelen R 1949 *The Geology of Indonesia*, The Hague

[3] Abdurrachman M and Yamamoto M 2012 Geochemical variation of Quaternary volcanic rocks in Papandayan area, West Java , Indonesia: A role of crustal component *The 12th Regional Congress on Geology, Mineral and Energy Resources of Southeast Asia* (*GEOSEA 2010*) pp 40-57

[4] Sabila F S N 2018 *Vulkanostratigrafi dan petrogenesis Gunung Raung dan sekitarnya, Kabupaten Jember dan Kabupaten Bondowoso, Provinsi Jawa Timur*, Skripsi, Institut Teknologi Bandung.

[5] Mader H M, Llewellin E W, and Mueller S P 2013 The rheology of two-phase magmas: A review and analysis *Journal of Volcanology and Geothermal Research* **257** pp 135-158

[6] Marsh B D 1988 Crystal size distribution (CSD) in rocks and the kinetics and dynamics of crystallization, *Contributions to Mineralogy and Petrology* **99**(3) pp 277-291

[7] Moitra P and Gonnermann H M 2015 Effects of crystal shape and size-modality on magma rheology, *Geochemistry, Geophysics, Geosystems* **16**(1) pp 1–26

[8] Klein J, Mueller S P, and Castro J M 2017 The Influence of crystal size distributions on the rheology of magmas: new insights from analog experiments, *Geochemistry, Geophysics, Geosystems* **18** (11) pp 4055–4073

[9] Klein J, Mueller S P, Helo C, Schweitzer S, Gurioli L, and Castro J M 2018 An expanded model and application of the combined effect of crystal-size distribution and crystal shape on the relative viscosity of magmas, *Journal of Volcanology and Geothermal Research* **357** pp 128-133.

[10] Mollo S and Hammer J E 2017 Dynamic crystallization in magmas, *Mineral Reaction Kinetics: Microstructures, Textures, Chemical and Isotopic Signatures* **16** pp 378-418.

Community preparedness against the threat of abrasion and tidal floods in Padang City

Putri Nilam Sari[1*], Yulina Wahyuningrum Purba[1], Azyyati Ridha Alfian[1]

[1] Faculty of Public Health, Andalas University, Padang, 25163, Indonesia

Corresponding Author: nilam@ph.unand.ac.id

Abstract. The tidal wave in Padang City resulted in 674 potentially eroded houses by sea abrasion over the next five years. Therefore, it is essential to improve community preparedness. This study aims to analyze the preparedness of coastal communities for the threat of abrasion and tidal flooding. This study used a cross-sectional study design, conducted from January to July 2020. The respondents were 74 inhabitants in Pasie Nan Tigo Village. The results showed that more than a half of the respondents were less prepared to face abrasion and tidal flooding (64.9%), having low knowledge of disaster preparedness (41.9%), low awareness of policies and guidelines concerning abrasion and tidal flooding (77.0%), having low emergency response plans (33.8%), having low disaster warning system (43.2%), and having low resource mobilization (44.6%). The statistical test results showed a relationship between knowledge, emergency response plans, disaster warning systems-resource mobilization, and preparedness. Meanwhile, there was no relationship between policies, guidelines, and preparedness. This study showed that the level of preparedness of coastal communities in the face of abrasion and tidal flooding is still unprepared, the society and government need to improve their disaster readiness by increasing their knowledge of abrasion and tidal floods preparedness in seminars, training, and providing sufficient facilities.

1. Introduction

One of the impacts of global warming is the increase of seawater volume and the slowly sinking of low-lying islands or coastal areas. A report from the Intergovernmental Panel on Climate Change (IPCC) in 2016 predicts that in the next 100 years, starting from the year 2000, sea level will rise by 15-90 cm [1]. Sea level rise causes coastal erosion or abrasion and tidal flooding. Tidal flooding occurs when maximum tidal conditions inundate areas lower than the mean sea level. Gravity helps seawater to flow towards low place and inundate these areas [2].

Indonesia is one of the countries that significantly influenced by the sea-level rise, which is estimated to lose a significant land area, of about 90,260 km^2. Based on the National Disaster Management Agency (BNPB) in 2017 risk assessment results, the total number of people exposed to the risk of extreme waves and abrasion in Indonesia is 9,825,782 people in all provinces, with a potential loss of IDR 259 trillion. BNPB noted that in 2019, there were 18 abrasion disasters caused by tidal waves throughout Indonesia. Abrasions are predicted to increase, in line with the climate crisis / global warming [3]. Flooding have been occurred in some areas in Indonesia since 2010. In January and February 2010, there was several tidal flooding in some areas of Surabaya, with inundation levels between 20 cm to 160 cm in 30 minutes to 6 hours flooding duration [4]. In 2011, tidal flooding occurred in the coastal area of Jakarta reaching a height between 30 cm and 50 cm [5]. In 2014, tidal flood occurred in the North Semarang sub-district, due to sea-level rise, covering an area of 823,545 ha or about 70.991% of the total area of the North

Semarang sub-district [6]. In Pariaman City, shoreline changes since 2010 caused residents' home damage [7].

Five factors that influence the disaster preparedness of the community are knowledge (and attitude) towards disaster risk, policies and guidelines, plans for disaster emergencies, disaster warning systems, and the ability to mobilize resources [8]. Disaster preparedness for families on the coast also depends on the perception of disaster risk. Previous research stated a significant and positive relation between perception of disaster risk and disaster preparedness. As many as 83% of high-risk country have a high level of disaster preparedness. A better perception of disaster risk means that the community believes that the place where they live, has a risk of natural disasters and ultimately feels that they have a high responsibility, control, acceptance, and response to this risk [9]. In that case, it will encourage the community to take preparedness in flood disaster event[10].

Padang City has the potential to experience inundation due to sea-level rise. Topographically, almost half of Padang City lying on coastal areas and lowlands. Rising sea level will cause some places on the coast of Padang City, that were predicted before, to be inundated [reference]. From 2008 to 2013, sea-level rise of about 1.1786 cm/year in Padang City. The height of inundation, in 2020, in Padang City is 89.671 cm, and cover an area of 0.382 km^2 (visualized on the tidal inundation map) [11,12]. Tidal flood damaged houses, road infrastructure, public facilities, such as schools and health services, sanitation, yards, fields, rice fields, and ponds [13]. In addition, tidal flood potentially causes various diseases too, even causes epidemics [14]. Some diseases often appear in floods, such as leptospirosis, diarrhea, digestive tract diseases, upper respiratory tract infections, typhoid, skin diseases, dengue fever, or malaria [15].

Pasie Nan Tigo Village (PNT) is one of the coastal areas in Padang City, which is often hit by disasters, especially coastal abrasion, tidal waves, floods, storms, seawater intrusion, and indicated to be prone to tsunamis. There are 674 houses inhabited by 1,164 families in the red zone and potentially eroded by abrasion over the next five years. This happens because the distance between the beach and the house is very close, only around 3-7 meters. In recent years, coastal scour has been high enough that many houses have been affected by abrasion. In fact, 12 houses have been destroyed by the waves, and other 24 houses will possibly be damaged because the waves often reach them. Therefore, it is crucial to determine the preparedness of coastal communities against the threat of abrasion and tidal flooding.

2. Method

The study was conducted in a quantitative research observational analytic approach using a cross-sectional design. This study was carried out by using data from 74 respondents, who live around coastal areas in the Village Pasie Nan Tigo, Tangah Koto subdistrict, Padang.

Through interviews with the respondents, data was collected in a questionnaire about preparedness as a dependent variable, knowledge, policies and guidelines, emergency response plans, disaster warning systems, and resource mobilization as independent variables [16]. First, the data was analyzed to see the frequency for each community preparedness indicators. Each variable was then divided into two categories (low, high) with a median cut-off point. The association between the dependent and independent variables was analyzed using the chi-square test with a significance level of 95%.

3. Result and Discussion

Pasie Nan Tigo Village is one of 13 villages in Koto Tangah District, Padang City. PNT village is located at 0.8406° S, 100.3203° E. The beaches width in this area are between 2m and 21m, and the coastline is 7.2 km long. The coastal area of PNT is located between 0 and 3m above sea level, with a land slope of 0-2%.

Table 1. Relationship between preparedness and the other five independent variables

No.	Independent Variables	P-value
1	Knowledge	<0.001
2	Policy and guidelines	0.580
3	Emergency response plan	<0.001

4	Disaster warning system	<0.001
5	Resources mobilization	<0.001

Table 1 shows the relationship between preparedness and the five independent variables. The statistical test results showed a significant relationship between knowledge, emergency response plans, disaster warning systems-resource mobilization, and preparedness (<0.001). Meanwhile, there was no relation between policies and guidelines with preparedness (0.580).

3.1 Knowledge

Family knowledge about abrasion and tidal flooding is necessary to understand how far the families, who live in PNT Village, is ready for this disaster. Based on the study results, only 13.5% of respondents knew that the tidal flood was a disaster that brought sources of disease such as diarrhea, typhus, DHF, malaria, etc. Table 2 shows the community knowledge.

Table 2. Distribution of knowledge

No.	Knowledge	True Answer	
		f	%
1.	Tidal flood is a natural event that disrupts human life.	41	55.4
2.	Human behaviour causes damage to nature.	19	25.7
3.	Tidal flood is a disaster caused by high tides that inundate the coastal area.	60	81.1
4.	Tidal floods bring sources of diseases such as diarrhea, typhoid, dengue fever, malaria, etc.	10	13.5
5.	Tides that inundate coastal areas cause tidal flood.	70	94.6
6.	During the tidal flood, we have to save ourself to a safer place.	44	59.5
7.	During the tidal flood, it is necessary to evacuate to an area other than the tidal flood area.	25	33.8
8.	During the tidal flood, we need to save ourself and invite family members.	40	54.1
9.	When the tidal flood occurs, we need to monitor the condition of the water level.	45	60.8
10.	During the tidal flood, we have to avoid the power line	19	25.7
11.	During the tidal flood, we (don't) have to save valuables things	40	54.1

Table 2 shows the high level of community knowledge only about "Tidal flood is a disaster caused by high tides that inundate the coastal area" (81.1%) and "Tides that inundate coastal areas cause the tidal flood" (94.6%). For other knowledge question answers, scores are still low. The results of the previous study show relationship between flooding and the incidence of diarrhea, water shortages, water quality (smelly dirty water, and the presence of insects and worms), as well as community resilience to help each other in water supply and treatment of water sources to avoid contamination [17].

However, people who live in coastal areas have a high level of knowledge because they have already lived for more than ten years in areas that are prone to abrasion and tidal flood. Therefore, they have more experiences with abrasion and tidal flood. In addition, some elders also passed their knowledge related to the tidal flood to the younger generation, for example that tidal flood takes place during the full moon every month.

It is better to re-educate the community, who can cooperate with related cross-sector such as schools, youth groups, and PKK groups, about tidal abrasion and flooding, to increase public knowledge. Furthermore, the education should cover all ages to accelerate the delivery of information to the community, especially people who live on the PNT coast.

3.2 Policies and guidelines

Policies and guidelines are also a benchmark for family readiness to face abrasion and tidal floods. Table 3 shows policies and guidelines regarding anticipation of tidal flood.

Table 3. Availability of policies and guidelines

No.	Policy and guidelines	Availability	
		f	%
1.	Dissemination of policies/programs related to tidal flood.	8	10.8
2.	Assistance from the government to overcome tidal flood.	59	79.7
3.	There is a tidal flood alert group.	12	16.2

The research shows that 10.8% of respondents stated that there was a socialization of village policies/programs related to abrasion and tidal flood. Furthermore, 79.7% of respondents said that there was assistance from the government for abrasion and tidal floods. Communities affected by abrasion and tidal flood said that disseminating policies/programs not evenly distributed. This study aligns with another research, which founds that 55.7% of respondents had unpreparedness in policies and guidelines for people in high-risk flood disaster areas [18]

To overcome flooding, government should give serious attention [19]. It is better if the local government also give an information related to disaster response policies/programs to the community, so the people in PNT, especially who live on the coastal areas will prepare themselves better in dealing with the possibility of abrasion or tidal flood. One of the efforts is by forming an abrasion or tidal flood alert group. The government should also provide an equal assistance to all communities affected by abrasion and tidal flood.

3.3 Emergency response plans

The community emergency response plan is illustrated in the following table (Table 4).

Table 4. Community emergency response plan

No.	Emergency response plan	Availability	
		f	%
1.	Have some prepared actions during tidal flood.	38	51.4
2.	Have a place for refuge.	21	28.4
3.	Prepare evacuation routes.	4	5.4
4.	Prepare long-lasting ready-to-eat food.	12	16.2
5.	Prepare first aid kit.	11	14.9
6.	Prepare important and valuable documents.	60	81.1
7.	Prepare clothes, cash, and special needs.	33	44.6
8.	Save family photos.	8	10.8
9.	Set up alternative communication tools.	43	58.1
10.	Save important addresses.	17	23.0
11.	Participate in tidal flood evacuation drills.	10	13.5
12.	Increase knowledge about tidal flood.	44	59.5
13.	Make an evacuation plan.	37	50.0
14.	Conducting family evacuation simulation before tidal flood.	10	13.5
15.	Build a house that prevents waterlog.	18	24.3
16.	Move from the riverside area.	10	13.5

In the emergency response plan, the highest score that exceeds 80% is preparing important and valuable documents (81.1%). Based on the research results, it is found that there are still at least 5.4% who are preparing for evacuation on the available routes. Low scores of emergency response plans can occur because the public awareness of the possibility of abrasion and tidal flooding, such as increasing knowledge about abrasion and tidal flooding, having an agreement on evacuation/evacuation places with family members, and conducting simulation for tidal flood evacuation, is still low. is study is in line with Damayanti's research in 2015, which states that only 14% of respondents have a good emergency response plan [20]. Therefore, the community should increase awareness and save themselves against the risk of abrasion and tidal flood. Improving emergency response plans can be done by increasing knowledge about abrasion and tidal flood, agreeing on evacuation/evacuation places, preparing important and valuable documents as well as storing numbers of emergency services such as hospitals, state electricity companies, disaster management agencies, and local police.

3.4 Disaster warning systems

The disaster warning system is a determining variable in the readiness of the families in PNT Village to face abrasion and tidal floods. The following table (Table 5) shows the disaster warning system owned by the community.

Table 5. Disaster warning systems

No.	Sources of disaster warning system	Availability	
		f	%
1.	City/village official.	6	8.1
2.	Police and security.	5	6.8
3.	Radio.	5	6.8
4.	Television.	22	29.7
5.	Printed media.	12	16.2
6.	Mosque.	12	16.2
7.	Community leaders.	60	81.1

Based on the research results, the source of information arrived to the respondents were received from the community leaders (81.1%). However, the disaster warning system from the local government is still low, 8.1%. The description of the community disaster warning system in PNT shows that it is essential to immediately give a socialization and training about disaster warning system by the government, mainly before the flood occurs. Community only relies mainly on the personal experience of observing the moon to know the conditions of the tides. This is in line with other research, in which the household preparedness in the disaster warning system is 43.3%. When disaster occurs, various prevention efforts including building and developing a non-traditional disaster warning system have to be carried out, even if it is only using a simple method [21]. Local governments should cooperate with the media to disseminate flood information to accelerate its delivery to the public.

3.5 Resources mobilization

Resource mobilization is the last variable in determining community preparedness to face abrasion and tidal floods. The following table (Table 6) shows mobilization of resources for tidal floods preparedness.

Table 6. Resources mobilization

No.	Resources mobilization	Availability	
		f	%
1.	Attend a tidal flood preparedness seminar.	11	14.9
2.	Have first aid skills.	18	24.3
3.	Have evacuation skills.	5	6.8
4.	Have scouting skills.	4	5.4
5.	Have clean water treatment skills.	5	6.8
6.	Have food processing skills.	9	12.2
7.	Have savings.	63	85.1
8.	Have life / property insurance.	14	18.9

9.	Have another house / place safe from disaster.	14	18.9
10.	Having relatives/friends ready to help during the tidal flood .	64	86.5

Based on this study, 14.9% of respondents attended training, seminars, or meetings related to preparedness to face tidal flood. This condition is due to their lack of first aid, preparedness, evacuation of victims, and clean water treatment skill. Only a small part of the community in PNT participated in the related training. The existence of savings and insurance as part of family resources not adequate because awareness, interest, access, or information related to these points are still limited to certain circles. However, most people claim to have relatives or friends who are ready to help in the event of abrasion and tidal flood. This will reduce the burden and risk of abrasion and tidal flood in this area.

The previous study shows that the resource mobilization index was only 66.5%, because not every families are prepared for disasters [22]. Almost the same as this study, where there are only a few scores that are more than 80%, the category of resources mobilization is still relatively low. It is advisable to hold training or seminars related to preparedness for abrasion and tidal floods. The community should also gain important skills such as first aid, victim evacuation, scouting, clean water treatment, and food processing that can be useful for helping all people after the floods.

4. Conclusions

The community preparedness to face tidal flooding is lower than expected, preferably more than 80%. The public should be well prepared to minimize the risks and losses caused by abrasion and tidal floods. The government also needs to provide seminars, training, and sufficient facilities to raise community preparedness against the threat of abrasion and tidal flood.

References

[1] D. Sutrisno 2012 The Model of Sea Level Rise in Indonesia *J. Shipp. Ocean Eng.* **2** 2
[2] P. L. Barnard *et al.* 2019 Dynamic flood modeling essential to assess the coastal impacts of climate change *Sci. Rep.* **9** 1 pp 1–13
[3] BNPB 2017 Data Informasi Bencana Indonesia
[4] T. B. Nur, A. N. Rusydi, dan S. A. Wicaksono 2018 Pengembangan Sistem Informasi Geografis Berbasis Website (WEBGIS) Untuk Simulasi Pemetaan Daerah Genangan Banjir Rob Menggunakan Metode Neighbourhood Analysis (Studi Kasus: Pantai Utara Kota Surabaya) *J. Pengemb. Teknol. Inf. dan Ilmu Komput. e-ISSN* **2548** 964
[5] Walhi 2011 Menanggapi Banjir Rob di Jakarta Utara
[6] D. Larasati, Johnny, dan Y. Fitrianingsih 2014 Perencanaan Sistem Penyaluran Air Buangan pada Komplek Perumahan Grand Sukati Kecamatan Sungai Kakap Kabupaten Kubu Raya
[7] A. P. Donal 2020 Coastal disaster mitigation through shoreline changes analysis in Pariaman city," in *E3S Web of Conferences* **156** 01007
[8] LIPI-UNESCO/ISDR 2006 *Kajian Kesiapsiagaan Masyarakat dalam Mengantisipasi Bencana Gempa Bumi dan Tsunami*. Deputi Pengetahuan Kebumian LIPI.
[9] A. Junaidi, N. Nurhamidah, dan D. Daoed 2018 Future flood management strategies in Indonesia *MATEC Web of Conferences* **229** 1014
[10] Y. T. Kurnianto 2019 Pengaruh persepsi risiko bencana terhadap kesiapsiagaan bencana pada keluarga di pesisir pantai kecamatan sumur, pandeglang, Universitas Negeri Jakarta
[11] D. Daoed, B. Rusman, B. Istijono, dan A. Hakam 2016 Predictions of Vulnerability Flood and Flood Prone Areas in Watershed West Sumatra Province using Arc-GIS and Category Value *International Journal of Earth Sciences and Engineering* **9** 3 pp 274-

279
[12] M. Rasyda, S. Widada, dan R. B 2015 Analisa Spasial Daerah Banjir Genangan (Rob) Akibat Kenaikan Muka Air Laut di Kota Padang *Oseanografi* **4** pp 379-385
[13] M. A. Marfai dan A. Cahyadi 2017 Dampak bencana banjir pesisir dan adaptasi masyarakat terhadapnya di kabupaten Pekalongan
[14] Kemenkes RI 2013 Buku Panduan Hari Cuci Tangan Pakai Sabun Sedunia (HCTPS) ke -6
[15] D. D. Saulnier, K. B. Ribacke, dan J. von Schreeb 2017 No calm after the storm: a systematic review of human health following flood and storm disasters *Prehosp. Disaster Med.* **32** 5 pp 568-579
[16] A. A. Shah *et al.* 2018 Schools' flood emergency preparedness in khyber Pakhtunkhwa province, Pakistan *Int. J. Disaster Risk Sci.* **9** 2 pp 181-194
[17] P. N. Sari dan Nofriya 2018 Flood Disaster, Diarrhea and Community Resilience in Water Provision: A Case Study in The City of Bukittinggi *J. Kesehat. Masy. Andalas* **12** 2 pp 23-29
[18] D. Sari dan C. Husna 2017 Kesiapsiagaan Bencana Banjir Pada Masyarakat Daerah Risiko Tinggi Dan Risiko Rendah Banjir *J. Ilm. Mhs. Fak. Keperawatan* **2** 3 pp 1-9
[19] B. Hidayat, B. Istijono, T. O. Irwan, dan A. Junaidi 2020 The Effects Of Batang Kandis River Flood Control In Padang City-Palapa Metropolitan Urban Are *Int. J.* **19** 71 pp 9-14
[20] H. N. Damayanti 2015 Kajian Kesiapsiagaan Individu dan Rumah Tangga Dalam Menghadapi Bencana Tsunami Di Kecamatan Grabag Kabupaten Purworejo. Universitas Negeri Semarang
[21] I. Murbawan, A. Ma'ruf, dan A. Manan 2018 Kesiapsiagaan Rumah Tangga Dalam Mengantisipasi Bencana Banjir Di Daerah Aliran Sungai (Das) Wanggu *J. Ecogreen* **3** 2 pp 59-69
[22] L. Hanifah, S. A. Susilowati, D. A. Sasmita, A. Demawan, F. Zain, dan H. F. Fitrianto, 2017 Tingkat Pengetahuan, Mobilisasi Sumberdaya dan Kesiapsiagaan Keluarga terhadap Bencana Gempa Bumi di Desa Tangkil, Kecamatan Sragen, Kabupaten Sragen

IOP Conf. Series: Earth and Environmental Science 1027 (2022) 012021 doi:10.1088/1755-1315/1027/1/012021

Farmers Perception of Climate Change on Pepper (Pipper nigrum L.) Productivity of In East Lampung District

Indah Listiana [1], Helvi Yanfika[1], Rinaldi Bursan [2], Habibbullah Jimad, Maya Riantini [1], R.A.D Widyastuti [1], Abdul Mutolib and Ali Rahmat [3]

[1] Faculty of Agriculture, University of Lampung, Indonesia

[2] Faculty of Economics and Business, University of Lampung, Indonesia

[3] Pusat Penelitian Limnology, Indonesia

Corresponding Author: mhmustofa@apps.ipb.ac.id

Abstract. This study aims to determine farmers' perceptions of climate change on pepper productivity in East Lampung Regency and Sukadana Baru Village, Marga Tiga District, East Lampung Regency, from March to April 2020. The research method used is survey method. The research population is 433 pepper farmers. in Sukadana Baru Village. The number of samples in this study were 108 people. The data analysis method used is descriptive statistical analysis. The results show that climate change causes changes in temperature, changes in rainfall, changes in production so that it has an impact on changes in pepper productivity. Changes in temperature that occur in the study area are indicated by the percentage of respondents of 51.85%. Meanwhile, 30.56% of respondents stated that there was no change in temperature in the study area. Rainfall experienced a change with the percentage of respondents amounting to 75.00%, while the rest considered it unchanged with a percentage of 7.41% and 17.59% stating they did not know. Most of the farmers stated that their production had decreased by 75.93%. farmers answered they don't know with a percentage of 63.89%, this situation occurs because most farmers rarely calculate the amount of productivity. Only 20.37% of respondents answered that there was a change in productivity due to climate change.

1. Introduction

Pepper is one of the most abundant spices among other spices (King of Spices), both in terms of its very distinctive uses and cannot be replaced with other spices [1]. Pepper has an important place in the world spice trade, where pepper was the first commodity to be traded internationally and opened trade routes for the western and eastern world [2]. The climate in the study area also greatly affects pepper productivity. During the rainy season, pepper is more easily damaged than during the summer. In the rainy season, pepper will be more susceptible to diseases, such as root rot, yellowing leaves and also decreased pepper productivity. Lampung Province is the second largest pepper centre in Indonesia after the Bangka Belitung Islands, with an area of 45,863 ha. Pepper production in Lampung Province also ranks second after the Bangka Belitung Islands with a total production of 14,860 tons/year.

Pepper centres in Lampung Province are North Lampung, West Lampung, Way Kanan, Tanggamus and East Lampung Regencies. East Lampung is one of the pepper centres in Lampung Province which has a relatively low productivity level (320 kg/ha), far below the average productivity in Lampung Province which reaches 493 kg/ha [3]. The largest black pepper producers in East Lampung Regency are Marga Tiga District, Melinting District, and Gunung Pelindung District. Marga Tiga District in 2016 was able to produce 434 tons of pepper with a planting area of 814 ha, an increase from the previous year which only produced 401 tons. To maintain productivity, pepper farmers in Marga Tiga village are required to know various kinds of obstacles that hinder the pepper production process, especially climate problems and weather conditions in the local environment. This is so that pepper plants can grow optimally, and pests or diseases due to climate change can be overcome. Based on the description above, the purpose of this study was to determine farmers' perceptions of the impact of climate change on pepper productivity in Sukadana Baru Village, Marga Tiga District, East Lampung Regency [4].

It is *vital* that you **do not add any headers, footers or page numbers to your paper**; these will be added during the production process at IOP Publishing (this is why the Header and Footer margins are set to 0 cm in table 1).

2. Materials and Methods

The research method used is a method survey. The research was conducted in Sukadanan Baru Village, Marga Tiga Subdistrict, East Lampung Regency, which was chosen purposively with the consideration that the area is a centre for pepper and has also received certification from MIGHL, for nursery production. The number of samples in this study was determined using the Slovin formula (Noor, 2012). [5].

$$n = \frac{N}{1 + N(e)^2}$$

n = Number of samples
N = Total population of lowland rice farmers
e = Error level (set at 10% with a confidence level of 90%)

The population in this study were 433 pepper farmers who were members of the pepper farmer groups in Sukadana Baru Village. The number of samples in this study were 108 people, with each farmer group consisting of three managers and three members of the farmer group [6].

The data collection techniques used were questionnaires and documentation. The technique of collecting data with a questionnaire was carried out by directly interviewing pepper farmers who were the research samples. Data collection techniques with documentation are carried out on data stored in materials in the form of documentation, such as daily notes, letters, reports and photos. The data analysis method used in this research is descriptive.

3. Result and Discussion

3.1. Age

The majority of farmer group in Sukadana Baru Village range from 25-40 years old, with a percentage of 50 percent, which means they fall into the productive age category. The farmer group members with age 41-55 years are 48.15 percent, which is categorized as still included as the productive age for working in farming.

3.2. Level of education

The education level of pepper farmers in Sukadana Baru Village is quite high with high school education (SMA) are 62%. And another member is junior high school.

3.3 Time period of Farming

The average duration of pepper farming is 16-29 years in pepper farming and as a percentage of 59.26 percent. Meanwhile, members of farmer groups are only 42.59 percent with a susceptibility to farming 16-29 years [7].

3.4 Land area

The majority of farmers have a land area of 0.25-1.5 hectares, which means that many pepper farmers still have small land for the size of their gardens or annual crops. This is because most of the farmers are continuing family farming that has been passed down from generation to generation.

3.5 The Impact of Climate Change on Pepper productivity

Indonesia is a tropical area that is crossed by the equator, so that in a year the sun crosses the equator twice. So, Indonesia has two seasons, namely summer and rainy season. When the sun is directly north of the equator, some parts of Indonesia will experience a dry season. On the other hand, if the sun is directly south of the equator, some parts of Indonesia will experience the rainy season. Climate change is something that cannot be avoided due to global warming. Climate change has an impact on various sectors and is very complex because it covers various aspects of human life [8]. Several studies have shown that climate change has had a negative impact on many aspects of life, especially farmers' lives, such as changes in rainfall patterns, an increase in the frequency of extreme climate events, and an increase in air and sea level temperatures. is a serious impact of climate change that affects the agricultural sector [9]. The climate recapitulation at the level of application of pepper cultivation techniques can be seen in Table 1.

Table 1. The climate recapitulation at the level of application of pepper cultivation techniques

Indicator of Climate Change	Classification	Farmers Group Member (%)
Temperature Change	Do not know	17,59
	Do not change	30,56
	Change	51,85
Average		**2,34**
Change in rainfall	Do not know	17,59
	Do not change	7,41
	Change	75,00
Average		**2,57**
Production change	Do not know	0,00
	Do not change	24,07
	Change	75,93
Average		**2,75**
Productivity chane	Do not know	63,89
	Do not change	15,74
	Change	20,37
Average		**2,16**

Table 1 shows that the climate that occurs at the level of application of pepper cultivation techniques has changed, both from changes in temperature, changes in rainfall to impact on changes in pepper production and productivity in the study area.

3.6 High Temperature

Temperature is the heat and coldness of air caused by the combination of work between the speed of the cooling and heating process of an area in the amount of water content and the earth's surface. Based on the results of the research in Table 1, it shows that there is a change in temperature that occurs in the research area with a percentage of respondents of 51.85. Meanwhile, 30.56 percent of respondents said there was no change in temperature that occurred in the study area. This shows that most residents can feel the temperature that occurs around the research area, which of course has an impact on the pepper farming they are doing.

Changes in temperature, which are the impact of climate change that occurs in the territory of Indonesia. This is in line with Bande's research [10]. This difference in temperature and humidity causes the transpiration rate of pepper in Konawe Regency to be higher, so the occurrence of plant wilting is faster. Likewise, in the research area the temperature changed, causing the leaves of the pepper plant to turn yellow.

3.7 High Rainfall

Rainfall is the amount of water that falls on a flat surface during a certain period measured in millimetres (mm) above the horizontal surface. Based on the results of the study, the rainfall that occurred experienced a change with the number of respondents as much as 75.00%. While the rest are considered unchanged, with a percentage of 7.41% and 17.59% stating they do not know. The higher the rainfall, the lower the pepper production will be. Plants rot more easily due to waterlogging, so pepper plant roots are more susceptible to rot. Precipitation with a greater amount usually causes erosion [11]. This situation will worsen if land cover vegetation and soil conservation efforts are not improved [12]. High rainfall during the flowering and fruiting phases of pepper plants can cause the loss of flowers and fruit seeds. In the process of fruit ripening, high rainfall accelerates fruit decay, thereby damaging fruit quality. Likewise, with pepper, if there is high rainfall it will rot more easily and in the flowering phase it can cause flower miscarriage [13,14].

3.8 Changes in Production

Climate change causes the productivity of pepper plants to be disturbed in the research area. Based on Table 1, most of the farmers said that their production decreased by a percentage of 75.93 percent, this indicates that climate change has a very large influence on the growing nature of pepper in the study area

3.9 Change in Productivity

The impact of climate change on the productivity of pepper plants is not known. Based on Table 1, most of the farmers answered they did not know with a percentage of 63.89%, because most of the farmers rarely calculated the amount of productivity. and 20.37% of respondents answered that there was a change in productivity due to climate change.

4. Conclusion

The impact of climate change on pepper plants that occurs is changes in temperature, changes in rainfall, changes in production and changes in pepper productivity. The temperature changes that occur in the research area are indicated by the percentage of respondents at 51.85% said there was changes.

Meanwhile, 30.56% of respondents said there was no change in temperature that occurred in the study area. Rainfall that occurred experienced a change with the percentage of respondents of 75.00% said there was changes, while the rest considered it unchanged with a percentage of 7.41% and 17.59% said they did not know.

The majority of farmers said that production had decreased by a percentage of 75.93%. the farmer answered that he did not know with a percentage of 63.89%, because most of the farmers rarely calculated the amount of productivity and 20.37% of respondents answered that there is a change in productivity that occurs as a result of climate change.

References

[1] Direkorat Jenderal Perkebunan. 2013. Statisktik Perkebunan Indonesia. Direktorat Jenderal Perkebunan. Jakarta

[2] International Papper Community (IPC) dan Agriculture Organization of The United Nations (FAO). 2005. Papper (Piper Nigrum L.) Production Guide for Asia and The Pacific. Indonesia Papper Community. Jakarta. www.ipcnet.org. Diakses pada 9 Maret 2020

[3] Direkorat Jenderal Perkebunan. 2017. Statistik Perkebunan Indonesia Lada 2015-2017. Direktoral Jenderal Perkebunan, Kementrian Pertanian. Jakarta. http://ditjenbun.pertanian.go.id/tinymcpuk/gambar/file/statistik/2017/Lada-2015-2017.pdf. Diakses pada 2 maret 2020 pukul 10: 27 WIB

[4] Badan Pusat Statistrik. 2017. Lampung Timur dalam Angka. BPS Kabupaten Lampung Timur. Lampung Timur

[5] Noor, J. 2012. Metodologi Penelitian. Kencana Prenada Media Group. Jakarta.

[6] Listiana I, Mutolib A, Bursan R, Yanfika H, Widyastuti R A D, Rahmat A. 2021. Institutional strengthening of farmer group to support sustainable agriculture and food security in Pesawaran regency. Journal of Physics: Conference Series **1796** pp 1-10

[7] Listiana I, Hudoyo A, Prayitno R T, Mutolib A, Yanfika H and Rahmat A. 2020. Adoption Level of Environmentally Friendly Paddy Cultivated Innovation in Pringsewu District, Lampung Province, Indonesia Journal of Physics: Conference Series Vol **1467** (Institute of Physics Publishing)

[8] Yanfika H, Viantimala B, Mutolib A, Listiana I, Widyastusi R A D, Rahmat A. 2021. Impact of Climate Change on Coastal Communities in Tanggamus Regency. Journal of Physics: Conference Series **1796** pp 1-9

[9] Mutolib A, Rahmat A, Yanfika H, Listiana I, Rudy, Haryanto Y. 2020. Level of income, knowledge, and impact of climate change on fishing household in Limau Subdistrict, Tanggamus Regency. IOP Conference Series: Earth and Environmental Science **739** pp 1-8

[10] Bande L.O.S, Bambang. H, Susamto. S, Bambang H.S. 2015. Peran Unsur Cuaca Terhadap Peningkatan Penyakit Busuk Pangkal Batang Lada di Sentra Produksi Lada Daerah Sulawesi Tenggara. Jurnal Manusia dan Lingkungan, Vol. **22**, No. 2, Juli 2015.

[11] Suroso. 2006. Analisis Curah Hujan untuk Membuat Kurva Intensity Duration Frequency (IDF) di Kawasan Rawan Banjir Kabupaten Banyumas. Jurnal Teknik Sipil Vol **3**

[12] Julismin. 2013. Dampak dan Perubahan Iklim di Indonesia. Jurnal Geografi. Vol **5**. No.1

[13] Rahayu, J, dan Muhandoyo. 2011. Dampak perubahan iklim terhadap usaha apel di Kecamatan Poncokusumo Kabupaten Malang. Laporan Penelitian. Faperta Univ. Wisnuwadhana

[14] Ruminta dan Handoko. 2011. Kajian risiko dan adaptasi perubahan iklim pada sektor pertanian di Sumatera Selatan. Laporan Penelitian. KLH Jakarta.

Analysis of Food Expenditures of Rice Farmers in Flooding Prone Region in South Lampung District, Lampung Province

Zainal Abidin [1]*, Dewangga Nikmatullah [1], Adia Nugraha [1] and Yuliana Saleh [1]
[1] Department of Agribusiness, Faculty of Agriculture, University of Lampung, Indonesia

* Corresponding Author: zainal.abidin@fp.unila.ac.id

Abstract. Flooding frequently happened in South Lampung District between January and February which caused loss of production significantly to rice farmers. This study aims at analysing model of food expenditures of paddy farmers using survey method. The study involved 120 farmers who were randomly selected from flooding prone area and was carried out in July-September 2020. Data were analysed using regression model where dependent variable is food expenditures, and independent variables include rice production, corn production, loss of production due to flooding, family income, number of owning livestock, price of rice, price of cassava, and number of family member. Before regression analysis was run, the study did standard test for ordinary least- squares (OLS) such as multicollinearity, heteroscedasticity, and autocorrelation tests. The data passed all the required tests. The study suggests that factors affecting food expenditures include rice production, corn production, family income, price of cassava, loss of production due to flooding, and price of rice. The model concludes R2 is 0.543 suggesting that there is still about 45% of other factors that were not included in the model. The t-test suggests that rice production, family income, and price of cassava have a highly significant influence in affecting the farmers' food expenditure in South Lampung District.

1. Introduction

The Food and Agriculture Organization in 2016 [1] has warned about the serious impacts of climate change, as stated in the FAO Report on the State of Food and Agriculture in 2016. In this report, all countries in the world have been and will be seriously affected, not only in countries. developing, but also in developed countries. Tropical developing countries, such as in Africa, Oceania and Asia, will experience the greatest losses [2]. This is exacerbated by the low economic capacity of these countries in dealing with the impacts of climate change independently. Furthermore, the impact of climate variability and change could reduce the production of food crops (serelia) in the Southeast Asian region by 2.5 percent to 7.8 percent.

Meanwhile, according to [3], variability and climate change with all its impacts have the potential to cause loss of food crop production of rice by 20.6 percent, corn by 13.6 percent, and soybean by 12.4 percent.

Lampung Selatan district is one of the districts that produce important food crops, especially rice, but it is vulnerable to floods and drought. On the other hand, according to [4], the South Lampung Regency Food Security Index is relatively low, which is ranked 172 out of all districts in Indonesia. The flood and drought disasters that occurred in South Lampung Regency resulted in a lot of agricultural land, especially rice fields, which were submerged in water during the rainy season and experiencing drought during the dry season, resulting in crop failure. On the other reason, decrease in rainfall intensity and changes in rainfall are the main reasons that cause a decrease in agricultural production, especially in dry land agriculture [5].

Several sub-districts in South Lampung Regency which are the main rice producers, namely Palas, Sidomulyo and Candipuro Districts, are very vulnerable to flood and drought disasters due to changes in rainfall. Palas, Sidomulyo and Candipuro Sub-Districts are areas that are centers of rice production in South Lampung District. Due to low rainfall in 2018, Palas District experienced a decrease in the harvested area of paddy fields covering 4,179 hectares and paddy fields covering 50 hectares [6]. But in 2019-2020, heavy rain has caused flooding in the study area So, even though the area is rice producer, due to various climate changes, the production of rice is uncertain. In second planting season of 2020, some farmers suffer a loss of production up to 60% due to combination of rat attack and insect of Wereng coklat. With production risk increases recently, it could seriously affect food expenditures of farmers living in the area. Further, it also seriously influences food security of farmers.

1.1. Previous study on food expenditure

Problem with food expenditure of rice farmers could also risks to future generation in terms of malnutrition as a study by [7] in Bangladesh has proved it. In their study, effect of price to food expenditure has cause malnutrition such as stunting.

Other study undertook by [8] in Ghana identified factors in determining vegetable expenditures which include farm income, non-farm income, and other socio-economic characters in rural food expenditure.

In South Sulawesi [9] examines the determinants of household food expenditure in the cassava development area. They use multiple linear regression model. The results showed that household income, household head education, rice prices, sugar prices, edible oil prices, kerosene prices, fish prices, cassava prices, household size, and cassava production affected household food expenditures. Food expenditure is close to 90% of household expenditure per year. The regression model used is not very robust because the R2 value is 0.669.

In Trinidad, [10] were using multiple linear regression analysis in order to analyse socio-economic factors on household food expenditure. The use of multiple linear regression analysis with independent variables, namely gender (dummy), marital status (dummy), educational level (dummy), Ethnicity (dummy), Monthly household income, food security index (ordinal), household size, and age. The results of the study show that all of these independent variables affect food expenditures for rural communities in North Trinidad.

A study organized by [11] in Iran further explored the impact of climate change on household food expenditures. The study uses secondary data such as weather data, food security, food expenditure throughout Iran. Data were analysed with Dynamic Model. The results of the study show that all variables, namely a decrease in rainfall, an increase in temperature, an increase in the retail price index of food and household income reduces the cost for food consumption but it is a trade off with food security.

From the exploration of the research mentioned above, this research has a novelty in terms of household expenditure issues on the main food producer, namely rice. So far there is an opinion that rice farmers, because they produce the main food ingredient, they are relatively secure in food expenditure. However, rice farmers are actually food consumers. Thus, this study intends to investigate rice farmers' expenditure model, especially in areas that are often hit by flooding.

This research aims at (1) identifying food family expenditure, and (2) analysing model in affecting food expenditure of the farmers living in prone flooding region.

2. Methodology

2.1. Type of study

This study uses survey method with study locations in Way Gelam Village, Candipuro Sub-District and Sidorejo Village, Sidomulyo Sub-District. All of them are in South Lampung District. The location determination was purposively chosen with the consideration that the two sub-districts are rice-producing areas and often experience flooding during the rainy season and drought during the dry season. Since the two villages shared the same lowland paddy field watershed, the study will refer both villages as study area.

2.2. Sample size

The sample or respondent is defined as farmers who reside in watershed of Way Gelam River in Candipuro and Sidomulyo sub-districts. The respondent is determined using Slovin method [12] with formula is as follows:

$$n = \frac{N}{Nd^2 + 1} \quad \ldots\ldots\ldots\ldots\ldots\ldots\ldots\ldots\ldots\ldots\ldots\ldots\ldots\ldots (1)$$

where:
n = sample size
N = population which was 4113 farmers who cultivate rice in the watershed of Way Gelam river
d = precision at 10% (d = 0,1)

Using the above formulae, the sample size is then computed at 99,64 respondent or 100 respondents. However, the study determined to expand sample size to 120 respondents for improving data distribution. The respondents consist of 60 respondents from Way Gelam and the 60 from Sidorejo village. Who were selectively using simple random sampling.

The research was carried out in July-September 2020. The respondents in this study were irrigated and rainfed rice farmers who experienced the impact of climate change as many as 120 people, namely 60 respondents from Sidorejo Village and 60 respondents from Way Gelam Village. Respondents were selected using simple random sample method.

2.3. Analysis

To estimate model of food expenditure of farmers, the study employs multiple regression analysis as formulated in the following

$$Y = \beta_0 + \beta_1 X_{1i} + \beta_2 X_{2i} + \ldots\ldots + \beta_n X_{ni} \ldots\ldots\ldots\ldots\ldots\ldots (2)$$

Where,
Y denotes to monthly food expenditures (rupiahs)
β_0 denotes to intercept
β_1, β_2 ....., β_n are regression coefficient
X_1 = rice production (kg/season)
X_2 = corn production (kg per season)
X_3 = decrease of rice production due to flooding
X_4 = family income (Rp/season)
X_5 = number of animal husbandry owned (unit)
X_6 = price of rice (Rp/kg)

X_7 = price of cassava (Rp/kg)
X_8 = number of family member (person)

3. Results and Discussions

3.1. Study area

The study site is located in Way Gelam Village of Candipuro Sub-district and Sidorejo Village of Sidomulyo Sub-District of South Lampung District, Lampung Province. Both villages are adjacent to each other and shared lowland paddy area surrounding Way Gelam watershed. Furthermore, the study site is located around 25 km from Kalianda, capital sity of South Lampung District and about 65 km from Bandar Lampung, capital city of Lampung Province.

Most families in the area are migrant from Jawa, back in around 1970s when the government-initiated transmigration program combined with land reclamation project in Rawa Sragi swamp land, just near of Way Gelam village.

The study area is considered accessible from outside. Road infrastructure is already well developed and since 2019, government has built access to road to this area that make the study site is growing rapidly in terms of population and land prices. It also has big market place nearby. Communication facilities are also well developed such as mobile network, internet, and cable and internet television.

3.2. Characteristic of respondent

The following table summarizes characteristic of farmers in the research site.

Table 1. Summary of respondent characteristic

Characteristics			
Age	Average: 42,4 years	Youngest: 34 years	Oldest: 73 years
Education	Elementary school (24%)	Senior high school (30%)	Junior high school (46%)
Size of paddy area (m2)	Average: 7125	Lowest: 1250	Highest: 35000
Ethnicity	Java (94%)	Lampungnese (4%)	Sundanese (2%)
Farm experience	Average: 23 years	Lowest: 20 years	Highest: 52 years
Members of family (person per family))	Average: 3,32	Lowest: 2	Highest: 5
Non-farm activities	No (94%)	Yes (6%)	

The above table suggests that in terms of education, the majority of farmers have earned junior high school and senior high school degrees. Those numbers suggest that farmers are quite well in terms of education. With regard to size of paddy land, there is discrepancies because some farmers owned large land (3,5 ha) and other farmer owned very small land (0,125 ha). Farmer who owned large land usually obtained from purchasing neighbour land while farmer who owned small parcel because they sell their property for various reasons, mostly due to economy reason.

In terms of ethnicity, majority of farmers in the study area are Javanese and only very small number of farmers are Lampungnese and Sundanese. This is, as explained above, because the majority of respondent are second generation of transmigration back in 1970s. In terms of farm experience, on average respondent already endured long experience in farming, especially paddy. On average, they have cultivated paddy for more than 23 years. In last, only 6% of farmers has non-farming activities such as: working in construction, owned small business, and transportation services. The most majority of respondent rely on farm activities.

3.3. Family expenditures

Household expenditure is divided into food and non-food expenditures. The structure of household expenditure is influenced by changes in expenditure according to time, differences in tastes, differences in income and the environment [13]. The components of food expenditure consisted of staple foods, side dishes, nuts, vegetables, fruits, sources of fat, beverages, sugar, spices, cigarettes, and fried foods. Non-food expenditure components consist of fuel, education, health, social gathering, recreation, public transportation, personal hygiene, beauty, social, communication, taxes, clothing, and jewellery expenditures. The following is summary of family monthly expenditures.

Table 2. Average family monthly expenditure

Description	**Expenditures** (Rp/month)	**Percentage** (%)
A. Food		
Staple food	302.292	13,59
Side dish	187.793	8,44
Beans	30.063	1,35
Vegetable	113.078	5,08
Fruits	15.879	0,71
Fat	36.692	1,65
Beverages	51.596	2,32
Sugar	43.213	1,94
Spices	214.430	9,64
Cigarette	262.383	11,79
Fried foods	13.125	0,59
Subtotal A	**1.270.543**	**57,11**
B. Nonfood		
Fuel	274.188	12,33
Education	268.557	12,07
Health	1.142	0,05
Community sharing	13.917	0,63
Tourism	5.625	0,25
Public transport	0	-
Body wellness	94.468	4,25
Cosmetic	44.026	1,98
Social	57.900	2,60
Communication	97.492	4,38
Tax	21.953	0,99
Clothing	41.515	1,87
Jewellery	33.289	1,50
Subtotal B	**954.069**	**42,89**
Total monthly expenditures	**2.224.613**	**100,00**

The above table suggests that food expenditures are bigger than non-food expenditures. That number is consistent with data of South Lampung District which suggests that food expenditures was in around 55% [14]. According to [15], for rural family or low-income family, food expenditures are usually larger than non-food expenditures. Furthermore, the table also suggests that expenditure for staple food contributes the highest. This is to suggests that fulfilling carbohydrate is the main contributor for food consumption. Purchasing cigarette is the second largest expenditure of paddy farmers. Indonesia Statistics Agency or [16] has included cigarette expenditure as food component. Recent study suggests that cigarette expenditure contributes to poverty in Indonesia. This is also consistent with [17] study which concluded that spending for cigarette is bigger than that of for vegetables, meat, milk, and other nutritious food.

Top three non-food expenditures included for fuel, education, and communication. This is to suggests that expenditure for mobility is very significant for family in the study area. In addition, during pandemic, communication expenditure for student in the area increases as school use online system for education.

3.4. Expenditures and poverty line

Poverty line can be determined by the level of expenditures in Indonesia using basic need approach. This approach suggest that poverty is defined as inability of people to satisfy basic necessities such as food and non-food needs. In short people is considered poor if he/she is unable to meet expenditure set by BPS. For South Lampung district, poverty line is drawn at monthly expenditure at Rp 389.236 per capita per month [18].

Using that benchmark, the study calculated further and found out the number of family living and above poverty line as suggested in the following table.

Table 3. Poverty status according to monthly per capita expenditure

No	Status	Number of families	Percentage (%)
1	Not ppoor (>Rp389.236)	108	90,00
2	Poor (<Rp389.236)	12	10,00
	Jumlah	120	100,00

The table suggests that around 10 percent of farmers are considered poor, the family who are unable to spend more than Rp. 389,236 per capita per month. This is also consistent with BPS Labupapten Lampung Selatan 2021.

3.5. Factors affecting food expenditures

As mentioned n methodology, we estimated 8 independent variables that are affecting food expenditure. The following is description of variables.

Table 4. Description of each variable

Variables	Description	Value
Food expenditures (Y)	Monthly expenditures spent by farmers for food consumption every month	Average: Rp 1,270,543.5 Minimum: Rp 704,000 Maximum: Rp 241,0000
Rice production (X_1)	Rice production in harvest season 1 and 2 of 2019/2020	Average: 9,460 kg Minimum: 2,900 kg Maximum: 42,000 kg

Corn production (X_2)	Corn production in dry season 2019/2020	Average: 41 kg Minimum 0 kg Maximum: 2500 kg
Production lost due to flooding (X_3)	Rice production lost as an impact of flooding that occurred in 2020	Average: 516 kg Minimum 20 kg Maximum: 7350 kg
Family income (X_4)	Family income obtained from various farming, off farming, and non-farming monthly	Average: Rp 4,323,517 Minimum: Rp 1,086,333 Maximum: Rp 12,723,750
Number of animal husbandry (X_5)	Number of animal husbandry owned by family	Average: 6 Minimum: 1 Maximum: 50
Price of rice (X_6)	Price of rice received by farmers per kg in planting season 1 and 2	Average: Rp 8,250 Minimum: Rp 8,000 Maximum: Rp 10,000
Price of cassava (X_7)	Price of cassava received by farmers per kg	Average: Rp 2,375 Minimum Rp 900 Maximum: Rp 3,000
Number of family member (X_8)	Number of family member living in respondent home (Person)	Average: 3 Minimum: 2 Maximum: 5

Before data was run in regression, the study team undertook several requirement tests, such as: multicollinearity, heteroscedasticity, and multicollinearity tests as required in Ordinary Least Square (OLS) procedure [19].

1. Multicollinearity test
   Multicollinearity occurs by analysing simple correlation between variables. Multicollinearity occurs ff VIF value is greater than 10 found in independent variables. The study found that there was no evidence of multicollinearity occurs in our data as all VIF values are less than 10 (Table 6). Hence, it passes the regression requirement.
2. Heteroscedasticity test
   Heteroscedasticity test is intended to investigate whether there is certain clear pattern of data. Using SPSS, the study analysed the scatterplot graph to check if there is a pattern in relation to data. Using scatterplot output as shown in the following graph, the study concluded there was no indication of heteroscedasticity violation within the data. Therefore, it passed regression analysis requirement.

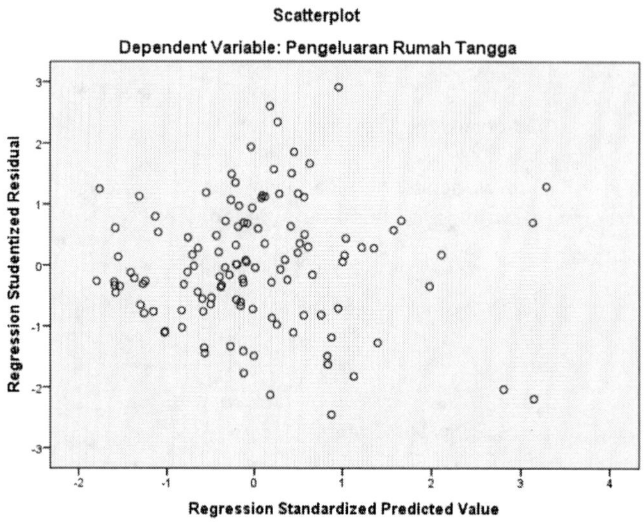

Figure 1. Analysis scatterplot of food expenditures

3. Autocorrelation test

Durbin-Watson test was employed to estimate if there is violation on autocorrelation. Using SPSS, the study obtained the DW value was 2,129. The DW table for α = 0,05 with n = 120 and k = 10 is as follows:

dL = 1,5262; 4 – dL = 2,4738
dU = 1,8852; 4 – dU = 2,1148

Since DW value is less dan dL table, 2,4738, henceforth there is no autocorrelation vilation within the data. The following figures summarizes the Durbin-Watson test of the data.

+ Correlation No correlation. - correlation

1,5162 2,1148 1,8852 2,1148
(dL) (4-dU) (dU) (4-dL)

2,129

Figure 2. Durbin-Watson test to investigate autocorrelation

Therefore, the data has passed all three requirement tests for OLS. Henceforth, the study proceeded with regression analysis model to estimate factors affecting food expenditures of farmers in the study area.

Using SPSS Statistic 22.0, the study reveals the regression summary as stated in the following table.

Table 5. Summary of regression analysis of factors affecting food expenditures in South Lampung District

Model	R	R Square	Adjusted R Square	Std. Error of the Estimate	Change Statistics					Durbin-Watson
					R Square Change	F Change	df1	df2	Sig. F Change	
1	.737[a]	.543	.510	262557.85924	.543	16.461	8	111	.000	2.129

The above table suggests that 54% of the model can be explained by the independent variables while 45% is explained by others not included in this model. Furthermore, however, F test suggests that all independent variables affect the food expenditures of farmers in the study site. This means that the model is enough to explain effect of all independent variables to dependent variables.

Furthermore, to investigate the power of each independent variable to dependent variables, the study uses t test where the results is summarized in the following table.

Table 6. Regression analysis of factors affecting family food expenditures in South Lampung District, 2020

Model		Unstandardized Coefficients		T	Sig.	Collinearity Statistics	
		B	Std. Error			Tolerance	VIF
1	Constant (B_o)	390714,679	444670,836	-,879	,381		
	Rice production (X_1)	-13,776	7,786	-1,769	,080**	,240	4,160
	Corn production (X_2)	109,742	89,081	1,232	,221	,964	1,037
	Production lost due to flooding (X_3)	45,524	31,702	1,436	,154*	,700	1,429
	Family income (X_4)	,124	,023	5,412	,000***	,265	3,773
	Number of animal husbandry (X_5)	2847,865	3118,308	,913	,363	,884	1,131
	Price of rice (X_6)	92,418	56,088	1,648	,102*	,812	1,232
	Price of cassava (X_7)	162,791	29,008	5,612	,000***	,886	1,128
	Number of family member (X_8)	44801,494	35688,393	1,255	,212	,820	1,219

Note:
***= significance at 99%
** = significance at 90%
* = significance at 85%

Of the eight factors, there are five variables that have significant impact on food expenditures including rice production (X_1), lost of production due to flooding (X_3), household income (X_4), rice prices (X_7), and price of cassava (X_8).

Rice production (X1) has a significant negative effect on family food expenditure (Y) with a regression coefficient of -13.776 at a 90 percent confidence level. This means that every 1 percent increase in rice production will affect a decrease in household spending by 13.776 percent, ceteris paribus.

The loss of rice production due to flooding (X_3) has a significant positive effect on family food expenditure (Y) at the 80 percent confidence level with a regression coefficient of 45,524. This means that every 1 percent increase in loss of rice production due to flooding will increase household food expenditures by 45.524 percent. This is because farmers must spend more money for food expenditures as they expect less rice production due to flooding. This will eventually threaten food security of rice farmers. In Nigeria, [20] study effects of flooding on rural food security. They suggest that flooding risks farmers to experience food insecurity.

Household income (X_4) has a significant positive effect on family food expenditure (Y) at the 99 percent confidence level with a regression coefficient of 0.124. This means that every increase in household income by Rp 100,000 will increase household expenditures by Rp 12,400. The results of this study are in line with the results of research by [21] that household income partially has a significant effect on household food expenditure.

The price of rice (X_7) has a significant positive effect on family food expenditure (Y) at a confidence level of 90 percent with a regression coefficient of 92.418. This means that every 1 percent increase in the price of rice will affect an increase in family food spending by 92.418 percent. This in line with [22] study which suggests that increase in price of rice will increase farmer's spending of substitute commodity hence it increases food expenditures.

The price of cassava (X_8) has a significant positive effect on household spending at the 99 percent confidence level with a regression coefficient of 162,791. This means that every 1 percent increase in the price of cassava will have an effect on an increase in household expenditure of 162,791 percent. Cassava is substitute commodity for rice in rural areas. Therefore, when the price of cassava is up, the spending on food will also go up too.

Food expenditure is important indicator to food security and welfare. Study by [23] suggests strong positive relationship between food expenditures and food security. The more farmers spend for food, the more secure their food security status. However, please keep in mind that proportion of food expenditure as compare to non-food expenditures must take into account. If food expenditures are higher than that of non-food expenditure, it suggests farmers welfare is low.

This study reveals that rice farmers are also consumers of rice. During the flood, they had to spend more money to buy rice due to crop failure. Thus, flood control mitigation will help rice farmers to maintain prosperity.

4. Conclusion

1. The average proportion of food expenditure to total household expenditure is 57.11% which is consistent with data of South Lampung District. This means that food expenditures still make up the majority of respondents' household expenditures. This shows that the food security status of the rice farmers is still classified as not prosperous.
2. Factors that significantly affect food expenditures (Y) include rice production (X_1), decreased production due to flooding (X_3), household income (X_4), rice prices (X_7), and cassava prices (X_8).

Acknowledgement
The study team acknowledge the contribution of the Faculty of Agriculture, University of Lampung in providing funding for this research.

References

[1] FAO 2016 The State of Food and Agriculture: Climate Change, Agriculture, and Food Security, FAO, Rome.
[2] Y. Apriyana, E. Susanti, F. Ramadhani and E. Surmaini 2016 Analisis Dampak Perubahan Iklim terhadap Perubahan Produksi Tanaman Pangan pada Lahan Kering dan Rancang Bangun Sistem Informasinya *Informatika Pertanian* **25** 1 pp 69-80
[3] I. Handoko, Y. Sugiarto and Y. Syaukat 2008 Final Report: Kajian Keterkaitan Perubahan Iklim dan Produksi Pangan Strategis, Partnership for Governance Reform, Jakarta.
[4] Badan Ketahanan Pangan Indonesia 2012 Road Map Diversifikasi Pangan Tahun 2011-2015. Edisi 2, Kementerian Pertanian Indonesia, Jakarta.
[5] S. Angles, M. Chinnadurai and A. Sundar 2016 Awareness on Impact of Climate Change on Dryland Agriculture and Coping Mechanism of Dryland Farmers *Indian Journal of Agricultural Economics* **66** 2 pp 365-372
[6] Selatan 2018 BPS Kabupaten Laampung, Kabupaten Lampung Selatan Dalam Anga, BPS Kabupaten Lampung Selatan, Kalianda.
[7] A. A. Campbell, S. de Pee, K. Sun, K. Kramer, A. T. Lyman, R. Moench-Pfamer, M. Sari, N. Akhter, M. W. Bloem and R. D. Semba 2010 Household Rice Expenditure and Maternal and Child Nutritional Status in Bangladesh *The Journal of Nutrition* **140** 1 pp 189S-134S

[8] T. Meng, W. J. Florkowsky, S. Kolavalli and M. Ibrahim 2012 Food Expenditures and Income in Rural Households in the Northern Region of Ghana [Online] Available: https://www.researchgate.net/publication/254384717_Food_Expenditures_and_Income_in_Rural_Households_in_the_Northern_Region_of_Ghana.

[9] M. Zani, H. Soediman, S. Abdullah, L. Daun and L. Yunus 2019 Determinants of Household Food Expenditure in a Cassava Growing Village in Southeast Sulawesi *Academic Journal of Interdisciplinary Studies* **8** 3 pp 302-310

[10] V. Rhamdanie, C. Pemberton and I. Granderson 2017 Socio-economic factors affecting household food expenditure in North Trinidad *Tropical Agriculture* **94** 1 pp 11-30

[11] A. Karbasi and C. Sayyadi 2016 Effects of Climate Change on Food Expenditures of Rural Households in Iran *International Journal of Agricultural Management and Development* **6** 2 pp 225-233

[12] H. Umar, Riset Pemasaran dan Prilaku Konsumen, Jakarta: Gramedia Pustaka, 2002.

[13] Sugiarto 2009 Analisis Tingkat Kesejahteraan Neayan Menurut Pola Pendapatan dan Pengeluaran di Pedesaan, Pusat Analisis Sosial Ekonomi dan Kebijakan Pertanian, Bogor.

[14] Badan Pusat Statistik, Kabupaten Lampung Selatan 2021 Indikator Kesejahteraan Rakyat Tahun, Badan Pusat Statistik, Kabupaten Lampung Selatan, Kalianda, 2021.

[15] S. Sukirno 2012 Makroekonomi: Teori Pengantar. Edisi ke 3, Jakarta: Raja Grafindo Persada.

[16] BPS Indonesia 2012 Profil Kemiskinan di Indonesia, Jakarta: Badan Pusat Statistik.

[17] E. Hernanda, Y. Indiani and U. Kalsum 2017 Pendapatan dan Ketahanan Pangan Rumah Tangga Petani Padi di Desa Rawan Pangan *Jurnal Ilmu-Ilmu Agribisnis* **5** 3 pp 283-391

[18] BPS Kabupaten Lampung Selatan 2020 Garis Kemiskinan 2017-2019, Badan Pusat Statistik Kabupaten Lampung Selatan [Online]. Available: https://lampungselatankab.bps.go.id/indicator/23/241/1/garis-kemiskinan.html. [Accessed 24 September 2020].

[19] D. Gujarati 2003 Ekonomika Dasar: Diterjemahkan oleh Zein, S., Jakarta: Erlangga.

[20] T. T. Akukwe, A. A. Oluoko-Odingo and G. O. Krhoda 2020 Do floods affect food security? A before-and-after comparative study of flood-affected households' food security status in South-Eastern Nigeria *Bulletin of Geography, Socio-Economc Series* **47** pp 115-131

[21] R. Faradina , Iskandarini and S. N. Lubis 2018 Analisis Faktor-Faktor yang Mempengaruhi Pengeluaran Konsumsi Pangan Rumah Tangga (Studi Kasus : Desa Karang Gading, Kecamatan Secanggang, Kabupaten Langkat) *Local Wisdom, Social and Arts (1)* pp 284-295

[22] D. Hafizah, D. B. Hakim, Harianto and R. Nurmalina 2020 The Role of Rice's Price in the Household Consumption in Indonesia *Agriekonomika* **9** 1 pp 38-47

[23] R. O. Babatunde, A. E. Omoniwa, A. 0. Adekunle and G. T. Oyeleke 2019 Effect Of Food Expenditure On Farming Households' Welfare In Osun State, Nigeria *Cercetări Agronomice în Moldova* **LII** 1 pp 79-90

IOP Conf. Series: Earth and Environmental Science 1027 (2022) 012023 doi:10.1088/1755-1315/1027/1/012023

The diversity of plant species in the proboscis monkey's habitat as a species reference for habitat restoration

T Atmoko

Research and Development Institute of Natural Resources Conservation Technology, Jl. Soekarno Hatta Km 38, PO BOX 578, Balikpapan 76112, Indonesia

Corresponding Author: three.atmoko@gmail.com

Abstract. Most of the Proboscis monkey (*Nasalis larvatus*) habitat is outside the protected area, so the potential for destruction is high. One of the efforts to improve its habitat is habitat restoration. This study aims to determine plant species for proboscis monkey habitat restoration based on reference habitats. The research was conducted at eight locations in East Kalimantan, Indonesia i.e. Kuala Samboja, Mahakam Delta, Suwi River, Ohong River, Berau Delta (mangrove and riverine), Sungai Wain, and Kutai National Park. Vegetation analysis with line-plot sampling method with a total sample of all locations of 11.7 ha. The results showed that the habitat included three main types: mangrove, riverine, and riverine-near coastal. A total of 152 tree species, 41 of which were potential proboscis monkey feed. The highest plant species diversity value was in the riverine forest-Berau Delta of 1.25, while the lowest was in the mangrove forest-Berau Delta of 0.54. Species that are a food source and have good adaptation in riverine habitat are *Vitex pinnata, Dillenia excelsa, Garcinia* spp, *Heritiera littoralis, Ilex cymosa, Cerbera manghas, Hevea braziliensis*, and *Sonneratia caseolaris*. Whereas in mangrove habitat are *Avicennia alba, A. officinalis, Bruguiera gymnorrhiza, B. parvifolia, Rhizophora apiculata*, and *S. alba*. They are recommended for habitat restoration.

1. Introduction

Kalimantan has broad tropical forests and is a habitat for diverse plants and animals. However, various economic interests cause damage and loss of animal habitats. The proboscis monkey (*Nasalis larvatus*) is one of the endemic primates of Borneo; its habitat is limited to riverine forests, mangrove forests, and swamp forests [1]. This monkey is a protected animal under Indonesian government regulations, an endangered species according to the IUCN red list, and is included in Appendix I of the CITES [2]. Most of their habitats are outside protected areas which are prone to damage and changes in function. The proboscis monkey's habitat has been highly fragmented, and many small populations are isolated. Small populations are very vulnerable to extinction [3].

One of the goals of habitat restoration is to connect fragmented habitats. These activities can be carried out through planting a wide ecological range of species, native species, and required proboscis monkeys [4]. The habitat restoration is in line with implementing the Proboscis Monkey Conservation Strategy and Action Plan 2013-2023, namely the river-bank restoration program.

The ecosystem restoration paradigm is oriented towards biodiversity-ecosystem functioning (BEF) [5]. One targeted function is restoring and repairing wildlife habitats that have been degraded, damaged, or destroyed, directed explicitly at improving habitat conditions for one or more target species [6]. Ecosystem restoration facilitates the re-establishment of the original structure, productivity, and species diversity close to the initial conditions [7]. Therefore, it is crucial to know the tree species composition and richness in the reference ecosystem as a basis for selecting plant species to be used in restoration activities [8]. Selection of the suitable species will determine the success and functioning of the built target animal habitat. The use of native species is prioritized because they are more effectively used to restore ecosystem function and biodiversity in restored forests [7, 9]. This study aims to determine plant species for proboscis monkey habitat-restoration activities based on reference habitats in East Kalimantan.

2. Methods

2.1. Study Site

The reference habitat locations used in this study were eight proboscis monkey's habitats in East Kalimantan. A total of six locations (Kuala Samboja, Mahakam Delta, Suwi River, Ohong River, mangrove-Berau Delta, and riverine-Berau Delta) are based on primary data collected from 2011-2019. While the other two in Sungai Wain Protected Forest and Kutai National Park are based on literature (Figure 1).

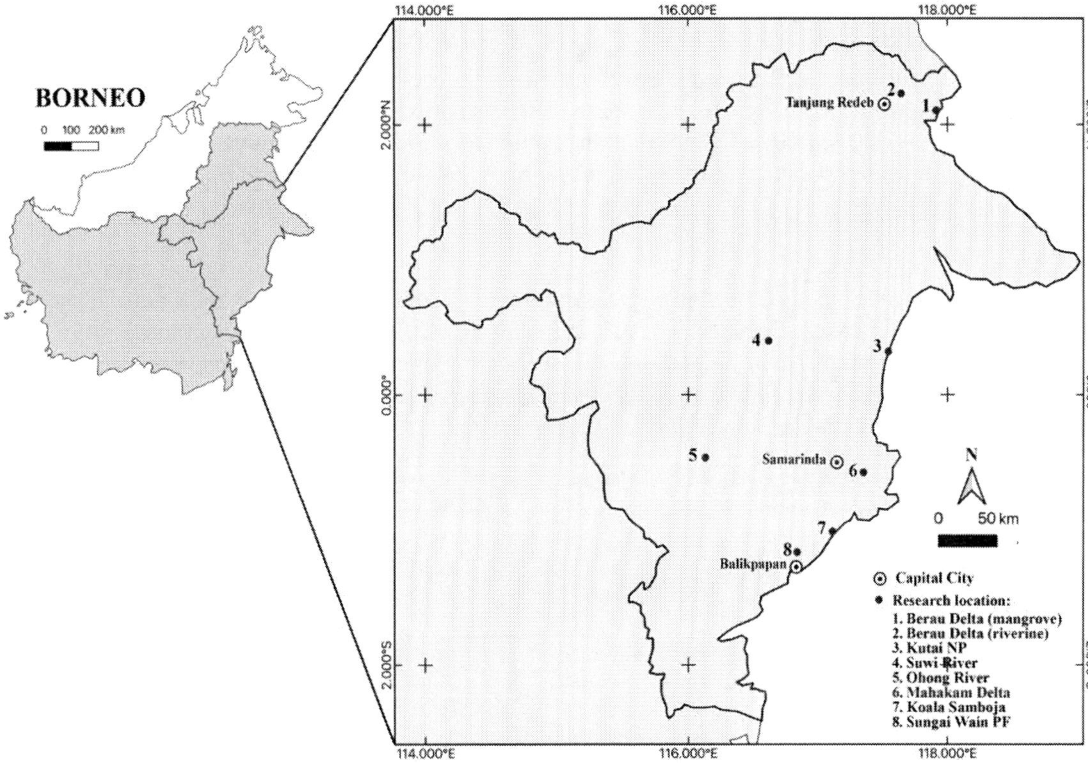

Figure 1. Map of study site in East Kalimantan

The description of the proboscis monkey's habitat is as follows:

- Berau Delta (BD), Berau District (2°10'N 117°44'E). The data was collected in 2019 on the forested mangrove and riverine habitat. Twenty plots of 20 x 20 m were created in mangrove forest (0.8 ha) and 40 plots in the riverine forest (1.6 Ha). Precipitation in the location is 1,845-2,258 mm/year, temperature 27.25-27.91 °C, humidity 82-92%, altitude range from 0-40 m asl, soil types are alluvial and red-yellow podzolic.
- Suwi River (SR), Muara Ancalong Sub District, East Kutai District (0°23'N 116°36'E). The data was taken in 2017 on the 60 plots size 20 x 10 m (1.2 ha). Yearly precipitation in this location is 2,488 mm/year with an average of 207 mm/month, temperature 23.00-31.00 °C, flat topography

with an altitude range from 0-20 m asl, and soil type is peat.

- Ohong River (OR), Tanjung Isuy Village, Jempang Sub District, West Kutai District (0°27'S 116°8'E). The data was collected in 2016 on the 17 plots size 20 x 20 m (0.7 ha). Yearly precipitation in this location is 2,151 mm/year with an average of 179 mm/month, temperature 26.40-32.90 °C, altitude range from 5-120 m asl, soil types are red-yellow podzolic and sandy.
- Mahakam Delta (MD), Sanga-Sanga and Anggana Sub District, Kutai Kartanegara District (0°34'S 117°21'E). The data was collected in 2019 on the 161 plots 20 x 10 m (3.2 ha). The yearly precipitation is 1,546 mm/year with an average of 128 mm/month, temperature 21.50-34.60 °C, humidity 70-92%, and soil type is alluvial.
- Koala Samboja River (KS), Samboja Sub District, Kutai Kartanegara District (1°0'S 117°6'E). The data was taken in 2011 on the 30 plots 20 x 20 m (1.2 ha). The precipitation is 2,290 mm/year (average 191mm/month), air temperature ranges from 23.34-34.70 °C, humidity 82-93%, sloping topography, red-yellow podzolic soil, clay, and sandy.
- Sungai Wain Protected Forest (SW), Sub District North Balikpapan Sub District, Balikpapan City (1°10'S 116°50'E). The data base on Noorhidayah *et al* [10]. The plots were done 63 in sizes 10 x 10 m (0.6 ha). The rainfall in this location is 2.283 mm/year (average 190 mm/month), temperature 24.08-31.24 °C, humidity 80-94%, the topography of 0-150 m asl, the soil types are red-yellow podzolic and alluvial.
- Kutai National Park (KNP), Sangkimah, East Kutai District (0°19'N 117°32'E). The location is a mangrove forest, and habitat data is based on Bismark [11]. One hundred plots were made measuring 20 x 20 m (4.0 ha). Rainfall at the habitat location is 1,543 mm/year, and the average is 128 mm/month, temperature 21.00-34.00 °C, humidity 67-90%, the mangrove area is flat, soil types are alluvial and muddy.

2.2. Methods

The vegetation analysis method used is a line plot transect (Figure 2). The vegetation criteria used are tree-level vegetation (trees; diameter at breast height/dbh $\geq$_10 cm) [12]. The data taken were plant species, branch-free height, total height, dbh measured $\pm$ 1.30 m from the ground. Plant species that could not be identified in the field were collected leaf samples and further identified at the Herbarium Wanariset (WAN) Samboja, Kutai Kartanegara, East Kalimantan. Identification of the proboscis monkey's food plant species is carried out directly in the field on an ad-libitum basis and based on the literature [11, 13-20].

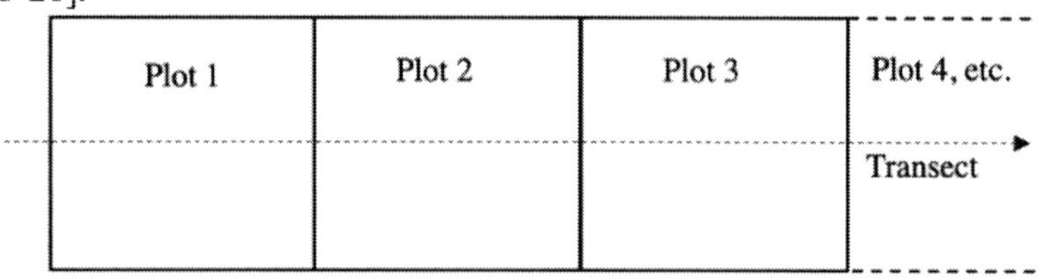

Figure 2. Design of line plot transect.

2.3. Data analysis

Plant species diversity was calculated using the Shannon index (H') $= -\sum_{i=1}^{s}\{(ni/N)\ log\ (ni/N)\}$, where *s* is the number of species, *ni* is the number of trees of species *i*, *N* is the total number of trees of all species, log natural (e) is 2,718 [21].

Vegetation analysis is done by calculating the value of the important value index (IVI), which is the sum of the relative frequency (RF), relative density (RD), relative dominance (RDo) [22].

$$RF_i = \frac{F_i}{\sum_{i=1}^{j} F_{ij}}\ x\ 100\%$$

$$RD_i = \frac{D_i}{\sum_{i=1}^{j} D_{ij}} x\ 100\%$$

$$RDo_i = \frac{Do_i}{\sum_{i=1}^{j} Do_{ij}} x\ 100\%$$

$$IVI_i = RF_i + RD_i + RDo_i$$

where RF_i is the relative frequency of species *i*, RD_i is the relative density of species *i*, RDo_i is the relative dominancy of species *i*, IVI_i is the important value index of species *i*. The frequency (F_i) is the intensity of species *i* was found at sample plots, density (D_i) is the number tree of species per sample plot area, dominancy (Do_i) is the basal area of species *i* per sample plots area.

Analysis of hierarchical clustering of habitats using the ward. D method. Cluster distance is calculated based on Jaccard's Similarity Index $(SI_{(x,y)}) = (|Ix \cap Iy|)/(|Ix| + |Iy| - |Ix \cap Iy|)$, where *Ix* and *Iy* are the numbers or IVI values of species found on locations *x* and *y*, respectively. The $Ix \cap Iy$ is the numbers or IVI values of species found in both *x* and *y* [23]. The values used in the analysis are the binary value of the presence of plant species in the habitat (1 = present; 0 = absent) and the IVI of plant species in the habitat; The analysis using software R version 3.4.4 (vegan package).

The approach to the adaptation of plant species in various habitats was analyzed by manually overlaying plant species on habitats in Venn diagrams. Plant species with the most habitat overlays have a high level of adaptation.

3. Results

3.1. The Number of Plant Species

In the sample plots with 11.7 ha in eight proboscis monkey's habitat locations, at least 152 plant species of woody plants species from 100 genera and 49 families. Of these, 41 species (25.8%) are plant sources of proboscis monkeys (Table 1, Figure 3). The highest species diversity was in the riverine habitat of the Berau Delta (H'=1.25), while the lowest was in the mangrove habitat of the Berau Delta (H'=0.05).

Table 1. Composition of the number of families, genera, plant species, and proboscis monkey's food sources on the habitat in East Kalimantan.

	All location	BD		SR	OR	MD	KS	SW	KNP
		Riverine	Mangrove						
Species	152	68	5	36	24	30	37	8	8
Food plant	41	18	2	17	7	10	11	3	1
Family	49	35	3	23	17	22	23	8	6
Genera	100	56	4	30	22	29	32	8	7
Sapling area (ha)	11.7	1.6	0.8	1.2	0.7	3.2	1.2	0.6	4
H'	0.54-1.25	1.25	0.54	0.94	1.21	1.14	0.97	0.73	0.76

Remark: DB = Berau Delta, SR= Suwi River, OR= Ogong River, MD= Mahakam Delta, KS=Koala Samboja, SW= Sungai Wain Protection Forest [9], KNP= Kutai National Park [10].

3.2. Clustering Habitat

The results of the comparison and grouping of habitats show that there are three clusters of habitat types. Clustering based on binary values is more representative of classifying habitat types than IVI values for plant species. The first clustering is mangrove habitat, the second is coastal riverine, and the third is riverine. Clusters using IVI data resulted in riverine forest clusters being closer to mangrove forests. However, based on the plant species, many species were the same as coastal riverine-type habitats (Figures 3a, 3b). However, IVI data and binary data resulted in the same three clusters (Figure 3c).

Mangrove habitat types include Kutai NP and the Berau Delta, Coastal riverine habitat clusters include those in the Mahakam Delta, Wain River, Koala Samboja, and the Berau Delta. While riverine habitats located far from the sea, namely the Suwi River and Ohong River, are 100 km and 125 km apart from the sea.

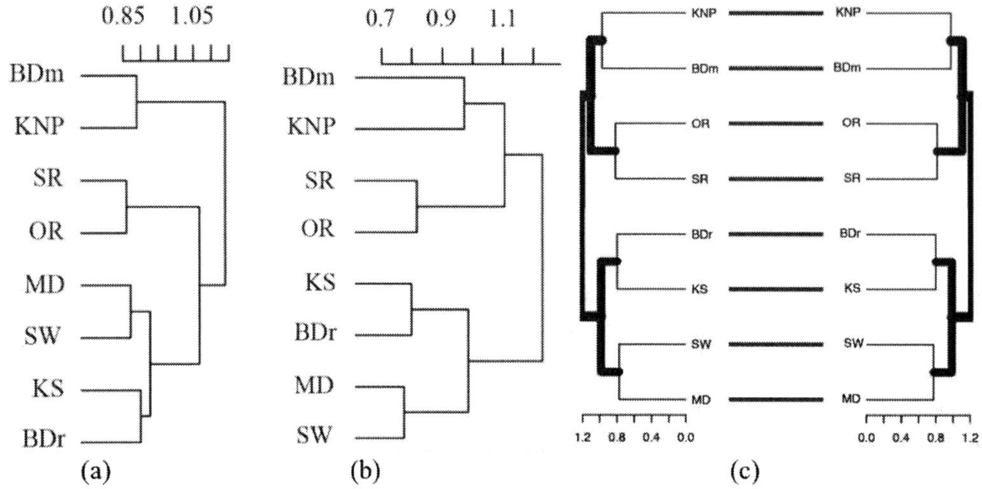

Figure 3. Clustering proboscis monkey's habitat-types in East Kalimantan is based on the binary species presence (a), the important value (b), and analysis of their dendrogram similarity (c).

3.3. Structure and Composition

The dominant species can be described by the IVI value. The species and highest IVI for riparian-coastal habitats were Berau Delta (*S. caseolaris;* IVI=82.33%), Mahakam Delta (*Alseodaphne* sp.; IVI=55.17%), Koala Samboja (*S. caseolaris*; IVI=78.03%), Sungai Wain (*Herritoera littoralis*; IVI=103.34%). The riverine habitat in Suwi River is dominated by *Malotus sumatranus* (IVI=106.97%) and Ohong River by *Gluta renghas* (IVI=58.50%). The mangrove habitats in the Berau Delta and Kutai NP are dominated by *Rhizophora mucronata* (IVI=141.52%) and *Rhizophora apiculata* (IVI=239.14%), respectively (Table 2).

Table 2. The top five tree species with the highest important values index (IVI) in several proboscis monkey's habitats in East Kalimantan.

Species	RF (%)	RD (%)	RDo (%)	IVI (%)	Species	RF (%)	RD (%)	RDo (%)	IVI (%)
Berau Delta (Riverine)					Koala Samboja (Riverine)				
Sonneratia caseolaris	11.88	11.88	58.56	82.33	*Sonneratia caseolaris*	14.94	22.77	40.32	78.03
Brownlowia argentata	11.04	11.04	6.97	29.04	*Vitex pinnata*	10.34	25.92	19.96	56.23
Syzygium lineatum	9.85	9.85	4.33	24.02	*Elaeocarpus stipularis*	11.49	16.49	6.86	34.84
Syzygium creaghii	7.64	7.64	1.96	17.24	*Heynea trijuga*	8.05	10.21	4.96	23.22
Nauclea officinalis	5.77	5.77	3.36	14.90	*Cerbera manghas*	3.45	5.50	4.13	13.08
Mahakam Delta (Riverine)					Sungai Wain PF (Riverine)				
Alseodaphne sp.	8.00	27.05	20.12	55.17	*Heritiera littoralis*	30.23	26.88	46.23	103.34
Sonneratia caseolaris	13.00	10.68	23.27	46.95	*Cerbera manghas*	16.28	35.48	32.04	83.81
Cerbera manghas	18.00	12.81	7.97	38.79	*Oncosperma horridum*	18.60	16.13	8.34	43.07
Ficus microcarpa	6.00	8.19	6.62	20.81	*Pouteria obovata*	13.95	8.60	5.62	28.18
Shirakiopsis indica	7.00	7.12	5.89	20.01	*Aglaia* sp.	13.95	7.53	4.21	25.69
Suwi River (Riverine)					Berau Delta (Mangrove)				
Mallotus sumatranus	15.18	42.44	49.35	106.97	*Rhizophora mucronata*	51.14	51.14	39.24	141.52

Barringtonia acutangula	11.28	13.07	8.73	33.09	*Avicennia alba*	23.80	23.80	36.30	83.89
Lagerstroemia speciosa	7.78	11.04	10.53	29.35	*Rhizophora apiculata*	16.96	16.96	11.58	45.51
Dillenia excelsa	12.84	7.39	8.47	28.70	*Sonneratia alba*	6.08	6.08	11.57	23.72
Gluta renghas	7.78	2.29	3.98	14.05	*Bruguiera* sp.	2.03	2.03	1.31	5.36
Ohong River (Riverine)					Kutai NP (Mangrove)				
Gluta rengas	9.27	11.69	37.54	58.50	*Rhizophora apiculata*	75.61	87.29	76.24	239.14
Mallotus sumatranus	8.61	27.64	10.95	47.21	*Avicennia officinalis*	7.32	5.09	17.91	30.31
Dillenia excelsa	8.61	8.35	3.94	20.89	*Bruguiera gymnorrhiza*	12.20	5.09	3.94	21.22
Barringtonia racemosa	6.62	7.61	4.31	18.54	*Bruguiera parviflora*	3.66	2.12	1.19	6.96
Crudia ripicola	3.31	2.23	12.01	17.55	*Aglaia cucullata*	1.22	0.42	0.72	2.37

3.4. Plant Adaptation

The level of species adaptation is seen from the presence of plant species in several locations of habitats. When the species are widely distributed in various locations, they are highly adaptive to various environmental conditions where they grow. The overlay results of plant species in eight habitat locations showed that species distributed in four habitat locations are *Gluta renghas* (BDr, SR, OR, DM), *Vitex pinnata* (BDr, SR, OR, KS), *Sonneratia caseolaris*, and *Cerbera manghas* (KS, SW, DM, BDr). Whereas *Dillenia excelsa, Garcinia parvifolia, Heritiera littoralis,* and *Ilex cymosa* grew in three habitats. While, *Avicennia alba, Carralia brachiate, Elaeocarpus stipularis, Ficus microcarpa, Garcinia littorale, Hevea braziliensis, Lepisanthes speciosa, Malotus sumatranus, Peternandra rostrata, Sonneratia alba*, and *Vatica rassak* grew in two habitat locations (Figure 4).

Most of these species are also sources of proboscis monkey's food plant. *Rhizophora apiculata* and *Bruguiera* sp. were found in three mangrove forest types with food plant species are *S. alba, A. alba,* and *A. officinalis.* Specifically, *S. caseolaris* is a true mangrove species commonly found in riparian areas because it adapts to low salinity levels.

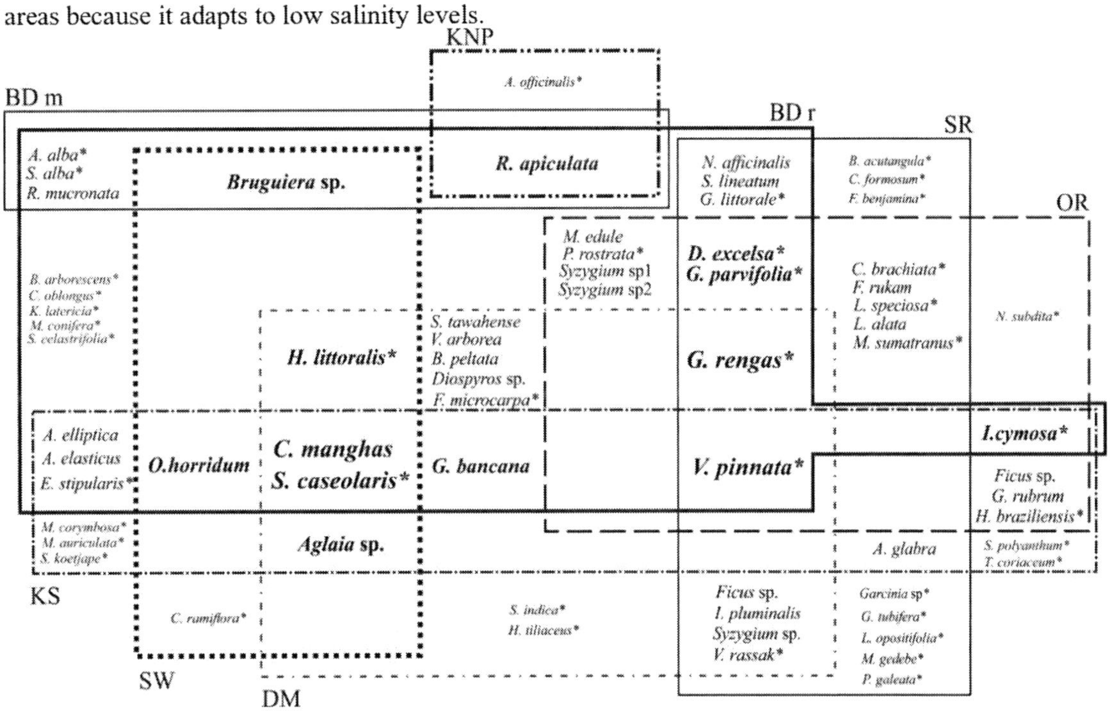

Figure 4. Overlay the plant on the proboscis monkey habitat on Berau Delta-riverine (BDr), Berau Delta-mangrove (BDm), Suwi River (SR), Ohong River (OR), Mahakam Delta (MD), Sungai Wain

Protection Forest (SW), Koala Samboja (KS), Kutai National Park (KNP) in East Kalimantan. Plants species presented in the figure are only found on the two or more habitat and food plant species. Asterisk mark shows that proboscis monkey food plant and font size represent overlapping plant species on the habitat.

4. Discussions

Proboscis monkeys have a wide range of coastal areas to inland forests up to 300 km from the sea [24]. This condition causes the proboscis monkeys to have several types of habitats. Generally, the proboscis monkey's habitat is in the forest, which is still related to water sources, rivers, lakes, seas, and swamps. The monkey's habitat types include riparian forest, mangrove forest, peat swamp forest [13], freshwater peat swamp [25], and galam swamp forest [26]. In addition, proboscis monkeys are also using mixed Dipterocarp forest habitat [27], heath forest [13], rubber forest, and limestone hill/karst forest [24], which are adjacent to the waters. The wide range of habitats makes the composition of plant species that form the habitat vary; even some habitats have different plant species.

The number of tree species in the proboscis monkey's habitat ranged from 5-68 species on all study sites. These species need to be selected to be used for habitat restoration. The selection of adaptive species to various soil and environmental conditions will increase the success of rehabilitating the habitat [28]. In addition, the selected plants are proboscis monkey's feed sources, grow fast, and have known propagation and planting techniques.

Mangrove species are characterized by their muddy habitat and are affected by high salinity. The number of mangrove plant species is relatively less, so that the proboscis monkey food plants are also limited in that habitat. Proboscis monkey's in mangrove habitats eat the leaves and/or fruit of *R. apiculata, A. officinalis, B. gymnorrhiza*, and *B. parvifolia* [11].

The genus *Sonneratia* (Family Lythraceae) is an essential plant for proboscis monkey. Apart from being a sleeping and activity tree, it is also a food source with high preference. *Sonneratia caseolaris* grows in areas with low salinity levels, often found in riparian areas relatively far from the sea. *Sonneratia alba* is more resistant to high salinity, so it is often found in areas directly adjacent to the sea or habitats with high salt concentrations. The pole-level vegetation of *S. caseolaris* and *S. alba* often become like bonsai because the shoots are eaten by proboscis monkeys continuously. After all, these species have a high level of preference as a food source.

Vitex pinnata is known as a species that has high adaptation to various types of habitats. This pioneer species can grow well at 400-1 000 m in dry regions in lowland forest habitats, riverbanks, and secondary forests [29]. This species is fast-growing and can survive in open land, so it is recommended to rehabilitate ex-coal mines [28]. Fast-growing species were chosen as one of the requirements for the restoration plant species to form the forest structure quickly. Growth in height and stem diameter at the age of four years of *V. pinnata* reached 5 m and 7.22 cm, respectively [28]. In addition, proboscis monkeys also have a high preference for using their young leaves as a source of food.

Gluta rengas (Familly Anacardiaceae) dominate the trees in the Ogong River habitat. When the local community cuts down other tree species for their wood, only *G. rengas* remains and becomes a proboscis monkey sleeping tree. Cutting down the *G. rengas* tree for local people will be fatal for them. The black sap can cause severe allergies to human skin. However, the effect of the sap is not harmful to proboscis monkeys, and they often eat the young leaves. Other Anacardiaceae families are also food sources for proboscis monkeys, such as *Buchanania arborescen* and *Mangifera gedebe*. Although this species can grow in several types of habitats and become a source of proboscis monkey food, the process of handling the propagation of seeds is hampered because the sap can cause skin irritation.

The selection of restoration plant species is based on species composition, diversity, and plant adaptation and must be easy to obtain, and the seed propagation technique is known. The dominant species indicated by a high IVI value and ability to grow in various locations showed a high degree of adaptation. While the high value of diversity in a habitat gives many choices of candidate species to choose the best restoration plant. A good plant propagation technique will provide sufficient, inexpensive, and quality plant seedlings. Plant propagation can be done generatively and vegetatively.

The success rate of producing *Vitex pinnata* seeds generatively reaches about 73% [30], while *Sonneratia caseolaris* reaches about 75% [31]. The stem cuttings propagation technique can be used for *Dillenia excelsa* species. Stem cuttings can be used for habitat restoration with extreme water level fluctuations in the swamp and riparian areas. Restoration plants that are submerged for a long time will die. For this reason, *Dillenia* stem cuttings as high as the highest tide are needed to survive.

5. Conclusion

Proboscis monkey habitat types in East Kalimantan include coastal riparian, riparian, and mangrove. The diversity index in mangrove forests (0.54-0.76) is lower than in riverine forests (0.73-1.25) because of the few species in the mangrove forest, so the limited for chosen for selected restoration plants. The selection of species for habitat restoration is based on its dominance in the habitat, height adaptability, food sources, and the formation of the structure and composition of plant species. The recommended species for riparian habitats are *Vitex pinnata, Dillenia excelsa, Garcinia* spp., *Heritiera littoralis, Ilex cymosa, Cerbera manghas, Hevea braziliensis,* and *Sonneratia caseolaris*. As for the mangrove habitat using *Avicennia alba, R. apiculata, S. alba, A. officinalis, B. gymnorrhiza*, and *B. parvifolia*.

References

[1] Phillips Q and Phillips K 2016 *Field Guide to The Mammals of Borneo and Their Ecology* (England: John Beaufoy Plubishing Ltd)
[2] Boonratana R, Cheyne S M, Traeholt C, Nijman V and Supriatna J 2021 *Nasalis larvatus* (amended version of 2020 assessment). The IUCN Red List of Threatened Species 2021 (https://www.iucnredlist.org/)
[3] Atmoko T, Mardiastuti A, Bismark M, Prasetyo L B and Iskandar E 2021 *J. Penelit. Kehutanan Wallacea* **10** pp 11-23
[4] Atmoko T 2020 *Habitat and Population Viability of Proboscis Monkey (Nasalis larvatus* Wurmb*) in Berau Delta, East Kalimantan* (Bogor: IPB University)
[5] Aerts R and Honnay O 2011 *BMC Ecology* **11** 29
[6] Borgmann K and Conway C 2015 *Wildlife Habitat Restoration*, in *Wildlife Habitat Conservation: Concept, Challenges, and Solutions*, ed M Morrison and H Mathewson (Johns Hopkins University Press: Baltimore, Maryland) pp 157-167
[7] Abella S R and Chiquoine L P 2019 *Restoration Ecology* **27** pp 343-351
[8] Pollock M M, Beechie T J and Imaki H 2012 *Ecosphere* **3** 11
[9] Lu Y, Ranjitkar S, Harrison R D, Xu J, Ou X, Ma X and He J 2017 *Plos One* **12** 1
[10] Noorhidayah, Sidiyasa K and Ma'ruf A 2007 *J. Penelit. Hutan dan Konservasi Alam* **4** pp 107-116
[11] Bismark M 1994 *Feeding Ecology and Behavior of Proboscis Monkey (Nasalis larvatus Wurmb) in the Mangrove Forest of the Kutai National Park* (Bogor: Institut Pertanian Bogor)
[12] Susilowati A, Rachmat H H, Elfiati D, Kholibrina C R, Kusuma Y S and Siregar H 2019 *Biodiversitas* **20** pp 1681-1687
[13] Salter R E, MacKenzie N A, Nightingale N, Aken K M and Chai P K P 1985 *Primates* **26** pp 436-451
[14] Yeager C P 1989 *In. J. Primatol.* **10** pp 497-530
[15] Boonratana R 1993 *The ecology and behaviour of the proboscis monkey (Nasalis larvatus) in the Lower Kinabatangan, Sabah* (Thailand: Mahidol University)
[16] Soendjoto M A, Alikodra H S, Bismark M, Setijanto H 2006 *Biota* **9** pp 101-109
[17] Atmoko T and Sidiyasa K 2008 *J. Penelit. Hutan dan Konservasi Alam* **5** pp 307-316
[18] Matsuda I, Tuuga A and Higashi S 2009 *Am. J. Primatol.* **71** 478-92
[19] Atmoko T 2012 *Bekantan Kuala Samboja, Bertahan dalam Keterbatasan* (Bogor: Pusat Litbang Kehutanan, Kementerian Kehutanan)
[20] Mukhlisi, Atmoko T and Priyono 2018 *Buku Flora di Habitat Bekantan Lahan Basah Suwi Kalimantan Timur* (Bogor: Forda Press)

[21] Chiarucci A, Bacaro G and Scheiner S M 2011 *Philos. Trans. R. Soc. Lond. B. Biol. Sci.* **366** 2426-37
[22] Bhadra A K and Pattanayak S K 2017 *Asian J. Sci. and Technol.* **8** pp 4304-4326
[23] Bag S, Kumar S K and Tiwari M K 2019 *Inf. Sci.* **483** pp 53-64
[24] Soendjoto M A, Alikodra H S, Bismark M, Setijanto H 2005 *Biodiversitas* **6** pp 40-44
[25] Yeager C P 1991 *Am. J. Primatol.* **24** 61-66
[26] Iskandar S 2017 *The Coexistency of Proboscis Monkey (Nasalis larvatus Wurmb. 1787) at Rawa Gelam, Tapin Region, South Kalimantan* (Bogor: IPB University)
[27] Bennett E L and Sebastian A S 1988 *In. J. Primatol.* **9** pp 233-255
[28] Adman B, Nugroho A W and Yassir I 2020 *Indonesian J. For. Res* **7** 2 pp 83-97
[29] Orwa C, Mutua A, Kindt R, Jamnadass R, Anthony S 2009 *Agroforestree Database:a tree reference and selection guide version 4.0* (http://www.worldagroforestry.org/sites/treedbs/treedatabases.asp)
[30] Hartina H, Kusuma R and Susanto D 2019 *Al-Kauniyah: J. Biologi* **12** pp 89-95
[31] Nugroho A W, Teguh and Atmoko T 2021 Perkecambahan rambai laut (*Sonneratia caseolaris*) secara generatif pada beberapa jenis media, in *Konservasi Kehati Skala Demo-Plot* (Bogor: IPB Press).

Conditions for Successful Local Collective Action in Mangrove Forest Management: Some Evidences from Eastern Coastal Area of South Sulawesi, Indonesia

S Suharti [1], D Darusman [2], B Nugroho [2] and L Sundawati [2]
[1] Forest Research and Development Centre, Jl Gunung Batu No 5, Bogor, Indonesia.
[2] Bogor Agricultural University, Kampus IPB Dramaga, Bogor, Indonesia.

Corresponding Author: suharti23@yahoo.co.id

Abstract. Resource management characterized by Common Pool Resources (CPRs) requires collective action for its sustainable management. CPRs with "rivalry" and "non-excludable" features face overexploitation problems because unlike the nature of pure public goods, the use of CPRs by one user will reduce the chances of other users taking advantage of it. This study aims to analyze the most appropriate local institutions and tenure arrangements for sustainable mangrove management in Eastern coast area of South Sulawesi, Indonesia. Data and information were collected through in-depth interviews involving key informants selected by using snow-ball method and continued by Focus Group Discussion. Design principles for sustainable management of common pool resources of Ostrom has been used in this study as an analytical framework which illustrate the structure of rules established and imposed by the local CPR's institutions. The findings show that collective action in sustainable mangrove management can be achieved through accepted rules and agreements that are participatory formulated. The agreed norms and rules have enacted as a benchmark for collective action in maintaining mangrove and to enforce sanctions for violators. The success of the community in sustainable mangrove resources management is mainly supported by the existence of several prerequisites that facilitate collective action. It is necessary to strengthen regulations both at the local and higher level through socialization to the community along with the development of incentive and disincentives system.

1. Introduction

Problems threatening the sustainability of ecosystems and the function of mangrove forests have been experienced by many countries, including Indonesia. Mangrove forest of Indonesia is the largest in the world reaching 22.4% of the world's mangrove area or around 3.22 million hectares [1]. Mangroves form distinct waterfront forest habitats with dense, undulating canopies in the wet and arid tropics of the world [2]. It is a typical forest type growing along the coast or river estuaries which is influenced by tides [3, 4]. Mangove has an important role and function in ecosystem, it also has unlimited real and intrinsic values in terms of ecology, economy and social [5, 6]. The various functions of mangroves generate tangible and intangible benefits and services, from the environment around/near mangrove resources and from far outside the mangrove forest [7]. This has resulted complexity in valuing the entire mangrove ecosystem and often has caused negligence in coastal area development plan. Therefore, it requires a deep understanding of the interrelations between social, economic, political, and environmental dimensions for effective management [5, 8]. As a result, lots of mangrove forests have been degraded and converted to other forms of land use such as fisheries, agriculture, industry and for housing/urban development [9, 10]. Since the last two decades, awareness on the value and importance of mangrove forests has been increasing and various conservation and rehabilitation efforts have been carried out in many countries including Indonesia. However, its success rates are still very limited. Many factors are thought to be the cause of the low achievement of mangrove rehabilitation and conservation efforts, including the lack of public awareness of the importance of mangrove forests [5, 7], sectoral management, conflicts of interest and the negative impact of activities on marginalized community groups. In addition, the lack of coordination between institutions and the lack of community involvement (using a top-down approach), one fits all policies and negating the potential role that the community can contribute to mangrove rehabilitation efforts are also the causes of this low achievement [11]. Policy initiatives and planning for mangrove forest management activities are highly dependent on the strength of local communities and their level of participation as well as recognition and support from the government. This is in line with [11, 12] who revealed that mangrove forest planning and management will be successful if the community can fully participate in the activities.

The success of the community in managing mangrove resources on the east coast of South Sulawesi, precisely in East Sinjai District, Sinjai Regency is an example of a success story that proves the ability of local communities to overcome various problems and complexities in natural resource management. However, despite having relatively similar condition (biophysical, socio-cultural and economic), the success of mangrove resource management did not occur broadly and evenly in all coastal village areas of East Sinjai. There are still many areas with poor mangrove forest performance and have not been able to function optimally both economically and ecologically. This condition raises questions about what factors might that cause these differences. This study aims to analyze the most appropriate local institutions and tenure arrangements for the success of sustainable mangrove management characterized by CPRs in the Eastern coastal area of South Sulawesi.

2. Materials and Methods

2.1. Data collection and analysis

This research is a qualitative research using the case study method. We used triangulation method to collect data and information. Triangulation in research is the use of more than one approach scrutinizing questions namely in-depth interviews with selected key informants using the snow-balling method, field observation and continued with focus group discussions. The aim of using more than one approach in data collection is to increase confidence in the findings through confirmation of propositions using two or more independent measures. Triangulation method is used to avoid the potential bias arising from the use of a single methodology. This technique is used to confirm the suggested findings, but can also be used to determine the completeness of the data [13].

In-depth interviews were conducted involving 17 key informants on behalf of community leaders, head of farmer group, traditional leaders and personnel from local Government of Sinjai Regency. Key informants were obtained through the snow-balling, a non-probability sampling method, to identify, select, and take samples in a network or chain of sustainable relationships. The information collected includes data on various norms and rules in mangrove resource management, both operational and collective rules, facts that occur, level of understanding and compliance with existing norms and rules. Collected data and information obtained from the interviews were then presented and discussed in a series of focus group discussions to confirm whether the data and information submitted by the key informants were valid.

We used qualitative techniques to analyze forest conditions in the study sites including the history of community's struggle to manage mangrove forests independently, community's perception and assessment on forest condition changes.

2.2. Research location

The research was conducted in the area of Sinjai District, South Sulawesi, with a focus on mangrove forest of Tongke-Tongke Vullage, East Sinjai Sub District (Figure 1). Of all mangrove forests in East Sinjai, Tongke-tongke is of two villages with good mangrove forest conditions that can function properly as a protector of villages along the coast as well as a source of community income. Mangrove forest in Tongke-tongke Village is a result of independent community management and it represent characteristics of mangrove forest in Sinjai Regency. The success of the community in managing forest area in the area is an example of a success story that proves the ability of local communities to overcome various problems and complexities in natural resource management.

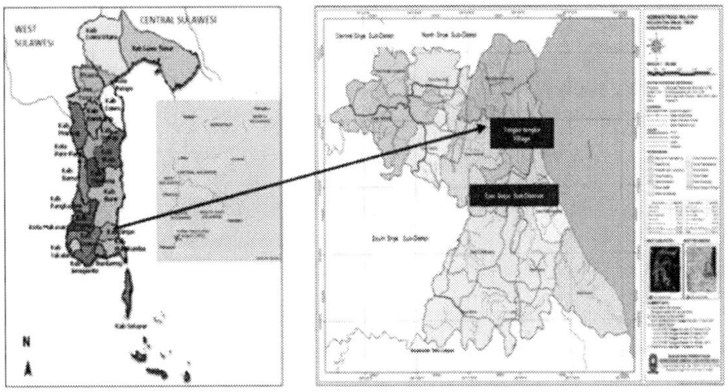

Figure 1. Tongke-tongke Village, in East Sinjai Sub District, Sinjai District, South Sulawesi

Tongke-tongke Village is 5 Km from the District Capital and 7 Km from the Capital of Sinjai Regency. The village has an area of 4.75 km^2, where the area extends from The North to The South following the coastline with an altitude between 0 - 25 m asl and a slope level of 0 - 8%. With flat and sloping morphological conditions, this area is influenced by sea tides and often inundated by sea water during high tide. Furthermore, the village location is directly expose to Bone bay. The border of Tongke-tongke Village to the north is Sinjai Utara District, to the south is Panaikang Village, to the West is Kaloling Village and Saukang Village and to the East is Bone Bay.

2.3. Theoretical Framework

Many factors are the key to successful management of natural resources by local communities. Maximizing benefits and minimizing uncertainty are some of the factors that become the keys for the success of community participation [14].

Mangrove forests in East Sinjai have two main characteristics of (CPRs) i.e. (i) difficulties to exclude or control potential user access, and (ii) each user can reduce the benefit of all other users (a matter of unity). Based on these two characteristics, some resources are referred to as shared (or common-pool) resources/CPRs and are defined as a group of resources whose exception is problematic, and shared use involves reduction/substractability. These types of resources include fish, wildlife, forests, grazing land, irrigation and groundwater [15]. Most wild lands, parks, and public space also reveal the same property characteristics, while most agricultural land and mineral resources do not. Resources that share the above factors are prone to depletion and degradation.

Apart from risk factors and resource scarcity, the success of the community in managing mangrove resources sustainably in Tongke-tongke is also supported by the existence of several conditions that facilitate collective action in natural resource management. Local institutions role for forest management, especially those related to maintenance, monitoring and product harvesting were evaluated qualitatively on the basis of their existence of agreed rules, rule enforcement effectiveness, and community's compliance with the applicable rules [15,16]. An evaluation of the overall institutional strength of the forest governance systems was done by using Ostrom's design principles as the theoretical and evaluative framework (Table 1).

Table 1. Eight design principles describing group efficacy that facilitate effective CPRs management

No.	Principles	Explanation
1.	Define clear boundary of resource and,	The identity of the group having rights to CPRs and the boundaries of the shared resource are clearly determined
2.	Relevance of the rules governing resource use with local needs and conditions.	There must be a balance between benefit sharing and contribution rules. Unfair inequality may harm collective efforts
3.	Collective and participatory rule-setting	There should be an inclusive decision-making process regarding adjustments to the rules for the use and management
4.	Effective monitoring of users and resource conditions	Managing commons is inherently vulnerable to free-rider behavior. There should be a responsible monitoring system for the management of CPRs and ensures its protection
5.	Determination of strict sanctions for each violation	Each violation of the rule for CPR management or extraction face graduated sanctions depending on the seriousness of the violation or repetition of violations
6.	Mechanisms for conflict resolution between parties	There must be an efficient (low-cost) and fast conflict resolution system that is fair by all group members
7.	Recognized rights to manage CPRs	Government allow groups to self-organize CPRs management by forming own internal rules of conduct. Externally imposed rules are unlikely to be adapted to local circumstances
8.	Building responsibility for managing CPR's in levels from the lowest level to the entire interconnected system.	The allocation, provision, monitoring, enforcement, conflict resolution, and government activities are regulated in various layers of nesting companies

3. Result and Discussion

3.1. Characteristic of Mangrove Forest in Tongke-Tongke Village

Mangrove forest in Tongke-Tongke Village, East Sinjai District, Sinjai Regency, South Sulawesi, covers an area of 173.5 hectares. The condition of mangrove forests in Tongke-tongke Village, East Sinjai Sub District is similar to mangrove forests in other areas, which are characterized by a strong root system that can bind/trap mud carried by high tide [17, 18]. The condition of the soil can be categorized into two types, namely sandy beaches and muddy beaches. Sandy beaches are dominated by gray sand and a little black sand, while muddy beaches contain a lot of humus resulting from sedimentation and sedimentation from river flows in addition to silt from the sea [19].

In areas with a high level of mangrove thickness, the soil tends to have a sandy clay texture class, this is due to the decomposition of litter which also determines the soil texture class and the binding of dust and mud particles by plant roots [20]. Whereas in areas without mangrove vegetation the soil texture class tends to be sandy because there is no vegetation that binds the mud particles. Mangroves spread all over the world and Indonesia is one of the countries having the largest mangrove forests in the world. Indonesia has the most diverse mangrove in the world with 202 mangrove species [21]. Among those 202 species, 43 species (including 33 species of trees and some shrubs) are classified as true mangrove species, while the rest are found around mangrove forests as associated mangrove species [22]. In the research site, there are fifteen mangrove species found (Table 2), but there are three the most dominant namely bakau (*Rhizophora* sp), api-api (*Avicennia* sp) and tancang (*Bruguiera* sp).

3.2. Dynamics of Mangrove Forest Development in Tongke-tongke

Mangrove forest is one of the few pristine ecosystems in the world and it offers a large number of ecological, economic, and protective functions and services [5, 6, 23]. Due to the recognized importance of mangroves and the ongoing threat to its sustainability, action has been taken for the conservation and sustainable use of mangroves. Unfortunately, several management and policy efforts have not been fully successful in ensuring the conservation and sustainable use of mangrove resources, due to a variety of reasons both technical and non-technical [5,7, 24, 25]. The mangrove forest ecosystem in Sinjai Regency which is the result of coastal rehabilitation carried out by the community independently is one example of the success stories of local communities in maintaining and managing mangrove resources sustainably. The development of mangrove forests in Tongke-tongke has gone through a long and winding process. The stages of the community's struggle to succeed in rehabilitating the coastal areas of the East Sinjai coast of South Sulawesi can be classified into two main periods namely institutional inertia period and institutional working period.

Table 2. Mangrove Plant Species Diversity in the Research Site

No.	Spesies Mangrove	Local name	Family
1	*Rhizophora mucronata*	Bakko / Bakau	Rhizophoraceae
2	*Rhizophora apiculata*	Bakko / Bakau	Rhizophoraceae
3	*Avicennia alba*	Api-api	Avicenniaceae
4	*Avicennia marina*	Api-api	Avicenniaceae
5	*Avicennia officinalis*	Api-api	Avicenniaceae
6	*Bruguiera cylindrica*	Bakko Panda	Rhizophoraceae
7	*Bruguiera gymnorrhiza*	Bakko Panda	Rhizophoraceae
8	*Sonneratia alba*	Padada	Sonneratiaceae
9	*Sonneratia caseolaris*	Padada	Sonneratiaceae
10	*Aegiceras corniculatum*	Otti-otti	Myrsinaceae
11	*Ceriop sp.*	Cokke	Rhizophoraceae
12	*Hibiscus tiliaceus*	Haru	Malvaceae
13	*Nypa fruticans*	Nipa	Palmae
14	*Terminalia catappa*	Ketapang	Comretaceae
15	*Pandanus tectorius*	Pandan	Pandanaceae

3.2.1. Institutional inertia period 1930s – 1980. The long history of mangrove forest development dates back to the 1930s. Initially, mangrove tree population in the area was very small. Coastal communities have made small efforts to plant mangroves themselves behind their homes. Mangrove seeds/seedlings were only collected from surrounding coastal areas when people go to the sea to look for fish. Therefore, the number of collected seeds/seedlings was still very limited. However, even though they are only planted on a small scale, mangrove trees have started to function a little as protection for the surrounding area from the waves and sea breezes.

Unfortunately, the initial efforts ended in vain when in the early 1940's. Economic pressure pushed many people to convert mangrove forests into fish ponds. As a result, the area became completely unprotected and when it was hit by the tidal waves it was badly damaged. Although there were still people who continued to try to rehabilitate the coastal area, but due to the very small and without support from the Government, the efforts have not been able to offset the ongoing damage. This went on for nearly 40 years. The impact was an increase in water salinity which made water not suitable for consumption, a lot of damage to settlements, significant decrease in fish catches and eventually the destruction of community ponds. This severe damage of the area raised community's awareness and then became their turning point of the people's to improve the situation and save the environment.

Although the people already realized the imminent danger from the sea, most of them did not understand well the potential benefits from mangrove planting efforts as coastal area protection, hence the appreciation for mangrove trees was still very low. This is quite reasonable considering that mangroves are slow growing species with relatively low market price. The basic question that arises is if they planted mangroves, what benefits will they obtained? Moreover, planting mangroves in coastal areas is not easy, it requires persistence and continuous replanting efforts until mangrove can grow well.

Therefore, apart from individual internal factors/characteristics (age, education and land area owned), factors that also affect people's willingness to participate in planting mangrove are understanding about Potential benefits that will be gained by planting mangrove. [26] revealed that knowledge and information factors also affected the level of community participation in efforts to overcome environmental problems.

During this period, efforts to plant and maintain mangrove plants were carried out independently without any coordination between community members. As a result, mangroves that have been planted are often damaged due to being hit by fishing boats passing along the coast, being exposed to nets that have been spread or being cut back for planting. The absence of rules that are mutually agreed upon by community's members have resulted in mangroves not growing properly. The period of mangrove development between 1930 -1980's (Figure 1) were a stage where the local institutions had not yet worked well (institutional inertia period).

Until 1980's, planting mangrove to protect coastal settlement area was not priority. Even, in the 1980s, the people of Tongke-tongke tried to cope with natural conditions by constructing embankments for breaking waves made of coral rock. The stone embankment that was built for almost two years is expected to protect the settlement area during the highest tide season. However, it turned out that the rock embankment did not last long, in just a few months the embankment was crushed by the waves so that two years of hard work were useless.

3.2.2. Institutional working period 1984 – now. The failure of making embankments in 1984 was a turning point in people's awareness that rock embankments could not protect coastal areas at all. Meanwhile they saw that even though the mangroves were small in number they were still able to withstand the waves and winds during high tide. This awareness then encouraged some community leaders to initiate and ask others to plant mangrove together to save their settlement and surrounding environment. Some of these initiators then initiated the formation of the farmer group called "Aku Cinta Indonesia" (I Love Indonesia/ACI farmer group). The cohesiveness among the community cannot be separated from the growing trust between group members and between the group and the initiators who initiate the planting mangrove. The element of mutual trust according to [27, 28] is an important factor in realizing collective action in an activity which in this case is saving the coastal areas in East Sinjai. This group then formulates various agreed norms and rules aiming of ensuring that mangroves that have been planted can be properly maintained. In 1984 Local Government of Sinjai District started to put attention on the condition of coastal settlement in the area. Institutional development and counseling programs from related agencies were provided for the community, and in 1988, ACI Farmer Group then was enacted to become the Natural Resources Conservation Group I Love Indonesia (KPSA-ACI).

The success of the Tongke-tongke community in planting mangroves independently has had a real positive impact, among others, the release of the village from coastal erosion/abrasion, crashing waves and tidal waves; decreased water salinity; abundance of aquatic biota such as shrimp fry, and crab which can be easily caught thereby increasing the income and welfare of coastal communities. Similar case also happened in the Philippines where the conservation of mangrove ecosystems involving local community institutions was able to restore mangrove function optimally both ecologically and economically so that the benefits (tangible and intangible) far exceeded the benefits obtained if mangroves were converted into shrimp pond areas which was previously thought to provide enormous benefits [29]. This shows that in an activity development, the existence of a stable institution is the first step for economic growth [30] as well as the improvement of the welfare of the people [27]. Local institutions that can play an effective role are also motivated by factors of respect and "trust" or mutual trust among community members who believe in a reciprocal relationship, namely that other people will not behave detrimentally if they also do not harm others [31; 28]. The period in which the community has succeeded in formulating various mutually agreed upon norms and rules shows that during that period the local institutions have begun to operate (institutional working period).

The success of planting mangroves in Tongke-tongke further inspired other areas around it so that mangrove plants have now spread to three sub-districts. Mangrove development continues to be carried out by the community both independently and with assistance from the Sinjai Regency Government and related agencies from both the provincial and central levels. The form of public awareness is to keep planting and developing mangroves. Area of Sinjai mangrove forest in 2013 reached 1,157 hectares.

Thus, the ability of the community to formulate various rules to protect mangrove plants so that they can function optimally as a protector of coastal areas and at the same time as a source of community income proves that the community has been able to manage mangrove resources independently in East Sinjai (institutional self-governing period). Local institutions that are able to play an effective role as a benchmark for action are also the key to success in managing mangrove resources sustainably as is the case in Nepal [32] as well as in other Southern Asia areas [33].

In order for mangrove forests to continue to develop and function optimally, community in Tongke-tongke village needs support from various parties, especially the local government which has authority over the status of mangrove forests in East Sinjai. If the government, through its authority, actually comes up with various formal policies and regulations that are not in line with local regulations that have been running so far, it can become a disincentive for the independence of mangrove resource management in East Sinjai.

3.3. Forest Governance Arrangements in Tongke-tongke Village

One of the local institutions that have been established and played a major role in the conservation of mangrove forests in Tongke-tongke is the KPSA-ACI group (the conservation group for natural resources - I love Indonesia). Although it has been established in 1984 (during the early stages of planting mangroves in Tongke-tongke), the KPSDA-ACI group was just formally inaugurated in 1988 with support from Local Government and its related agencies. The ACI group that has received formal legal support already has a statute/bylaws (*AD/ART*) and has an organizational structure with several sections in day-to-day management. For daily operational of mangrove management, community formulate rules that cumulatively influence the form of activities to be taken and the impacts that will be faced as a consequence which are formulated in operational rules (rule in use) and collective choice [15; 24].

3.3.1. Operational and collective rules. Operational rules exist at the implementation level which are formulated based on collective rules agreed upon by group members. Operational rules regulate when, where and how to use every resource unit, who and how to monitor its use, what information should be conveyed or stored and what forms of reward and sanctions will be given for the use of a resource unit and the impact it causes. Hence, the operational rules agreed upon in mangrove forest management in Tongke-tongke arrange the use of mangrove resources for group members. Utilization includes catching marine biota (fish, shrimp, crab) by setting traps in the plots traps, collecting firewood, and catching bats. Other operational rule is related with designing layout of the plots. The mangrove plots have an alley between 10-15 m wide that enables fishing boats to pass and enter the village. The distance in the hallway is adjusted according to the needs.

Those everyday settings are not in the form of written rules, but the form of an oral agreement has been agreed and adhered to by all members. All members have the same rights and obligations regarding mangrove resource management. If there are differences of opinion about the implementation, completion is done in groups through regular monthly meetings. The operational rules for mangrove management in East Sinjai also addressed the issue of monitoring and the obligation to preserve the area as well as law enforcement/imposing sanctions if a member commits a violation. Each member of the group becomes an overseer for the other members. If there are members who violate the group rules, then they will be given sanctions in the form of social warnings or if someone cuts the mangrove trees without permission and/or for clear reasons, they will be given sanctions to plant as many as the number of mangrove trees cut and ensure that the planted seeds grow well (replanted repeatedly until the mangroves grow well).

The collective rules in mangrove forest management are more related to agreements/policies about the direction and strategy of resource management that will be carried out. Collective rules concern the arrangements regarding the involvement of resource users and user preferences in resource management [34]. The formulation of this collective rule involves stakeholders related to mangrove forest management in East Sinjai, namely the Village Head, Local Government and local NGOs.

3.3.2. Conditions for Successful Local Collective Action in Mangrove Management. There are many factors became the keys for the successful management of natural resources by local communities. Maximizing benefits and minimizing uncertainty are key factors to success [28]. Apart from risk factors and resource scarcity, the success of the community in managing mangrove resources sustainably in Tongke-tongke is also supported by the existence of several conditions that facilitate collective action in natural resource management. Several conditions that enable sustainable CPRs management [15] are used to evaluate the success of mangrove resource management in East Sinjai as follows:

a. Clarity of management area boundaries
Several previous research results show that cooperation in society increases when user groups obtain certainty of rights in resource management and have exclusion rights (the authority to prohibit outsiders from entering their managed areas) [15]. This is usually done by clearly define the boundaries of the management area and user groups. Clarity of resources and groups entitled to use resources can reduce the chance of conflict and increase their commitment to established rules, reduce uncertainty in resource tenure and clarify who will bear the costs of collective resource management. In Tongke-tongke Village, boundaries are clearly defined using boundary markers and through face-to-face discussions with external entities entering the area. This clarity of territorial boundaries ultimately eliminates conflicts between local communities and outsiders.

b. Relevance
The distribution of benefits in accordance with/proportional to the costing rules is one of the conditions that facilitates collective community action to carry out sustainable resource management in Tongke-tongke. Synchronization of rules is carried out in accordance with local conditions regarding the time, place, and the availability of managed resources. Mutually agreed rules are ensured to run well and are must be obeyed by all members of the community. Although the rules exist, if not well implemented it will lead to failures, such as the ineffectiveness of the Watershed Forum ~~(DAS)~~ in watershed management causing the number of watersheds categorized as critical in Indonesia continues to increase instigated sedimentation rates and flood frequency continues to increase.

c. Collective and participatory rule-setting
In Tongke-tongke, operational rules for the use and management of mangrove resources were formulated by the KPSDA ACI group and decided jointly by group members in accordance with local customs and prevailing social norms. These rules can be changed by consensus among members of the KPSDA ACI group in the group meeting. Although there may be some external influences during the formation and modification of operational rules, all decisions were taken by consensus among all group members.

d. Monitoring effectively
Effective monitoring is absolutely necessary in the independent management of natural resources because there are always conditions that tempt some individuals to cheat to the detriment of others [15]. The residents of Tongke-tongke village who manage mangrove resources live and work in the area around the mangrove forest, both as fishermen, farmers, firewood collectors and bat seekers. This condition is very beneficial because they can monitor as well as act as direct guardians of the condition and utilization of mangrove resources. The willingness to carry out joint monitoring and the dependence of the community on mangrove resources for the life and livelihoods of the community creates a strong sense of belonging to mangrove resources.

e. Determination of strict sanctions for each violation
Giving strict sanctions for any violation of the use and management of mangrove resources is also a principle developed by local institutions in Tongke-tongke. According to [15], the imposition of sanctions is necessary because mutually agreed regulations have been socialized beforehand and every violation committed will reap sanctions and if it is repeated, it will result in heavier sanctions. Examples of violations are cutting down mangrove trees without prior notification and using it for unclear purposes. In cases like this, that person is obliged to replant mangrove trees and maintain them until it grow well. The results of the study also indicated that rule breakers never repeat violations again because they realize that the sacrifices that must be borne as sanctions for violations are greater than the benefits received. Finally breaking the collective rules becomes an option that doesn't appeal to them at all.

f. Mechanism for conflict resolution
The tenets presented by [15] assume that several types of conflicts in the use and management of natural resources will still occur in the field, even though various rules have been clearly formulated and in detail. Conflict is unavoidable because it is a consequence of team work. Conflict is the result of behavior and it is an integral part of human life [35]. Conflict emerge may be due to differences in the interpretation of the rules among users. If these conflicts are not resolved at low cost and in an orderly manner, it will become a disincentive for users to comply with mutually agreed rules. Conflict resolution in Tongke-tongke Village is set in stages. Internal conflicts related to the extraction and distribution of benefits from mangrove resources are resolved between conflicting members in a family manner. Conflict resolution actually is seen as an adjustment process, which itself can be subject to procedures to contain and regulate conflict behavior and ensure a fair outcome. This is also in accordance with the results of studies on local communities in the Arau watershed in managing protected forest areas where conflicts in management between residents are resolved through family negotiations [36]. Meanwhile, more complicated conflicts will be resolved in the assembly through a meeting of all members groups as well as with facilitation by local forestry staff where necessary (if conflict arises due to external factors).

From the description above, it is evident that the ability to identify the capacity of individuals involved in a situation to design their own regulations based on their knowledge and then formulate rules to protect the preservation of existing resources are the main key to successful mangrove management in Tongke-tongke. For the success of collective action in mangrove management, integrating local community knowledge and local adhered norms in program activities implementation especially related with natural conservation, will become the initial important step for its success [37]. The success in formulating mutually agreed rules also proves that the rational consideration of individual users to always maximize utility as feared by [38] did not always occur [15]. Eventually, as the structure and function of mangroves is site-specific, there is no "one-size-fits-all solution". all efforts for mangrove conservation and rehabilitation should be formulated with adequate site-specific information rather than global generalizations and should be combined with price-based instruments, such as carbon credits, payments for ecosystem services, taxes on deforestation and certified green products. , which would increase the value of maintaining and protecting mangroves rather than converting them for other uses (Lee et al. 2019).

4. Conclusion
The dynamics of the mangrove forest development process in Tongke-tongke Village show that even without the support of the government, collective action to plant mangroves can be realized through various rules and agreements formulated collectively by the community, although it requires a long and winding process. The success of the community in managing mangrove resources is also supported by the existence of several conditions that facilitate collective community action, namely; clarity of area boundaries; relevance of rules, collective and participatory rule-setting; determination of strict sanctions; effective monitoring; and conflict resolution mechanism. Those six conditions guide residents in interacting with mangrove resources. The research results strengthen the theory of self-governing

institutional development where operational rules (rule in use) and collective choice can be used as a benchmark for action for each member of the community in treating mangrove resources.

References

[1] Ilman M, Dargusch P, Dar P and Onrizal 2016 A historical analysis of the drivers of loss and degradation of Indonesia's mangroves. *Land Use Policy.* **54** pp 448–459.

[2] Duke NK and Schmitt K 2015 *Mangroves: Unusual Forests at the Seas Edge.* Tropical Forestry Handbook DOI 10.1007/978-3-642-41554-8_129-1# Springer-Verlag Berlin Heidelberg

[3] Hartshorn GS 2013. *Tropical Forest Ecosystems* In: Levin SA(ed) Encyclopedia of biodiversity, 2nd ed. Elsevier Inc., Nederland. p 5504

[4] Spencer T, Möller I and Reef R 2016 *Mangrove Systems and Environments.* In book: Reference Module in Earth Systems and Environmental Sciences 10.

[5] Suharti S, Darusman D, Nugroho B and Sundawati L 2016 Economic valuation as a basis for sustainable mangrove resource management: A case in East Sinjai, South Sulawesi. *Journal of Tropical Forest Management* **22** 1 pp 13-23

[6] Neumann B, Ott K and Kenchington R 2017 Strong sustainability in coastal areas: a conceptual interpretation of SDG 14 *Sustain Sci* **12**.

[7] Baderan DWK 2013 Modal valuasi ekonomi sebagai dasar untuk rehabilitasi hutan Mangrove di wilayah pesisir Kecamatan Kwandang, Kabupaten Gorontalo Utara, Provinsi Gorontalo [disertasi]. Yogyakarta (ID): Program Studi Geografi, Universitas Gadjah Mada

[8] Newton A and Elliott M 2016 A Typology of Stakeholders and Guidelines for Engagement in Transdisciplinary, Participatory Processes. *Front. Mar. Sci.* **3**:230

[9] Nicholls RJ, Hanson S, Herweijer C, Patmore N, Hallegatte S, Corfee-Morlot J, Cha^teau J and Muir-Wood R 2007 Ranking port cities with high exposure and vulnerability to climate extremes—exposure estimates. *OECD environmental working paper* no. **1**, Organisation for Economic Co-operation and Development (OECD), Paris

[10] Ahmed N and Glaser M 2016 Coastal aquaculture, mangrove deforestation and blue carbon emissions: Is REDD+ a solution? *Marine Policy.* **66** pp 58-66.

[11] Suharti S 2017 Development of Bio-rights incentive scheme for participatory restoration and conservation of mangrove resources *"BIODIVERSITAS" Journal of Biological Diversity.* **8** 1 pp 121-128

[12] Hakim L, Siswanto D and Nakagoshi N 2017 Mangrove Conservation in East Java: The Ecotourism Development Perspectives. *The Journal of Tropical Life Science.* **7** 3 pp 277 –285

[13] Heale R and Forbes D 2013. Understanding triangulation in research. *Evid Based Nurs* Vol **16** Number **4**

[14] Davenport MA, Bridges CA, Mangun JC, Carver AD, Williard KW and Jones EO 2010 Building local community commitment to wetlands restoration: A case study of the Cache River Wetlands in Southern Illinois, USA. *Environmental Management* **45 (4)** pp 711-722

[15] Ostrom E 1990 *Governing the Commons: The Evolution of Institutions for Collective Action* (Political Economy of Institutions and Decisions) 271 pp Cambridge: Cambridge University Press

[16] Ostrom E 2000 Collective action and evolution of social norms. *The Journal of Economic Perspectives* **14** 3 pp 137-158

[17] Chaudhuri P, Chaudhuri S and Ghosh R 2019 The Role of Mangroves in Coastal and Estuarine Sedimentary Accretion in Southeast Asia. 10.5772/intechopen.85591.

[18] Murdiyarso D, Hanggara B, Lubis and Ali A 2018 Sedimentation and soil carbon accumulation in degraded mangrove forests of North Sumatra, Indonesia (Preprint DOI: 10.1101/325191)

[19] Hossain MD and Nuruddin AA 2016. Soil and mangrove: A Review. *J. Environ. Sci. Technol* **9**: pp 198-207 DOI: 10.3923/jest.2016.198.207

[20] Dewiyanti I, Darmawi D, Muchlisin ZA, Helmi TZ, Imelda I and Defira CN 2021 Physical and chemical characteristics of soil in mangrove ecosystem based on differences habitat in Banda

Aceh and Aceh Besar. IOP Conf. Ser.: *Earth Environ. Sci.* **674** 012092 DOI:10.1088/1755-1315/674/1/012092

[21] Baderan DWK, Utina R and Lapolo N 2018 Vegetation structure, species diversity, and mangrove zonation patterns in the Tanjung Panjang Nature Reserve Area, Gorontalo, Indonesia. International Journal of Applied Biology, **2 (2)** 2580-2119.

[22] Noor YR, Khazali M and Suryadiputra INN 2012 *Panduan Pengenalan Mangrove di Indonesia.* PHKA/WI-IP, Bogor. 228 pp

[23] Halder NK, Merchant A, Misbahuzzaman K, Wagner S and Mukul SA 2021 Why some trees are more vulnerable during catastrophic cyclone events in the Sundarbans mangrove forest of Bangladesh. *Forest Ecology and Management* **490** (2021) 119117

[24] Romañach S S, Deangelis, D L, Koh H L, Li Y, The S Y, Sulaiman RRB and Zhai L 2018. Conservation and restoration of mangroves: Global status, perspectives, and prognosis. *Ocean and Coastal Management.* **154** pp 72-82. 10.1016/j.ocecoaman.2018.01.009.

[25] Richards DR an Friess D A 2016. Rates and drivers of mangrove deforestation in Southeast Asia, 2000–2012. *Proc. Natl. Acad. Sci.* **113** 2 pp 344–349

[26] Dimitrakopoulos P G, Jones N, Iosifides T, Florokapi I, Lasda O, Paliouras, F and Evangelinos K I 2010 Local attitudes on protected area: evidence from three Natura 2000 wetland sites in Greece. *Enviromental Management.* **91** pp 1847-1854

[27] Hermanto, NFN and Swastika D 2016 Penguatan Kelompok Tani: Langkah Awal Peningkatan Kesejahteraan Petani. *Analisis Kebijakan Pertanian.* **9** p 371 (In Indonesia)

[28] Suharti S, Darusman D, Nugroho B and Sundawati L 2016. Strengthening social capital for propelling collective action in mangrove management. *Wetlands Ecology and Management.* **24 3** pp 13-23

[29] Farley J, Batker D, Torre I and Hudspeth T 2009 Conserving Mangrove Ecosystems in the Philippines: Transcending Disciplinary and Institutional Borders. *Environmental management.* **45** pp 39-51

[30] Bosma, N, Content J, Sanders M 2018 Institutions, entrepreneurship, and economic growth in Europe. *Small Bus Econ.* **51** pp 483–499

[31] Tabellini G 2010. Culture and institutions: Economic development in the regions of Europe. *Journal of the European Economic Association* **8** 4 pp 677-716

[32] Baral N and Stern MJ 2011. A comparative study of two community-based conservation models in Nepal *Biodiversity Conservation* **20** pp 2407-2426

[33] Datta D, Chattopadhyay RN and Guha P 2012 Community based mangrove management: A review on status and sustainability. *Journal of Environmental* Management. **107** pp 84-95

[34] Varughese G and Ostrom E 2001 The contested role of heterogeneity in collective action: Some evidence from community forestry in Nepal *World Development* **29** 5 pp 747-765

[35] Thakore D 2013. Conflict and Conflict Management. *OSR Journal of Business and Management* (IOSR-JBM) **8** 6 pp 7-16 www.iosrjournals.org www.iosrjournals.org

[36] Nursidah, Nugroho B, Darusman D, Rusdiana O and Rasyid Y 2012. Pengembangan institusi untuk membangun aksi kolektif lokal dalam pengelolaan hutan kawasan lindung SWPDAS Arau, Sumatera barat. *Jurnal Manajemen Hutan Tropika* **XVIII (1)** pp 18-30

[37] Dharmawan B, Böcher M, Krott M 2016. The failure of the mangrove conservation plan in Indonesia: weak research and an ignorance of grassroots politics. *Ocean & Coast. Manag.* **130** pp 250–259

[38] Hardin G 1968 The Tragedy of The Commons. *Science.* **162** pp 1243-1248

The potential of *kabuyutan* sacred natural site towards a sustainable landscape management in Indonesia

M Z Dahlan[1], **K Fukamachi**[2] **and S Shibata**[2]

[1] School of Architecture, Planning, and Policy Development, Institut Teknologi Bandung, Ganesa No. 10, Bandung, West Java, Indonesia
[2] Graduate School of Global Environmental Studies, Kyoto University, Yoshida-Honmachi, Sakyo, Kyoto, Japan

Corresponding Author: mzd19@itb.ac.jp

Abstract. Sacred natural sites have been recognized by International Union for Conservation of Nature (IUCN) as a landscape heritage that contains biological and cultural diversity due to conserved over generations and sacred by indigenous people. This study identified the characteristics and ecological significance of *kabuyutan*, a sacred natural site perceived by Sundanese people in West Java Province, Indonesia. The observation participant method was used to gain a close and intimate relationship between Sundanese people and their natural and cultural environment, particularly *kabuyutan*. We identified 33 *kabuyutan* and found a typical structure compared to the surrounding landscape which is *dominantly* covered by dense vegetation and marked by sacred trees, springs, ancestor's graves, or stones. *Kabuyutan* were perceived to have a vital role in conserving land and water, which is proven by their specific land variables, and proposed a watershed-based management concept. However, we found that changes in the buffer zone of *kabuyutan* into the agricultural or residential area have potentially affected their existence due to the land ownership status. We suggested that empowering custodians and transferring information about the critical role of sacred natural sites are necessary for landscape management to prevent the negative impact of landscape dynamic.

1. Introduction

Thousands of sacred natural sites remain globally, and many are little known and ignored [1, 2]. These places have been considered important in maintaining and enhancing the quality of the environment through their function as a safeguard of biocultural diversity [1]. In the last two decades, the topic of sacred natural sites has been widely discussed by scholars [2]. However, few case studies have been conducted in Indonesia [1]. To expand the geographical range of studies and enrich the repertoire of knowledge, we conducted a study on *kabuyutan* sacred natural sites, which is closely related to Sundanese people in West Java, Indonesia. Further, like other sacred natural sites, *kabuyutan* is facing challenges maintaining the balance along with social, cultural, and spatial dynamics, such as land-use change [3].

Historically, the term of *kabuyutan* was used to define the sacred place allocated for the center of knowledge, settlement of priests, or conservation areas [4]. This term was occasionally used by philologists and archeologists who concern Sundanese study. The evidence of its existence can be found in the ancient Sundanese manuscript such as *Amanat G*alunggung [5].

In the manuscript of *Amanat Galunggung*, for instance, *kabuyutan* have considered as an essential place for the Sunda Kingdom that is strongly protected from the destruction. The King Darmasiksa

(1175 to 1297 A.D.) mandated *kabuyutan* to be kept and he despised anyone who ruined it even more despicable than sable in the trash [5]. This shows the important role of *kabuyutan* in a landscape as well as the strong suggestion to ensure its sustainability.

As for physical features, *kabuyutan* landscape can be easily distinguished within their surrounding landscape because it is mainly covered by dense vegetation or marked by the particular sacred element that shaped their structure. The sacredness of *kabuyutan* has been understood over generations and its existence has been protected by the customary rule that perceived as taboo [4]. Sundanese people argued that restriction of accessibility is the main rule in managing *kabuyutan*. Thus, there are two types of status of *kabuyutan*, namely, restricted access and limited access. Common people might be allowed to enter the limited access with permission from custodian, whereas they are strictly prohibited from entering the restricted ones.

Kabuyutan located in Ciomas Village were known as *kabuyutan* with limited access. This provides an opportunity to explore their characteristic as a basis for deeper understanding of their role in a landscape management. As an interdependent system, the characteristic of *kabuyutan* could not be disassociated from a whole landscape [2]. Therefore, understanding *kabuyutan* landscape characteristics is important to highlight a type of whole Sundanese landscape [4]. Concerning the vital role of *kabuyutan* and its challenges towards sustainability, the objectives of study were set (1) to identify the structure, function, and change of *kabuyutan* landscape and (2) to analyze its significance for landscape management.

2. Methods

2.1. Study area

This study was conducted in 33 *kabuyutan* distributed in Ciomas Village, Panjalu District, Ciamis Regency, West Java Province (latitude 07^{0}07'00" to 07^{0}12'00" S, longitude 108^{0}15'00" to 108^{0}19'00" E). Ciomas Village is located in the northern part of Ciamis Regency dominated by the hilly area as part of Mt. Sawal, ranging between 512-1762 masl. The average daily temperature is about 25 ^{0}C, humidity 85%, and the average annual precipitation is about 2.000 mm with rainy season almost all year long except in June, July, and August. Dryland farming, paddy field, and forests dominate the land-use type to indicate the prominent local people's activity as farmers (Figure 1).

2.2. Data collection and analysis

Data were obtained by observation participant through in-depth interviews with four key informants selected by snowball sampling method, starting from the Custodian then continuing to the eligible person until data saturation. In total, ten sessions (5-8 participants per session) within three Focus Group Discussions were conducted with key informants and 63 purposively selected local people living close to *kabuyutan*. The questions covered three main topics that focused on identifying the landscape structure, their major function within a whole landscape, and the changes during the last decades. A field survey also has been conducted to verify data spatially.

Two topographical maps from 2000 and 2014 were collected from the National Land Authority of Ciamis Regency to better understanding the spatial characteristic of *kabuyutan* within a whole landscape. These maps were used to derive spatial data of six particular land variables that support the existence of *kabuyutan*, namely elevation (ELEV), aspect (ASPC), slope (SLPE), approximate distance to a river (DTRV), approximate distance to a road (DTRD), area, and land-use (LU). Information about the main element for each *kabuyutan* and the status of land (LS) were also collected to analyze the significance of *kabuyutan*.

A participatory GIS-based analysis method [6, 7] was conducted to identify the structure, function, and change of *kabuyutan*. Analytical process [8, 9] was conducted by selecting important land variables to determine their relative influence and then overlaying all variable maps to determine their significance. Finally, results were discussed qualitatively to describe the phenomena.

3. Results and discussion

3.1. Landscape structure of kabuyutan

The findings from both the in-depth interview and the FGDs revealed that the most of respondents have a good understanding according to the condition of *kabuyutan* and the surrounding landscape. They perceived that landscape has been divided into three types regarding to the traditional forest's function. Basically, a whole landscape was designated as a protected forest (*leuweung larangan*) and *kabuyutan* exist within these areas. Next, people utilized the landscape as a production forest (*leuweung baladahan*) while perceiving the area between them as a conservation forest (*leuweung tutupan*).

Over generations, protected forest has been gradually changed to other uses and apart into small patches. Other studies have reported this condition as a landscape fragmentation [3]. Currently, *kabuyutan* landscape were distributed within a whole landscape in the small patches and perceived as an anchor which confers resistance to the surrounding landscape. However, the existence of sacred sites, such as *kabuyutan*, which mainly fragmented within a whole landscape were revealed an endangerment to damage unless it is adequately managed. This study found that about 33 *kabuyutan* were found and their characteristics were identified based on six land variables and other attributes (Table 1).

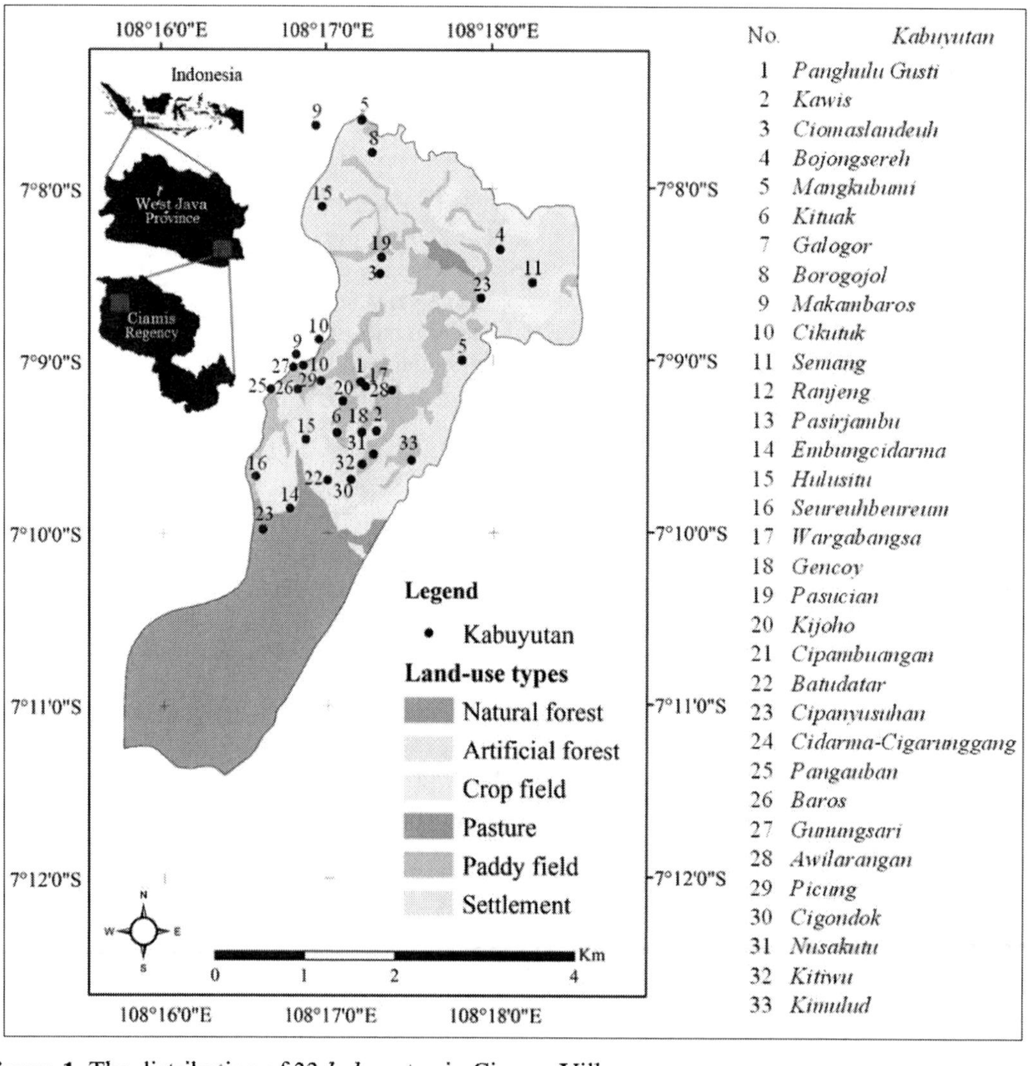

Figure 1. The distribution of 33 *kabuyutan* in Ciomas Village

Table 1. The characteristic of the 33 *kabuyutan*'s landscape structures.

No.	ELEV (m asl)	ASPC	SLPE	DTRV (m)	DTRD (m)	Area (ha)	LU	LS	Element
K01	730	W	VGS	180.0	84.9	6.6	AF (PF)	G	Spr, Plt, Grv
K02	748	E	VGS	150.0	84.9	3.3	AF (PF)	G	Grv, Plt
K03	703	E	VGS	67.1	94.9	2.2	AF (S)	G	Grv, Plt
K04	680	E	VGS	174.9	67.1	2.1	AF (CF)	G	Grv, Plt
K05	693	E	VGS	134.2	216.3	1.7	AF (S)	G	Grv, Plt
K06	784	N	GS	84.9	284.6	1.2	AF (PF)	P	Spr, Plt
K07	718	E	VGS	60.0	417.9	1.1	AF (PF)	P	Grv, Plt
K08	665	E	NL	127.3	90.0	1.1	AF (CF)	P	Spr, Plt
K09	794	E	VGS	90.0	60.0	1.0	AF (CF)	G	Grv, Plt
K10	760	N	VGS	30.0	60.0	1.0	AF (PF)	P	Spr, Plt
K11	645	E	VGS	283.0	67.1	0.8	AF (CF)	P	Grv, Plt
K12	728	E	VGS	0.0	0.0	0.7	AF (PF)	G	Plt, Riv
K13	731	E	VGS	42.4	180.0	0.7	AF (CF)	G	Grv, Plt
K14	987	E	GS	0.0	150.0	0.7	CF (CF)	P	Spr, Plt
K15	861	W	GS	0.0	123.7	0.6	AF (CF)	P	Spr, Plt
K16	929	W	GS	94.9	84.9	0.5	CF (PF)	G	Spr, Plt
K17	748	W	NL	123.7	60.0	0.4	DS (DS)	G	Grv, Plt
K18	755	W	VGS	247.4	60.0	0.3	AF (PF)	P	Spr, Plt
K19	741	E	NL	161.6	42.4	0.3	CC (DS)	P	Plt
K20	729	E	VGS	84.9	108.1	0.3	PF (PF)	P	Grv, Plt
K21	736	E	NL	42.4	67.1	0.2	AF (PF)	P	Spr, Plt
K22	928	E	MS	201.2	247.4	0.2	CF (CF)	P	Plt, Stn
K23	1043	W	NL	60.0	390.0	0.2	NF (CF)	G	Spr, Plt
K24	591	E	VGS	0.0	0.0	0.2	PF (S)	P	Plt, Riv
K25	794	E	VGS	0.0	0.0	0.2	PF (S)	P	Plt, Riv
K26	788	E	VGS	84.9	30.0	0.2	DS (DS)	G	Plt
K27	764	E	NL	0.0	30.0	0.1	PF (CF)	P	Spr, Plt
K28	712	E	NL	134.2	30.0	0.1	PF (CF)	P	Spr, Plt
K29	756	E	VGS	0.0	60.0	0.1	PF (CF)	P	Spr, Plt
K30	815	E	MS	30.0	84.9	0.0	PF (CF)	P	Spr, Plt
K31	734	W	VGS	42.4	120.0	0.0	PF (CF)	P	Plt, Stn
K32	762	E	VGS	42.4	42.4	0.0	PF (CF)	P	Spr, Plt
K33	772	E	GS	161.6	0.0	0.0	PF (CF)	P	Grv, Plt

Abbreviation: Number of Kabuyutan (K00); Elevation (ELEV); Aspect (ASPC) with area towards northern (N), western (W), eastern (E), and southern (S); Slope (SLPE) with area nearly level of 0-1% (NL), very gentle slope of 1-3% (VGS), gently slope of 3-5% (GS), and moderately sloping of 5-15% (MS); Approximate distance to river (DTRV) and road (DTRD); Land-use (LU) with primary forest (PF), artificial forest (AF), crop field (CF), pasture (P), paddy field (PF), and settlement (S); Land status (LS) with land belongs to government (G) and private (P); and main element of *kabuyutan* such as grave (Grv), spring (Spr), plant (Plt), river (Riv), and stone (Stn).

According to the elevation (ELEV), most *kabuyutan* are located on an average of 718 masl. They are mostly located around the highest and lowest point of the surrounding area. This showed the role of *kabuyutan* in conserving land and water resources. *Kabuyutan* located around the highest point more functioned as water catchment areas, while those in the lowest point serve as a water reservoir. This important role corresponds to the area located above 600 m asl that is recognized as the mountainous region and designated as a protected area by the formal rules. This condition might support the existence of *kabuyutan* landscape that play important role in managing a whole landscape.

In term of slope (SLPE), core zone of *kabuyutan* mostly located on average 1.22% of slope. However, they are surrounded by buffer zones that range from nearly to steep slopes. This area potentially causes damages ranging from splash erosion to material movements such as landslides and land erosion [10]. The instability of the potentially disastrous land has been well-understood and they arranged *kabuyutan* within this area to prevent and reduce the negative impact of slope-related disasters.

Among aspect (ASPC) and land-use (LU), most *kabuyutan* are located on the area towards the whole direction except areas without direction (flat) where functioned for developed areas (settlement). The increase in demand of settlement areas in line with population growth has been well-considered by local people. This has a potential impact on land-use change [3] that will also affect the existence of *kabuyutan* and the surrounding area. Therefore, people set *kabuyutan* in other directions to control the development of settlement areas. Furthermore, *kabuyutan* located on areas toward the east and west were set for protecting agriculture lands, while the north and south direction were set to support the hydrological system due to understanding of water cycle that flows from the north (forested areas) to the south (developed and cultivated areas).

According to approximate distance to river (DTRV), most *kabuyutan* were found close to water bodies such as springs, rivers, tributaries, or lakes. The existence of water bodies in *kabuyutan*, indicated the important role of customary rules (taboo) in protecting them from the disturbance. According to spatial arrangement, findings also showed that *kabuyutan* has been distributed along the river system from upstream to downstream. This indicates the main function of *kabuyutan* in conserving land and water resources in a landscape.

The reason can be shown that traditionally distribution of *kabuyutan* were set to mark a boundary of area. The position of *kabuyutan*'s symbol, for example spring, indicates the benchmark to set the covered area within the *kabuyutan* as their buffer zone. Then, the edge of buffer zone would be connected to determine the boundary of the whole landscape. This means that Sundanese landscape has been managed according to hydrological system to conserve water resources originating in watershed as well as land of watershed for more efficient and sustained production. This finding is similar to a previous study that stated the function of watershed management in dealing with land and water resources utilization in efficient ways [10].

Understanding the importance of these areas, the Indonesian government set the area around 200 m as the protected area for springs. In addition, the area within 100 m along the river is designated as protected areas. Moreover, such of rules would be more potential to ensure the existence of *kabuyutan* by combining with customary rules. According to the proximity to road (DTRD), most *kabuyutan* located more than 100 m from the road. The existence of the road for some *kabuyutan* perceived for ease of access and management, such as in *Kabuyutan Panghulu Gusti* which is often visited for pilgrimage. Also, for regular activities held on this *kabuyutan*, the existence of the road was helpful in management. Nevertheless, the existence of roads may potentially cause an instability of *kabuyutan*. This condition was strengthened by the development's actual condition, which follows the road (strip development). Prevention effort, especially revitalizing the function of buffer zone is important to be implemented.

Regarding those land variables and associated cultural values, informants reported that several *kabuyutan* were arranged to protect the vital area to ensure sustainability. For example, five *kabuyutan* were gradually set to protect land and continuity of the hydrological system (Figure 2).

Figure 2. The arrangement of five *kabuyutan* which bounded by a myth to unsure the continuity of hydrological system (a), the condition in *Kabuyutan Batudatar* (b), *cigondok* (c), *kitiwu* (d), *nusakutu* (e), and *kawis* (f) (source: documentation of author).

In addition, *kabuyutan* perceived as a part of watershed-based landscape management that borders the region by flowing water. This can be shown by the arrangement of four *kabuyutan* in order to bound the Cidarma Watershed. *Karantenan* was set in the peak of Mt. Sawal as the main *kabuyutan* in the upper-stream. Then, *kabuyutan tembong, jagakaya*, and *kimulud* were set to mark the boundary in the west, south, and east, respectively (Figure 3). *Karantenan* was symbolized by a sacred spring, while ancestral graves symbolized others.

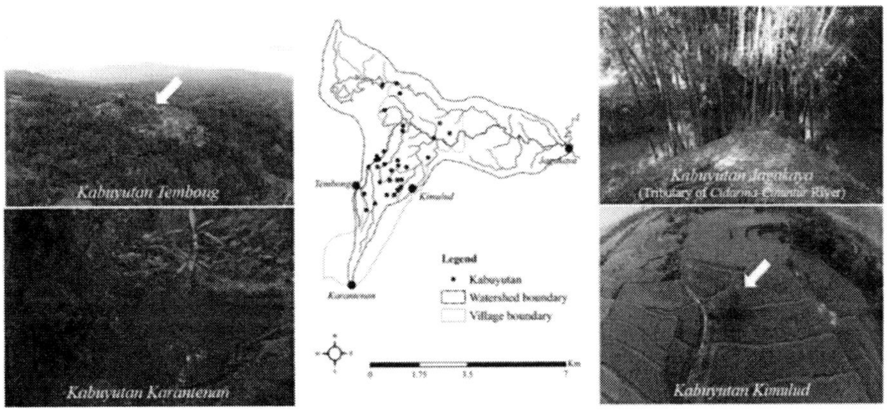

Figure 3. The arrangement of *kabuyutan* to mark the boundary of Cidarma Watershed (source: documentation of author).

3.2. *Dynamic of kabuyutan landscape*

Findings showed that major changes have occurred in cultivated pasture areas into non-natural woody areas located in the center of the village (Figure 4). Economic reason is perceived as the main reason due to higher income earned from tree cultivation. This condition also showed the evidence of successful implementation of one billion tree planting program at the village level. These reasons are consistent with the Regional Planning of Ciamis Regency Year 1999-2104 which stated that the northern part was planned as a protected area by encouraging people to plant trees through tree planting program which promoted since 2010.

In general, the current changes did not significantly affect the existence of *kabuyutan*, especially in their core zone. The existence of main symbol in each *kabuyutan* indicated the core zones are well-protected. However, fragmentation in *kabuyutan* which led their landscape into smaller patches has affected the existence of buffer zone (Figure 4). Historically, these *kabuyutan* have been fully covered by standing trees. But along with the increasing of people's needs and decreasing of their awareness due to lack of knowledge about the significance of *kabuyutan* are considered the reasons for fragmentation. Respondent reported that managing buffer zones has been traditionally recognized as an appropriate rule in managing *kabuyutan*. According to this function, previous study stated that a protected area, particularly small patches, is supported by their buffer zone [1, 2, 3].

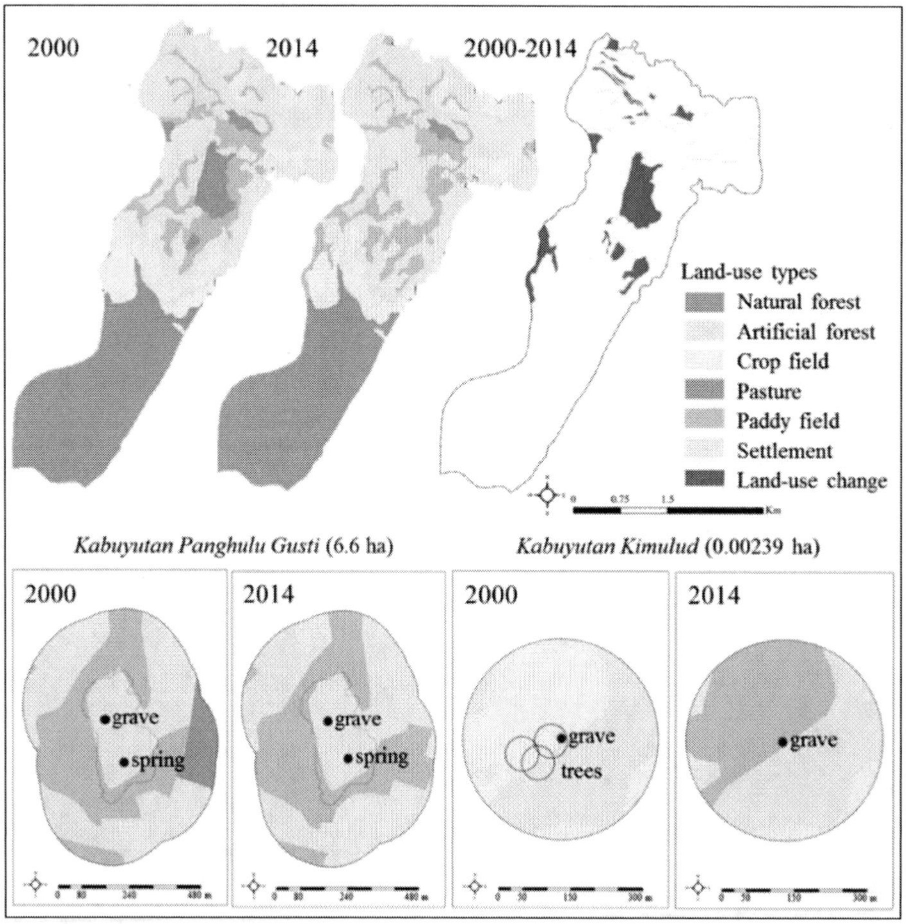

Figure 4. Changes in land-use during 14 years in Ciomas Village which led to loss of main elements and buffer zone of *kabuyutan*.

Moreover, it is found that most of *kabuyutan* located on personal lands indicated more vulnerability due to landowner's interest. Unpredictable change of landowner's need leads *kabuyutan* landscape more susceptible to damage and even loss of their area. Figure 4 presented the vulnerability of *kabuyutan* located in personal land. These *kabuyutan* have lost their buffer zones due to land-use changing. However, the sacredness of these places has succeeded in preventing extensive damage to the core zone. Therefore, revitalizing land status according to improvement the awareness of local people related to the importance of *kabuyutan* and other sacred natural sites in general is essential as a preventive effort.

4. Conclusions

This study concluded that local knowledge has provided significant information in defining the landscape as well as the changes and its reasons. About 33 *kabuyutan* were found and identified to have a unique structure compared to surrounding landscape, specifically covered by plant species and marked by springs, graves, or stones. *Kabuyutan* varied in characteristic depending on their physical features. *Kabuyutan* in Ciomas Village were mostly found in higher than 600 m above sea level, located on a wide range of slope, close to water bodies, far away from the road, and distributed in various types of land-use with a variety of its functions. This showed that *kabuyutan* greatly functioned as a landscape element to conserve land and water resources and proposed a landscape management based on watershed. However, *kabuyutan* located on personal land were vulnerable to damage and even loss of their areas. This study suggested that knowledge sharing related to imporant role of sacred natural sites is essential to increase local people's understanding that may affect their perception and attitude more wisely. This is a potential to be reinforced by customary and formal rules through revitalization of the core zone and its buffer zone. Further, the sustainability of landscape would be realized by protecting these natural sacred sites.

References

[1] Dudley N, Bhagwat S, Higgins-Zogib L, Lassen B, Verschuuren B and Wild R 2010 Conservation of biodiversity in sacred natural sites in Asia and Africa: a review of the scientific literature (London & Washington, DC: Earthscan) pp 19-32

[2] Verschuuren B, Wild R, Mcneely J and Oviedo G ed. 2010 Introduction: Sacred Natural Sites the Foundations of Conservation (London&Washington, DC: Earthscan) pp 1-13

[3] Daye D D and Healey J R 2015 Impact of land-use change on sacred forest at the landscape scale *Glob. Ecol. and Cons.* **3** pp 349-358

[4] Wessing R 1999 The sacred grove: founders and the owners of the forest in West java, Indonesia *L'homme et la foret tropicale* pp 59-74

[5] Danasasmita S 1987 Sewaka Darma (Kropak 408), Sanghyang Siksakandang Karesian (Kropak 630), Amanat Galunggung (Kropak 632). Transcription and translation (Bandung: Department of Education and Culture

[6] Hessel R, Van den Berg J, Kabore O, Van Kekem A, Verzandvoort S, Dipama J and Diallo B 2009 Linking participatory and GIS-based land use planning methods: A case study from Burkina Faso. *J. Land Use Pol.* **26** pp 1162-1172

[7] Malczewsk J 2004 GIS-based land-use suitability analysis: a critical overview. *Prog. in Plan.* **62** 3-65

[8] Saaty T L 1990 How to make a decision: The analytic hierarchy process. *Euro. J. Operat. Res.* **48(1)** 9-26

[9] Bello-Dambatta A, Farmani R, Javadi A and Evans B 2009 The analytical hierarchy process for contaminated land management *Adv. Eng. Inf.* **23** pp 433-441

[10] Lalika M, Meirea P and Ngaga Y 2015 Exploring watershed conservation and water governance along Pangani River Basin, Tanzania. *J. Land Use Pol.* **48** 351-361

Acknowledgments

This research was supported by The Indonesia Endowment Fund for Education (LPDP).

Level of gender equality in salted fish agro-industrial production at Tulang Bawang

H Yanfika[1], Rudy[2], I Listiana[1], R A D Widyastuti[3], M Riantini[4] and A Mutolib[5]

[1] Study Program of Agricultural Extension, Faculty of Agriculture, University of Lampung, Sumantri Brojonegoro Street 1, Bandar Lampung, Lampung, Indonesia, 35145
[2] Constitutional Law Department, Faculty of Law, University of Lampung, Sumantri Brojonegoro Street 1, Bandar Lampung, Lampung, Indonesia, 35145
[3] Agrotechnology Department, Faculty of Agriculture, University of Lampung, Sumantri Brojonegoro Street 1, Bandar Lampung, Lampung, Indonesia, 35145
[4] Study Program of Agribusiness, Faculty of Agriculture, University of Lampung, Sumantri Brojonegoro Street 1, Bandar Lampung, Lampung, Indonesia, 35145
[5] Postgraduate Program, University of Siliwangi, Siliwangi Street No.24, Kahuripan, Tawang District, Tasikmalaya, West Java 46115

Corresponding Author: helviyanfika81@gmail.com

Abstract. This study aims to find out how the level of gender equality in the production of salted fish agro-industry in Tulang Bawang Regency. The data analysis method used in this study is descriptive with quantitative approach. This research was conducted in East Menggala Subdistrict and Menggala Subdistrict of Tulang Bawang Regency. The number of respondents in this study was 39 samples. The data used is primary and secondary data. The results of this study showed that the activities carried out by the saltfish processing community in Tulang Bawang Regency both in reproductive, productive and social activities show that women dominate in playing a leading role in their daily lives. The existence of gender in terms of access and control of farmers to resources and benefits shows the dominant role of women to be the first reference in the life of salted fish processing society. Thus, women with this role in salted fish processing can already be productive in helping support the family economy, the average receipt per producer in one month amounted to Rp10,891,025.64 and the average total cost per producer in one month amounted to Rp8,757,081.00 so that the average profit per producer in one month was Rp2,143,944.64. The profit obtained by each producer varies due to the difference in total revenue received and the total cost incurred by each producer, and the main obstacles experienced by fishermen is the weather that is difficult to predict so as to decrease the productivity of salted fish processing.

1. Introduction

Indonesia is the world's largest archipelago and maritime nation. Indonesia's ocean area reaches 3.5 million km2 and land area of 1.9 million km^2 and has a coastline of 104,000 km with a total island of 17,504 islands [1]. Indonesia, which consists mostly of this sea, makes Indonesia has a very large and diverse fishery potential.

Lampung Province is one of the potential areas producing fisheries and has supported national fishery production as much as 40 percent, ranging from aquaculture, catches, to ponds. Lampung Province, especially Tulang Bawang Regency is a potential district that has the third largest fishery potential. The area of catching/maintaining fisheries in Tulang Bawang Regency amounted to 525,981.86 ha with a total production of 48,436.20 tons. The area of fishery is dominated by fishery production in the seawater area of 498,240.00 ha with the amount of production reaching 19,079.20 tons and freshwater fishery production of 27,741.86 ha with the amount of production reaching 29,357.00 tons per year 2017 [2].

The effectiveness of fishery sector development in Tulang Bawang Regency is not only limited to industrial development, but can give birth to agro-industrial development model with regional approach, which is based on potential natural resources. The approach of natural resources potential referred to so far is in Tulang Bawang Regency in large quantities, namely tilapia, patin, bandeng, and others. Most fishery production is obtained through traditional means to meet consumption demand [1]. This is important in order to ensure continuity and stability of agro-industry, given the nature of easily damaged fish and the availability of abundant fish in the harvest season, it is necessary to develop agro-industrial efforts based on potential natural resources. But from various commodities fish can not be known the most potential commodity to be developed [3]. The processing training and mentoring are given to produce good fish processing product qualit y, requiring fresh fish raw material and good processing capabilities [4].

The problem can then be solved by post-capture processing for the purpose of producing food and industrial raw materials, including preservation, canning, drying, smoking, shipping, and other processing activities so that fishery products can last longer as foodstuffs and can then be distributed from production centers to consumption centers [5]. The traditional fish processing business in Lampung Province has not been performed optimally and sustainably [6]. Basically these efforts initially only make use of natural processes that are done traditionally. Along with the development of science and technology, mechanical equipments are also developed that can accelerate and improve the quality of its products. Qualified technical and managerial competencies could improve the quality of products, resulting in sustainable traditional fisheries processing businesses [7]. Fishery products that have been processed and preserved include a variety of salted fish, smoked fish, pindang fish and others [5].

The traditional fishery processing business in Lampung Province is a small-scale business which have problems; ranging from the use of raw materials, technology, facilities and infrastructure, and food additives to the handling and processing of raw materials [8], related to processed products that are not safe for consumption. Related to this it is necessary to increase the competence of traditional fish worker to produce quality processed products. It is conceivable if traditional worker are not able to produce quality products then the processed products do not sell and the effort made will be a loss and close [9].

Women's participation in economic support activities is nothing new. Women work outside the home to earn income due to several things, one of which is the need to increase the family income. The widening of employment opportunities for women is also a factor for women to work. Labor participation reflects the relationship between wage levels and the number of hours worked offered, the outpouring of working hours can be affected by wage levels. Increased wages can provide increased income thus increasing the consumption of goods. So the increase in wages will make someone willing to go the extra mile to increase their consumption [10].

The potential role of women in the household to get additional income in the household economy, one of which is agro-industry. Because activities in agro-industry can be done easily anywhere, anytime, the size of the scale of the business anything, and all of it can be arranged freely according to the will of someone who does his business.

Based on the background description of the problem, it is necessary to conduct research on the role of women in the saltfish agro-industry in Tulang Bawang Regency which aims to analyze gender based on the profile of activities carried out by the saltfish processing community, identify gender based on the profile of access and control of farmers to resources and benefits, analyze the agro-industrial efforts of salted fish and know the obstacles experienced in the saltfish agro-industry.

2. Research Methods

This research will be conducted in Tulang Bawang Regency. The location of the research is determined intentionally (Purposive), this is done with consideration of salted fish agro-industry in East Menggala Subdistrict and Menggala Subdistrict is a salted fish agro-industry center in Tulang Bawang Regency. The research was conducted in November 2020. The population in this study is relatively small therefore all members of the population were sampled by 39 people divided from 21 people in East Menggala Subdistrict and 18 people in Menggala Subdistrict, so the method used in sampling is saturated sampling method or census.

The types of data used in research are primary data and secondary data. Primary data is obtained through direct interviews with respondents using questionnaires (list of questions) that have been prepared. Secondary data was obtained from agencies related to research such as the Central Static Agency, the Ministry of Marine Affairs and Fisheries, and the Fisheries Office of Tulang Bawang Regency.

The method used in this research is quantitative research with descriptive approach. Descriptive approach is used to describe the picture of social phenomena of the community in processing fish in Tulang Bawang Regency. This is related to the definition of a descriptive method approach that clearly shows an object's situation without affecting the object that is the focus of its research.

3. Result and Discussion

3.1. *Activity profile*

Profile of activities commonly carried out by men and women in the household is one of the factors that can be analyzed the existence of gender equality or not in its implementation. It is quite clear how men and women divide their roles in the household. This effort continues to be made to find out the extent to which people's understanding of gender equality, which is expected to be realized [4].

Household activities in salted fish processing societies, just like households in general include reproductive, productive and social activities. Household activities are carried out by men (husbands) and women (wives) with different portions of roles the difference is seen from the production activities carried out in accordance with managed efforts, namely salted fish [11]. Here is a more detailed explanation of each activity in the household saltfish processor about gender equality in it.

The role of women in the production of salted fish consists of several activities, in the processing of salted fish male labor and women's labor both have a role in it but in the proceedings of its activities women's labor plays more role by doing more work compared to the work of the male workforce.

3.1.1. Reproductive division of work. Reproductive work in a saltfish processing household includes domestic activities that are usually done in the home to meet all the needs of family members in the form of domestic services and facilities. The reproductive activities of salted fish processor can be seen in Table 1.

Table 1. Division of reproductive activities of salted fish processing in Tulang Bawang Regency.

Reproductive Activities	Men	Together Dominant Men	Together	Together Dominant Women	Women
	%	%	%	%	%
Cleaning the house	0	0	0	12.82	87.18
Parenting	0	0	35.90	41.03	23.07
Cook	0	0	0	0	100
Wash clothes	0	0	0	0	100
Washing dishes - Home furnishings	0	0	0	0	100
Accompanying children to learn	0	0	53.85	46.15	0
Child pick up and drop off	23.07	71.78	5.13	0	0

Shop for kitchen supplies	0	0	5.13	20.51	74.56
Total	**23.07**	**71.78**	**100.01**	**120.51**	**484.81**
Average	**2.88**	**8.96**	**12.5**	**15.06**	**60.6**

Table 1 above shows that the average reproductive activity is dominated by women with a percentage contribution of 60.60 percent, including cleaning the house, entering the child, cooking, washing clothes, washing dishes and household furniture, accompanying the child to study and shopping for kitchen needs. The role of men as a whole only contributes 2.88 percent of the child shuttle. Other activities were also carried out jointly by both men and women by 12.5 percent. Based on the above information shows that gender equality in productive activities of salted fish processing households is still widely borne and worked by women.

When likened to a team of army squads, women and men in households tend to describe the functions of women as nurses and men as gun holders. Therefore, women are commonly identified with domestic service work. In fact women and men have the same opportunities and life opportunities, so it would be fairer if all roles and functions can be divided in a good way the existence of consent between male or female students, no one is discriminated against from one side [6].

3.1.2. Productive division of work. The productive role in the household is usually identical to the work that provides financial benefits in order to need the needs of the Family [12]. Similar to salted fish processing households, there are activities that are part of the process to produce a product to meet the needs of family life by processing salted fish. The management of salted fish processing requires relatively little manpower, including the involvement of men and women in the family to manage this business so that it is able to produce. The production activities of salted fish processor in Tulang Bawang Regency can be seen in Table 2.

Table 2. Division of productive activities of salted fish processing in Tulang Bawang Regency.

Productive Activities	**Man**	**Together Dominant Men**	**Together**	**Together Dominant Women**	**Woman**
	%	**%**	**%**	**%**	**%**
Procurement of raw materials	20.51	35.90	30.77	0	12.82
Weeding	0	0	10.25	35.90	53.85
Sanctification	0	0	7.69	41.28	51.28
Salting	0	0	5.13	33.33	61.54
Drying	0	10.26	56.41	17.95	15.38
Packing	0	0	5.13	33.33	61.54
Storage	0	0	5.13	43.59	51.28
Sales	0	0	0	12.82	87.18
Total	**20.51**	**46.16**	**120.41**	**218.2**	**394.72**
Average	**2.57**	**5.77**	**15.05**	**27.27**	**49.34**

Table 2 above shows that the most average production activities carried out by women with a percentage contribution of 49.34 percent, there are production activities carried out by women entirely such as the sale of salted fish. Meanwhile, joint activities amounted to 15.05 percent. The smallest percentage is in salted fish processing activities conducted by men by 2.57 percent.

3.1.3. Social division. The social role in the household is usually related to how members of the household have contributions and roles in public, having extensive interactions in community social activities [13]. Salted fish processing households also have relatively the same social activities as other households, only there are some social activities that are still related to the processing of salted fish that it does. This social activity is very important in the role of the household, because basically human beings as social beings desperately need other human beings to meet the needs of their lives. The social activities of salted fish processing in Tulang Bawang Regency can be seen in Table 3.

Table 3. Distribution of social activities of salted fish processing in Tulang Bawang Regency.

Social Activities	Men	Together Dominant Men	Together	Together Dominant Women	Women
	%	%	%	%	%
Arisan	0	0	0	0	100
Celebration	0	0	100	0	0
Studies	0	0	35.90	10.26	33.33
Gather the village	89.74	10.26	0	0	0
Total	**89.74**	**10.26**	**135.90**	**10.26**	**133.33**
Average	**23.64**	**2.7**	**35.81**	**2.7**	**35.13**

Table 3 above suggests that the largest share of social work on average with a percentage of 35.81 percent of involvement is carried out by both men and women together. While the activities carried out by women are entirely 35.13 percent and those carried out by men by 23.64 percent. This suggests that male and female engagement does not experience much difference but based on the data is dominated by women.

3.2. Profile of access and control of resources and benefits

The concept of gender equality is considered very important in life. The roles and positions of men and women are equal partners who must have equal opportunities and opportunities to develop and have the same level of participation in different sectors of life. This is inseparable in the participation of men and women in the ability and control of their resources and benefits [14].

The access profile referred to here is something that shows a process of salted fish processing activities reviewed from the use or steps in its implementation. While the control profile is something that shows how decisions are taken from the course of activities. Access and control is indispensable for the implementation of an activity in order to run as expected. In this study, access and control will be reviewed based on the state of resources and how they benefit, this is to find out more about how gender is in society. Therefore, it is necessary to show some indicators of gender analysis as follows:

3.2.1. Equality level in access to resources. Equality is a common condition for men and women in obtaining their opportunities and rights as human beings, in order to be able to play a role and participate in political, legal, economic, socio-cultural, educational and defense and national security activities, as well as similarities in development. Based on the definition of gender equality, there are several kinds of indicators to see how the implementation of equality in society, has been successful or not.

The level of equality in access to resources means aimed at an indicator that sees equality between human rights and obligations in resource processing activities, where usually equality in the family or society can be understood through a variety of activities. In this case the access of resources in the agricultural sphere includes farming activities from upstream to downstream. So it is necessary for cooperation between men and women in its implementation. The level of equality in access to resources can be seen in Table 4.

Table 4. Equality level in access to resources.

Resource Access	Men	Together Dominant Men	Together	Together Dominant Women	Women
	%	%	%	%	%
Procurement of raw materials	20.51	35.9	30.77	0	12.8
Assisting production preparation	0	0	61.54	20.51	17.95

Water procurement	0	2.56	30.77	48.97	17.95
Fish drying	0	10.26	56.41	17.95	15.38
Fish packaging	0	0	5.13	33.33	61.54
Sales of processed products	0	0	0	12.82	87.18
Total	**20.51**	**48.72**	**184.62**	**133.45**	**212.7**
Average	**3.42**	**8.12**	**30.77**	**22.24**	**35.45**

Table 4 above shows that the implementation of salted fish processing activities that refer to the use of access shows the amount of female involvement is very high at 35.45 percent. While the involvement of men or roles of the two performed jointly showed a small number where the male role was 3.42 percent and the role together by 30.77 percent. Thus, it shows that gender equality between women and men in access to salted fish processing resources in Tulang Bawang Regency has not been equal, because the role of women still dominates the activities of both raw material procurement activities, helping production preparation, water procurement for production, fish drying, fish packaging, and the sale of processed products.

3.2.2. Equality level in access to benefits. The level of equality in access to benefits means intended to see the equality side of the indicators of utilization of resource processing results, especially in the processing of salted fish in East Menggala Subdistrict. The benefits obtained as a result of the achievement of farming activities must be felt by everyone in the family, not just certain people. So it is necessary to know whether equality in terms of benefits has been felt or not. The level of equality in access to benefits in salted fish processing in Tulang Bawang Regency can be seen in Table 5.

Table 5. Equality level in access to benefits.

Access Benefits	Men	Together Dominant Men	Together	Together Dominant Women	Women
	%	%	%	%	%
Proceeds from the sale of salted fish	0	0	79.49	0	20.51
Feel the money made	0	0	100	0	0
Household needs	0	0	20.51	33.33	46.15
Meet your eating needs	0	0	2.56	35.9	61.54
Total	**0**	**0**	**202.56**	**69.24**	**128.2**
Average	**0**	**0**	**50.64**	**17.31**	**32.05**

Table 5 shows that access to the benefits of salted fish processing activities is felt together with a percentage of 50.64 percent. But access to benefits was felt more by women at 32.05 percent. Indicators that become measurements are reviewed from the benefits of the sale of salted fish in the process of fulfilling household needs, to the benefits felt as other needs in the processing of salted fish. So it can be concluded that the level of equality in access to benefits is most widely owned and felt by women, so this has not reached an equal point.

3.2.3. Equality level in control of resources. The level of equality in the control of resources is a matter of how supervision of the role of men or women in making decisions on the implementation of activities, especially salted fish processing. Equality can be seen from the fairness between the participation of men and women in a balanced manner in determining a decision together. The decision made must be a thing that has been agreed together as a good step in achieving the results of salted fish processing, salted fish processing as a production activity is what will determine the level of economic contribution to the household. The level of gender equality in the control of resources in salted fish processing in Tulang Bawang District can be seen in Table 6.

Table 6. Equality level in control of resources.

Control Resources	Men	Together Dominant Men	Together	Together Dominant Women	Women
	%	%	%	%	%
Selling processed products	0	0	0	12.82	87.18
Deciding on production time	0	0	23.08	33.33	43.59
Deciding where to sell	0	2.56	48.72	18.46	30.26
Deciding on a sale time	0	0	10.26	31.03	58.97
Decided to buy a production tool	0	0	0	36.15	64.1
Total	**0**	**2.56**	**82.05**	**131.59**	**284.1**
Average	**0**	**0.51**	**16.41**	**26.31**	**56.77**

Table 6 shows that the level of equality in control of resources in decision making, women being the largest party to policy in a household or a society, it is shown that the percentage of female control is 56.77 percent, while men 0 percent. But in the decision-making process was also done jointly by men and women by 16.41 percent. Women's control in decision making for resource utilization activities is reviewed from how much is the decision to sell processed products, production time, where to sell processed products, sales time, to buy production tools. So this shows that the level of gender equality has not occurred in the processing of salted fish in Tulang Bawang Regency.

3.2.4. Equality level in control of benefits. The level of equality in control of benefits means how the control of men or women in determining decisions from the benefits of salted fish processing activities. One's control over these benefits shows how one's role and dominance or even demonstrates the creation of a group of people's life harmony. The level of equality in control of benefits in salted fish processing in East Menggala Subdistrict can be seen in Table 7.

Table 7. Equality level in control of benefits.

Benefit Control	Men	Together Dominant Men	Together	Together Dominant Women	Women
	%	%	%	%	%
Adjusting benefit allocations	5.13	35.90	48.72	0	10.26
Basic needs fulfillment	0	0	17.95	33.33	48.71
Fulfillment of education	7.70	28.21	56.41	5.13	2.56
Managing expenses	0	0	2.56	46.15	51.28
Total	**12.83**	**64.11**	**125.64**	**84.61**	**112.81**
Average	**3.21**	**16.03**	**31.41**	**21.15**	**28.2**

Table 7 shows that the level of equality in control of benefits shows that controls on the utilization of the results of fish processing activities are dominated by joint decisions with a percentage of 31.41 percent. This benefit control can be seen from how the control in managing the allocation of benefits, fulfillment of basic needs, fulfillment of education, to the management of expenditures. So it can be concluded that the level of equality in control of benefits is almost capable of being said to be equal but still dominated by women. The decision-making process in control of benefits is carried out jointly by men and women.

4. Analysis of salted fish processing business

4.1. Cost analysis

4.1.1. Fixed costs. Production costs consist of fixed costs and variable costs. Fixed costs are costs that do not depend on changes in production amount, such as equipment depreciation costs. Fixed costs in the saltfish processing business in Tulang Bawang Regency include labor costs and equipment shrinkage costs. The average fixed cost incurred by salted fish producers can be seen in Table 8.

Table 8. Average Fixed Cost on Salted Fish Processing Business in Tulang Bawang Regency.

No.	Type of Fixed Cost	Average (Rp/Month)	Percentage (%)
1	Labor	238,750.00	68.77
	a. Weeding	71,250.00	20.53
	b. Purification	83,750.00	24.12
	c. Salting	83,750.00	24.12
2	Equipment Depreciation	108,400.00	31.23
3	Investment Capital Interest	0,00	0.00
Total		**347,150.00**	**100.00**

Table 8 shows that the largest average fixed cost in a month in salted fish processing business is labor costs, Rp238,750.00 (68.77 percent) divided into labor in weeding activities amounting to Rp71,250.00 (20.53 percent), purification of Rp83,750.00 (24.12 percent), and refining of Rp83,750.00 (24.12 percent). Labor costs consist of labor costs outside the family. Usually the cost of labor in the family does not count. The wages of workers outside the family are given in the form of daily wages. The wage rate of female and male workers is the same ranging from Rp75,000.00 – Rp100,000.00. The highest average labor cost incurred in salted fish processing business in Tulang Bawang Regency in one month is the cost of labor in purification and salting activities.

The fixed cost of equipment depreciation, the average amount per month is Rp108,400.00 (31.23 percent). The cost of equipment shrinkage in fact does not need to be calculated because it is a cost that is not actually incurred by the manufacturer. However, because it uses the concept of profit depreciation costs equipment should still be calculated. Salted fish producers need equipment to carry out the production process. The equipment used is simple equipment, which has a fairly long economic life between 1-5 years. Equipment used includes knives, buckets, baskets, barrels, and widigs.

4.1.2. Variable costs. Variable costs are costs incurred by salted fish producers whose magnitude changes according to the amount of salted fish production produced. The variable cost of salted fish consists of the cost of the main raw materials, the cost of supporting raw materials, and the cost of others. The following data on variable costs in salted fish processing business in Tulang Bawang Regency can be seen in Table 9.

Table 9. Average variable cost on salted fish processing business in Tulang Bawang Regency.

No.	Type of Variable Cost	Average (Rp/Month)	Percentage (%)
1	Main Raw Materials	7,432,179.00	88.37
2	Supporting Raw Materials	865,983.00	10.30
3	Others	111,769.00	1.33
Total		**8,409,931.00**	**100.00**

Table 9 shows that the average variable cost spent per month is Rp840.931,00. The largest variable cost incurred in the saltfish processing business in Tulang Bawang Regency is the cost of the main raw material of fish, the amount per month averages Rp7,432,179.00 (88.37 percent). The type of fish used for raw materials for the manufacture of salted fish consists of a variety of various with varying prices. The high low cost of the main raw material of fish is strongly influenced by fluctuations in the price of fresh fish.

The supporting raw material used in the salt fish processing business is salt, with a ratio of 1:3 to the weight of fish. The average cost of supporting raw materials (salt) per month is Rp865,983.00 (10.30 percent). Salt as a supporting material is used to marinate fish. Salt used in the salt fish processing business in Tulang Bawang Regency is salt purchased at a price of Rp8000.00/Kg. Miscellaneous costs are the lowest variable costs incurred by producers. The average miscellaneous cost per month is Rp111,769.00 (1.33 per cent). This type of miscellaneous cost includes packaging used by manufacturers consisting of a wide variety of plastics, cardboard, and newspapers. This type of packaging was chosen because it is practical and able to maintain the quality of salted fish to keep it dry and not easily damaged.

4.1.3. Total cost. The total cost is the cost incurred by the producers of salted fish as a whole. The average total cost per month on salted fish processing business in Tulang Bawang Regency can be seen in Table 10.

Table 10. Average total cost of salted fish processing business in Tulang Bawang Regency

No	Type of Total Cost	Average (Rp/Month)	Percentage (%)
1	Fixed Cost	347,150.00	3.96
2	Variable Cost	8,409,931.00	96.04
Total		**8,409,931.00**	**100.00**

Table 10 shows that the average monthly total cost incurred by salted fish producers is Rp8,757,081.00. The largest cost incurred by salted fish producers was variable costs of Rp8,409,931.00 (96.04 percent), while the average fixed cost per month incurred by salted fish producers was Rp347,150.00 (3.96 percent). The variable cost incurred by respondents is due to the increase in the price of fresh fish raw materials as the main raw material due to the availability of fresh fish that began to decrease from the results of acquisition.

4.2. Acceptance analysis

The acceptance received by salted fish producers is a multiplication between the amount of salted fish production produced and the price of salted fish. The acceptance received by salted fish producers in Tulang Bawang Regency comes from the acceptance of various salted fish products produced, among others, namely cork fish, layis, tilapia, baung, tembakang, seluwang, lomow, sepat, and eel. The following data on the acceptance of salted fish processing business in Tulang Bawang Regency can be seen in Table 11.

Table 11. Acceptance according to the type of salted fish in the salted fish processing business in Tulang Bawang Regency.

Description	Total Production(Kg)	Unity Price (Rp/Kg)	Total Revenue (Rp)
Cork Fish	1,373.00	100,000.00	137,300,000.00
Fish Layis	280.00	70,000.00	19,600,000.00
Tilapia	1,098.00	60,000.00	65,880,000.00
Baung Fish	593.00	50,000.00	29,650,000.00
Gunfire	653.00	50,000.00	32,650,000.00
Seluwang	815.00	70,000.00	57,050,000.00
Lomow Fish	508.00	40,000.00	20,320,000.00
Sepat Fish	850.00	50,000.00	42,500,000.00
Eel	165.00	120,000.00	19,800,000.00

Total	**6,335.00**	**610,000.00**	**424,750,000.00**
Average	**703.88**	**67,777.77**	**10,891,025.64**

Table 11 shows that the total receipts in the saltfish processing business amounted to Rp424,750,000.00 with a total production of 6,335 kilogram and a total price of Rp610,000.00. Meanwhile, the average amount of receipts amounted to Rp10,891,025.64 with an average amount of production per type of fish in one month of 703.55 kilogram and an average price per type of fish of Rp67,777.77. The highest total receipt was at the cork salted fish receipt of Rp137,300,000.00 and the lowest total acceptance of layis salted fish at Rp19,600,000.00. This is influenced by the total production value produced by all respondents and the selling price of salted fish per unit. Each type of salted fish has a varying amount of production, unity price and amount of receipt. This is because each producer produces different types of salted fish with the price of each type of salted fish that also varies.

4.3. Profit analysis

The profit obtained by salted fish processing business in Tulang Bawang Regency is the difference between total receipt and the difference in total cost. In knowing the amount of profit can be seen in Table 12.

Table 12. Average profit on salted fish processing business in Tulang Bawang Regency.

No	Description	Average per-manufacturer (Rp/Month)
1	Acceptance	10,891,025.64
2	Total Cost	8,757,081.00
Total		**2,143,944.64**

Table 12 shows the average revenue per producer in a month of Rp10,891,025.64 and the average total cost per producer in a month of Rp8,757,081.00 resulting in an average profit per producer in one month of Rp2,143,944.64. The profit obtained by each manufacturer varies due to the difference in total revenue received and the total cost incurred by each manufacturer. This calculation uses the concept of profit, then the actual cost is not really incurred by the manufacturer, namely the cost of equipment depreciation.

5. Obstacles Faced in the Implementation of Salted Fish Agro-Industrial Production

Every business must have obstacles that can interfere with the smoothness in the production process so that every obstacle must be faced so that the business can continue to run and develop in the future. Similarly, salted fish processing business in Tulang Bawang Regency, also has constraints that can hinder the smoothness in the production process.

The obstacle faced in the saltfish processing business in Tulang Bawang Regency is the weather (sunlight). Salted fish processing business in Tulang Bawang Regency is still traditionally done by relying on sunlight in the drying process. If the weather is cloudy or rainy season, salted fish processing business faces obstacles in the drying or drying process. In the rainy season or cloudy weather the drying or drying process of salted fish will take a relatively longer time and the resulting product can also be at risk of damage. The step that has been done by manufacturers to overcome these obstacles is to stack the widig and cover it with a tarpaulin. Although, in reality the drying process of salted fish can be done mechanically by using a modern dryer that is a dryer oven. However, for the reason that the production of dried salted fish using the oven is not perfect, such as uneven dry so that fish producers in Tulang Bawang District do not use dryer ovens and still make use of sunlight in the drying process of salted fish. Another reason why manufacturers do not use dryer ovens is also because the costs incurred will increase because the price of dryer ovens are quite high.

Another obstacle faced by salted fish producers in Tulang Bawang Regency is the availability of fish raw materials that depend on the season. Fresh fish is the main raw material in the salted fish processing business in Tulang Bawang Regency, but its availability is highly dependent on the season. Such a situation will have an impact on the business of processing salted fish, which results in the continuity of salted fish production can not last well throughout the year. Some manufacturers even choose not to produce due to the low availability of raw materials. However, there are also those who overcome the scarcity of raw materials by still producing using existing raw materials or adequate raw materials by using other types of fish raw materials or from outside the city when there is an urgent order.

Besides the above constraints there are other obstacles faced by respondents salted fish processing business in Tulang Bawang District, namely the absence of organizations that accommodate respondents salt fish processing business. This causes the attention of the local government to the saltfish processing business has not been poured evenly to all respondents, for example on local government assistance given in the form of facilities and infrastructure has not been evenly distributed and the procurement of counseling / coaching that does not exist for the development of salted fish processing in Tulang Bawang Regency. In addition, with the establishment of an organiasi so that no communication between respondents salted fish processing business, as a result of which the respondents can not exchange information about the business information of salted fish processing.

Another obstacle experienced by respondents processing salted fish in terms of marketing is the difficulty of access to consumers and business rivals are quite numerous. Respondents who still sell salted fish with consumers who have to come directly to buy salted fish, so there needs to be a development of skills for respondents especially in the management of information systems. Some respondents already use information systems with social media in promoting their salted fish products.

6. Conclusions

Based on the results of the field can be known that activities carried out by salted fish processing communities in Tulang Bawang Regency both in reproductive, productive and social activities show that women dominate in holding the lead role in their daily lives, while men also still have a role in all three activities. This indicates that there are still gender differences in society, where between men and women have not achieved the word equality.

The existence of gender in terms of access and control of farmers to resources and benefits shows the dominant role of women to be the first reference in the life of the saltfish processing community, although there is also in access and control to benefits is still said to be balanced, but in other aspects women who hold full access and control in salted fish processing activities. Thus, women with this role in the processing of salted fish can already be productive in helping support the family economy.

Average revenue per producer in one month was Rp10,891,025.64 and the average total cost per producer in one month was Rp8,757,081.00 so that the average profit per producer in one month was Rp2,143,944.64. The profit obtained by each manufacturer varies due to the difference in total revenue received and the total cost incurred by each manufacturer. This calculation uses the concept of profit, then the actual cost is not really incurred by the manufacturer, namely the cost of equipment depreciation.

There are several obstacles faced by salted fish processors in Tulang Bawang Regency. These constraints are the process of processing salted fish is still done traditionally that relies on sunlight, the availability of raw materials that depend on the season that results in raw materials are difficult to obtain, and the absence of institutions/ organizations that accommodate salted fish processors in supporting their business.

References

[1] Junianingsih I 2013 Analysis of profitability of traditional processing business of smoked cobs in Situbondo Regency Anchor Village. *Samakia: Journal of Fisheries Science.* pp 86-93.

[2] Central Bureau of Statistics 2018 Tulang Bawang District in 2018 Figures BPS Bone Onion Onion Bone

[3] Tauhid M, Sri H, and Tirza 2017 Analysis of Agro-Industrial Types and Feasibility of Fish-Based Agro-Industrial Establishment in Tulang Bawang Regency. *Journal of Industrial Technology and Agricultural Products.* **22**

[4] Kusumastanto T 2014 Indonesia's Development Strategy as a Maritime Country Bogor Agricultural University Bogor

[5] Sari, M K 2011 Analysis of Salted Fish Processing Business in Cilacap Regency Thesis University of Eleven March Surakarta
[6] Laisa DD, Sayekti WD and Nugraha A 2013 Analysis of The Cost of Production and Development Strategy of Dry Rice Anchole Processing Industry in Pasar Island, West Betung Bay District, Bandar Lampung City. *Journal of Agribusiness Sciences*, pp 111–117
[7] Yanfika H, Listiana I, Mutolib A, and Rahmat A 2019 Linkages between extension institutions and stakeholders in the development of sustainable fisheries in Lampung Province. *Journal of Physics: Conf. Series 1155 012014.*
[8] Yuliana E, Suhardi DA, and Susilo A 2011 Level of use of harmful chemicals in salted fish processing. Cases in Muara Angke and Cilincing; Jakarta. *Indonesian Fisheries Products Processing Journal.* pp 14-21
[9] Yanfika H, Siti A, Anna F, Pang SA 2018 Worker Performance from Perspective of Profit, Quality and Work Accuracy in Traditional Fishery Business in Lampung Province. *International Journal of Social Science and Economic Research.* Pp 578-589
[10] Reyka 2013 Influence of Wage Factors, Age, Husband's Income, Last Child's Age, and Household Expenditure on The Outpouring of Working Hours of Married Women in Magelang City Thesis Diponegoro University Semarang
[11] Pawitasari E 2015 Women's Education between Gender Equality and Islam. *Journal of Islamic Civilization.* **II** pp 249-272
[12] Martiany Dina 2012 Implementation of Gender Mainstreaming (IGM) as a Strategy for Achieving Gender Equality (Study in North Sumatra and Central Java Provinces). *Aspiration Journal.* **2** pp 122-136
[13] Handayani T A 2015 Realizing Gender Justice through Legal Protection of Women. *Journal of Reschtstaat Nieuw.* **1**
[14] Manaf M 2007 Analysis of Space Utilization in Coastal Areas of Bontoharu Subdistrict of Salayar Islands Regency. *Plano Madani: Journal of Regional and Urban Planning*, pp 10–21.

Regulatory Impact Analysis on Mangrove Forest in the Coastal Area of the Bandar Lampung

N Tresiana[1], N Duadji[1], IG Febryano[2], MK Maharani [2], A Rahmat[3]

[1]Departemen of Public Administration, Faculty Social and Politic, University of Lampung, Indonesia
[2] Department of Forestry, Faculty of Agriculture, University of Lampung, Indonesia
[3] Research Center for Limnology, Indonesian Institute of Sciences, Indonesia.

E-mail: novita.tresiana@fisip.unila.ac.id

Abstract. Kota Karang Mangrove is one of the mangrove forests located in urban areas with a limited existence in Indonesia. Mangrove forests have economic, physical, and ecological potential. Several policy designs have been issued to protect and develop the potential of mangrove forests, but they have not been optimal in protecting and managing mangrove forests. The purpose of this study was to assess the quality of the mangrove forest policy in Karang City. The research method uses a qualitative case study design, the research period is September to December 2020. The research location is in the coastal area of Karang City Village, Teluk Betung Timur, Bandar Lampung City, Lampung Province, Indonesia. The data collection technique used in-depth interviews, the determination of the informants was purposive, consisting of the Ministry of Maritime Affairs and Fisheries, the Lampung Province Maritime Affairs and Fisheries Service, the Lampung Provincial Forestry Service, the Bandar Lampung City Marine and Fisheries Service, the Bandar Lampung City Housing and Settlement Service, NGOs (Walhi , Mitra Bentala, and Hands) and the Karang City Community. Data has been collected through desk review and stock-taking using regulatory impact analysis. The results showed at the problem analysis stage; Mangrove forests have not become the main issue, the location of the authority is limited, it contradicts national policies, there are obstacles in the implementation and implementer processes. Only 3 policies are used as a basis for policy development, integration, and collaboration of 3 key stakeholders in policy development. The choice of ecotourism and green open space policy development becomes smart regulation that can improve policy quality, transparency, accountability, and costs.

1. Introduction

Mangrove Kota Karang Mangrove is one of the mangrove forests located in urban areas which is very limited in existence and is found only in five regions in Indonesia, including DKI Jakarta, Central Java, Yogyakarta, East Java, North Kalimantan, and Bandar Lampung City [1]. The existence of mangrove forests has economic, physical, and ecological potential and functions [2]. Mangroves function physically as carbon sequestration and protection for the coastal zone to eliminate tidal wave energy [3]. Economically, mangroves can be used as ecotourism so that they can create job opportunities, increase alternative income, and become business opportunities for the community [4]. The ecological function of mangroves as a habitat for marine biota and other types that live in mangrove ecosystems, provides soil fertility for the surrounding land [5], including mangrove forests which have various functions in line with the function of green open spaces in urban areas [6].

The study was based on the issue of neglect and minimal protection of mangrove forests in urban areas in several parts of Indonesia. This has relevance to public policy as well as a means of success in achieving two international standards, in the form of: first, the sustainable development goals (SDGs) item 14, which contains the preservation and utilization of marine and oceanic resources for sustainable development. Marine resources in coastal areas are generally characterized by mangrove forests [7]. Second, the Ramsar Convention on Wetlands, which contains an international work plan on cooperation in the conservation and sustainable use of mangroves [8]. The two agreements form the basis for various national policy implementations in all countries with mangrove forests, including Indonesia.

This study focuses on the protection and management of mangrove forests in urban areas. The picture of the mangrove forest ecosystem that is still under threat, is prone to change its function, whether it is conversion to ponds, gardens including residential areas, even though several protection policies have been established and the implementation of policies is carried out by cross-sectoral agencies. Several roles of state institutions with strategic mangrove policies have been established in Indonesia, including: first, the Ministry of Forestry through the Forestry Law and Law No. 5/1990 on Conservation of Living Natural Resources and Ecosystems views mangroves as forests; second, the Ministry of Marine Affairs and Fisheries has duties and functions regarding coastal resources, including mangrove forests; The Ministry of Environment is involved because mangrove damage is the standard criterion for ecosystem damage. Several laws related to mangrove forests have been issued, including Law No. 41/1999 on Forestry, Law No. 26/2007 on Spatial Planning, Law No. 27/2007 on Management of Coastal Areas and Small Islands, and Law No. 32/2009 on Environmental Protection and Management [9, 10]. At the local government level, the Lampung Provincial Government has issued Lampung Provincial Regulation No. 1 of 2018 which regulates the zoning plan for coastal areas and small islands, and the Governor of Lampung Province Regulation No. 20 of 2019 which regulates the development of water areas and coastal conservation areas and small islands in Lampung Province. However, the protection of mangrove forests is still not optimal. Some of the causes are overlapping policies on mangroves and mangrove protection on paper [11], wrong perceptions of the role and function of mangrove forests [12], several cases of criminalization of fishermen and community groups who conserve coastal resources by replanting mangroves due to lack of law enforcement. and weak enforcement of crimes against cutting and destroying mangrove forests [13].

This study is important because several studies related to the current conditions and the existence of mangrove forests in various parts of the world describe opportunities as well as threats. A study using an environmental economics perspective that has been carried out previously describes the contribution of US$ 200-900 thousand of the economic benefits contained in one hectare of mangrove forest area [14]. These benefits are obtained from the production of wood, charcoal, fish, shrimp, and crab products, as well as other economic products. Contribution is obtained from the revenue from the natural tourism sector (ecotourism) because it can attract local and international tourist visits. Some of the threats include garbage dumped in the forest, oil and toxic chemicals, human encroachment, development reasons, changes in seawater temperature due to weather changes, and global warming. Meanwhile, another study [15], describes a decrease in the area of mangrove forests in the world. From 18.8 million hectares to 15.2 million hectares in 2018. The decline was due to massive damage that occurred in several countries, including Brazil, Mexico, and Indonesia. The FAO study states that the damage to mangrove debt is due to the conversion of environmental functions into salt ponds, shrimp ponds, overexploitation for economic purposes, as well as industrial waste that is carelessly dumped in river estuaries [16]

Several previous studies have linked the protection and management of mangrove forests, especially in cities, with a lack of awareness, lack of knowledge, and education about the importance of preserving the environment [17]. Another study relates to preservation methods by keeping forest ecosystems by their natural habitats [18], law enforcement and processes, conflicts over authority over mangrove forests [19]. However, it is still seldom related to how the policy-making process, including the assessment of several alternatives and consultation with stakeholders, requires encouragement of policy-making that provides the greatest benefits to the environment, the existence of forests and communities, policies become strong and justified. The importance of policies or regulations for the protection and management of quality mangrove forests because the applicable policies/regu must have reasons, the policy chosen is the best alternative, the benefits and costs are calculated, public consultation (stakeholders) is carried out and most importantly no adverse impacts of the policy are felt. Policy development through structured, systematic, tested steps with a rigorous assessment of various choices nations is a key instrument for achieving policy quality [20, 21].

The research objective was to analyze the quality of the mangrove forest policy in Karang City using Regulatory impact analysis. The resulting policy recommendations will be useful for policymakers and stakeholders in making policies that are more efficient, effective, transparent, and accountable so that they have a better impact on the welfare of society and the state. Besides that, it is an important part of the effort or movement to create smart regulations.

2. Methods

This study uses a qualitative case study design [22]. The research period was from September to December 2020. The research location was in the coastal area of Kota Karang Village, Teluk Betung Timur District, Bandar Lampung City, Lampung Province, Indonesia. Data collection techniques using in-depth interviews. The technique of determining informants was purposive, consisting of the Ministry of Maritime Affairs and Fisheries, the Lampung Province Maritime Affairs and Fisheries Service, the Lampung Provincial Forestry Service, the Bandar Lampung City Marine and Fisheries Service, the Bandar Lampung City Housing and Settlement Service, NGOs (Walhi, Mitra Bentala, and Hands).) and the Urban Community at Kota Karang.

Apart from qualitative design, the study also used desk reviews and stock-taking / stock talking for data collection. Desk reviews were conducted to map various regulations, the development of policy impact literature from various aspects (social, economic, political, environmental), environmental and sustainable development standards and objectives. Desk reviews are useful in enriching the concept of specific mangrove and environmental policies and by the objectives, international and national standards. Stock-taking / stock talking focuses on research results and current publications on the impact of regulations in the form of iissues/problems popolicies/regulationsstakeholders, costs and benefits, alternative policies [23].

The data collected through qualitative design, desk review a,nd stock taking/stock opname were reviewed using regulatory impact analysis [24]. Regulatory impact analysis, can explain the success of the protection of urban mangrove forests through design and policy reforms through the following stages: 1) Problem analysis: describes the main issues / problems of urban mangrove forests, their causes and impacts; 2) Mapping of regulations: identifying mangrove forest regulations, whether vertical, horizontal, national or regional;3) Stakeholder analysis: mapping actors and their interests in mangrove forest policy issues; 4) Policy development options: analyze the costs and benefits of policy options

3. Result and Discussion

3.1. Research Location Conditions

The study was conducted in Kota Karang Kelurahan, located in Teluk Betung Timur District with an area of 35 ha and has a coastline (coastal area). Topographical conditions in the mangrove forest area are flat, both in forest and land areas. The soil condition in the mangrove area has muddy soil. Rainfall is relatively low and there are no freshwater sources in the mangrove area.

Based on data from the Central Statistics Agency (BPS) of Bandar Lampung City, in 2020 the Ward of Kota Karang has a population of 10,186 people consisting of 5,440 male residents and 5,180 female residents. The number of household heads in the Kota Karang Kelurahan is 2,642 households. The population density of Kota Karang Kelurahan is 30,349 people / km2. Its strategic location makes this area the center of general trade, services, and the center of the economy. The total population of Karang City is 10,186 people consisting of 5,440 male residents and 5,180 female residents. The number of household heads in the Kota Karang Kelurahan is 2,642 households. The population density of Kota Karang Kelurahan is 30,349 people / km2. The population in this kelurahan has various religions, Islam 9,676 people, Catholicism 117, Hindu 213, and Buddhist 53. The heart of the gate of the Teluk Betung

Timur Subdistrict is the Kota Karang Village. Its strategic location makes this area the center of general trade, services, and the center of the economy. This economic activity is supported by the existence of traditional markets, processing centers for salted fish and anchovies on Pasaran Island. Most of the population makes a living as fishermen, construction workers, and entrepreneurs / trading.

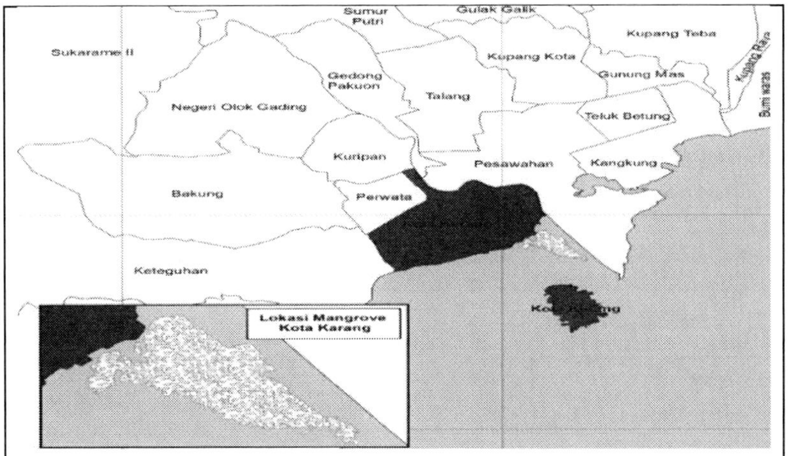

Figure 1. Map of Kota Karang

3.2. Regulatory impact analysis

3.2.1. Problem analysis. Currently, the area of mangrove forest in Bandar Lampung is only 5,478 ha from 59.35 ha [1]. One of the contributing factors is the conversion of mangrove forests on the urban coast to ports, settlements and other activities caused by human activities and illegal encroachment from local communities. The low level of income of the Head of Family in the Karang City Village is accompanied by the level of education of the people in this area which tends to be low, resulting in people living in this area, especially those on the banks of the river, unable to get good jobs. This affects the settlement pattern in the form of settlement expansion towards the riverbanks towards the mangrove location. In the early 1990s, the coast of Bandar Lampung City had a mangrove ecosystem that almost covered the entire coastline of Bandar Lampung City. The increase in population and development causes the mangroves in Bandar Lampung City to experience degradation. Degradation causes the mangrove ecosystem to remain only a few hectares and in certain locations. Meanwhile, the potential for the utilization of mangrove forests in Bandar Lampung was found to be very minimal. The government does not get the potential of mangrove forests in urban areas. This attention is one of the keys to management and protection, given the existence of mangrove forests that are only found in a few areas in Indonesia. The following is the problem analysis in the form of a problem tree, which can be seen in Figure 1.

The analysis in Figure 1 shows that the problem that has occurred in the mangrove forests of Karang City is that the area has decreased since 1993. This can be seen from the absence of an increase in the area of mangrove forests in the city of Bandar Lampung. Another factor is the conversion of mangrove forests on the urban coast to ports, settlements, and other activities in Bandar Lampung. The potential for utilization of mangrove forests in the city of Bandar Lampung is minimal. The analysis (Figure 2) illustrates that the main problem is that the protection and management of mangrove forests in urban areas is minimal and unsustainable. The success of determining mangrove forests for further development is seen from the authority of local government institutions with reference to the regional autonomy law. The position of mangrove forests in urban areas makes the authority limited to the OPD of the Housing and Settlement Service, Bandar Lampung City area, while the forestry OPD, which has the main mandate of protecting and managing mangrove forests, is in the provincial government. This in governance policies has relevance to authority, policies, programs, and budgeting. The government's understanding (Housing and Settlement Service), regarding the presence of mangroves in urban areas is a problem that has an impact on reducing mangrove area significantly. The government's focus on managing environmental issues is low. Implementers are less focused on environmental issues, one of which is due to high rotation, weak data and information, all of which have an impact on the planning of the potential for various sustainable policies and programs not yet a priority. Governance in the form of allowing the conversion of mangrove forests to various human activity facilities results in the destruction of the forest, which implies an inability to maintain the existence of urban mangrove forests, contributing to the loss of one of the urban mangrove forests in Indonesia [25].

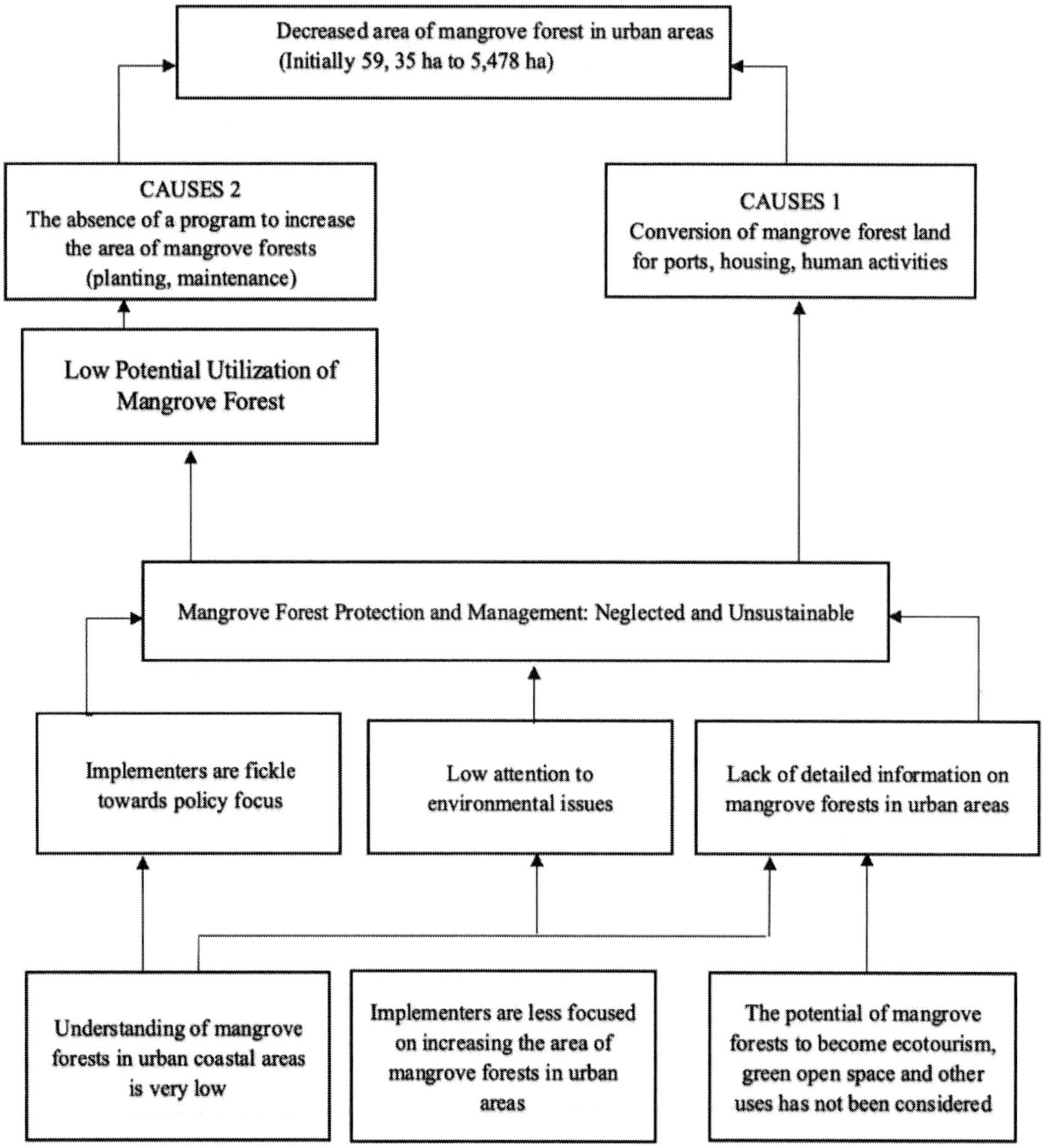

Figure 2. Problem analysis

3.2.2. Regulatory mapping. Mapping of regulations can be done to review a regulation in order to improve the quality of existing regulations, so that the policies taken can hit the target [26]. Regulations related to mangroves in Karang City are regulated in national to regional policies. The governing regulations are Law No. 27 of 2007, Regional Regulation of the Province of Lampung No. 1 of 2018, and Regional Regulation of the City of Bandar Lampung No. 10 of 2011. The success of the mangrove forest policy in Karang City can be seen in table 1.

Table 1. Mapping of 3 Relevant Policies for Management and Protection of Mangrove Forest in Karang City

No	Policy	Substance / Analysis
1	UU. No. 27 Tahun 2007 tentang Pengelolaan Wilayah Pesisir dan Pulau-Pulau Kecil (UU. No. 27 of 2007 concerning the Management of Coastal Areas and Small Islands)	1. Formation of the Zoning Plan for Coastal Areas and Small Islands (RZWP3K) with the aim of protecting, dividing, utilizing coastal areas and small islands for the ecosystem in a sustainable manner. 2. Strengthen the role of the community and government institutions for the management of coastal areas and small islands. 3. The basis for local governments to make policies related to zoning for coastal areas and small islands.
2	Peraturan Daerah Provinsi Lampung No. 1 Tahun 2018 (Lampung Provincial Regulation No. 1 of 2018)	Provide confirmation that Kota Karang mangroves are included in a coastal park area, and their development is used for the development of science, research, education and ecotourism.
3	Peraturan Daerah Kota Bandar Lampung No. 10 Tahun 2011 (Bandar Lampung City Regional Regulation No. 10 of 2011)	Contains spatial plans for mangroves including other protected areas, coastal border areas, and Green Open Space (RTH)

Analysis of the implications of three policies: Policies at the national level related to mangroves serve as guidelines for determining zoning plans in coastal areas and small islands. Derivative national policies are regional regulations that are stipulated by each region in Indonesia and are guided by national-level policies. The zoning determination of areas on the coast and small islands in each area is based on the characteristics of each zoning. Coastal park zoning is a stipulation in the zoning plan at the provincial level. Coastal parks have the attraction of living natural resources so that their existence is in the conservation status of coastal areas and small islands. The status of mangroves as a conservation area is in line with the rules of the RTRW of Bandar Lampung City which classifies mangroves as other protected areas. The determination of the status of the area has not been written in the RTRW, so that mangroves in Karang City are classified as "coastal border areas and green open spaces". Mangroves as coastal parks in urban areas can also become ecotourism so that their existence supports the increase of green open space in Bandar Lampung.

Policies related to the determination of mangroves nationally are regulated in Law Number. 27 of 2007 concerning the Management of Coastal Areas and Small Islands. This national policy has accommodated the interests of the parties who are the implementers of the policy. The purpose of a national policy regulation is to protect mangroves and empower communities. The policies in place include the prevention of mangrove conversion. The first attempt is made by prohibiting the use of methods and methods that damage the mangrove ecosystem in the utilization of coastal areas. The second is to prohibit the conversion of mangrove ecosystems in cultivation areas or zones that do not take into account the sustainability of coastal ecological functions. This regulation is a regulation that is national in nature, so that in its implementation there are derivative regulations.

The mangrove ecosystem in the regulations is classified as a Coastal Conservation Area and Small Islands (KKP3K) and is included in the type of coastal park. Kota Karang Teluk Betung Timur has the code KKP3K-TPM-1. This provincial-level regulation contains the main programs and priority activities regarding the plan for coastal conservation areas and small islands. This plan contains several programs and activities. The programs carried out include strengthening institutions, drafting supporting regulations for coastal areas and small islands, empowering communities, as well as monitoring and evaluating the management of coastal conservation areas and small islands. The program implementation matrix is scheduled to be implemented in the 2019-2032 period. This program receives funding from the APBN (National Revenue and Expenditure Budget), APBD (Regional Revenue and Expenditure Budget), and the private sector. The implementing agencies for the coastal and small island conservation plan based on the Perda are the Marine and Fisheries Service, the Public Works and Spatial Planning Service.

The policy at the regional level, especially Bandar Lampung City, has not specifically included Kota Karang mangroves as other protected areas. The Bandar Lampung regional regulation classifies other protected areas as mangroves, mangroves, and seagrass beds. Other protected area management regulates area management direction. Management directives include prohibitions against damaging mangrove ecosystems. The activities allowed in the area are tourism and research. The spatial planning carried out by the city government of Bandar Lampung regarding the mangrove area is not spatially determined.

All of three regulations are the basis for the implementation and management of the Kota Karang's mangrove forest, but based on the results of interviews and research on various programs, it is found that the capacity and commitment of the city government is still low. Some of the findings include: The City Government does not yet know its duties as an implementer and does not see mangroves as the focus of ecological development, mangrove development programs through the potential of Green open space (RTH) and tourism have not been implemented. On the other hand, the Regional Government of Lampung Province actually shows the opposite condition. The government's commitment has been realized by formulating policies related to mangroves as derivatives on a national to regional scale. Policies that regulate mangroves in Lampung province related to mangroves and coastal areas are realized with policies up to development directions. The success of planting carried out in several mangrove locations showed an increase in area. The capacity of the provincial government in carrying out the authority of mangrove management in macro is high.

3.2.3. Stakeholders analysis. Stakeholder analysis is carried out to understand the scope of the institution, the roles of relevant actors and key policy holders that determine the course of a policy at the formulation and implementation levels, starting from the central, provincial, to local governments [27] [28]. Their influence and importance will have an impact on the management of mangrove forests in Kota Karang. Stakeholder analysis can be seen in Table 2.

Table 2. Mapping the roles of stakeholders.

Actors	The Role of Formulation	Implementation Role	Interest	Action
Key stakeholders				
Public service unit of Marine Affairs and Fisheries (DKP), Lampung Province	Conducting mangrove forest policy formulation in Lampung Province	Conduct monitoring, work coordination, work activities related to mangrove forest management in Lampung Province	Making policies regarding the allocation of funds and the location of mangrove forests in Lampung Province	Planting mangroves in several mangrove forest locations in Lampung Province
Lampung Provincial Forestry Service	Making suggestions regarding the location of mangrove forests in Lampung Province	Coordinating work with the DKP of Lampung Province	Carry out rehabilitation related to mangrove forest damage in Lampung Province	Planted mangroves in several locations designated as mangrove forests in Lampung Province
Public service unit of Marine	Planning mangrove forests for tourism in	Coordinating work with Disperkim for	Establishing mangrove management activities as ecotourism in the	Making ecotourism work plans in the

Affairs and Fisheries, Bandar Lampung	the city of Bandar Lampung	mangrove objects as ecotourism	city of Bandar Lampung	mangrove forests of Bandar Lampung City
Housing and Settlement Service of Bandar Lampung	Arrange a location in the city of Bandar Lampung which is included in the mangrove forest according to the RTRW	Determine the mangrove forest area in the city of Bandar Lampung	Sufficient and mapping the location of mangrove forests in the city of Bandar Lampung	To build and repair facilities, facilities and infrastructure for mangrove forest locations
Community		Establishing a management team at the mangrove forest location in Karang City	Acting as the target group at the location so that the presence of mangroves can support income	Maintain the location so that the mangroves are not converted into settlements, ports, and other activities.
Supporting stakeholders				
Marine and Fisheries Ministry	Prepare guidelines for the use of mangrove forests	Conduct monitoring related to mangrove forest management	Make decisions regarding how the location can be used as a mangrove forest	Encourage the development of mangrove forest facilities
NGO (Walhi, Mitra Bentala, andTangan)	Contributing to making proposals related to policies to determine the expansion of mangrove forests	Carry out activities related to saving mangroves in collaboration with the government	Provide information to the surrounding community regarding the importance of mangrove forests	Contributing to counseling and planting mangrove forests in Karanh City

Based on the table above, stakeholders play different roles. Stakeholders who have influence and interests are divided into 2, namely:

1. The main stakeholders are those who make policy actors, groups who are directly affected by the positive or negative impacts of the mangrove management policy of Kota Karang, including the Lampung Province Marine and Fisheries Service, the Lampung Provincial Forestry Service and the Bandar Lampung City Housing and Settlement Service and the community;
2. Supporting stakeholders are intermediaries in assisting the process of realizing the Mangrove Policy in Karang City. Among them are the Ministry of Maritime Affairs and Fisheries and several NGOs, including: Walhi, Mitra Bentala, and Tangan.

The existence of guidelines for the preparation, determination, and utilization of mangrove forests nationally is regulated by the Ministry of Marine Affairs and Fisheries, which is a supporting stakeholder, so that this plays a major role in the formulation of national mangrove forest policies. These guidelines form the basis for local governments to implement derivative policies and their implementation. Meanwhile, non-government NGOs run outreach programs to the community about the importance of the existence of mangrove forests, so that people have awareness to protect the mangrove forests in these locations.

Meanwhile, Key Players include: the Department of Maritime Affairs and Fisheries of the Province of Lampung, the Forestry Service of the Province of Lampung and the Department of Housing and Settlements of the City of Bandar Lampung, holding the key policies for the success of the protection and management of the city's mangrove forests. Collaboration between stakeholders is a priority in successful implementation. Some of the collaborative actions among key stakeholders are as follows:

1. Department of Marine Affairs and Fisheries of Lampung Province, with the authority to manage mangrove ecosystems in Lampung Province, including mangrove locations in Karang City. Important role as implementer, facilitator, and conduct cost planning related to advanced management.
2. The Lampung Provincial Forestry Service acts as a facilitator and collaborates to plant and utilize mangroves in urban areas. Collaboration with DKP can save the remaining mangroves

in the city of Bandar Lampung. Collaborative policies that involve the Forestry Service in managing mangrove forests need to be done so that the Forestry Service has the responsibility to protect mangrove forests on urban coasts.

3. Collaboration between the Bandar Lampung City Marine and Fisheries Office and the Bandar Lampung City Housing and Settlement Service is key to policy. Planning related to mangrove forest management in Karang City has been prepared by the DKP of Bandar Lampung City to become an ecotourism location whose existence can be directed towards increasing Green Open Space (RTH) in Bandar Lampung City. The Housing and Settlement Service must play a role in establishing a Regional Regulation that the location of mangrove forests in Karang City is included in the RTRW so that the existence of mangrove forests does not experience a decrease in area because their status is recognized by RTRW. The RTRW designation of mangrove forests will support the policy of mangrove forests as protected areas and can be managed by local agencies where mangrove forests are located. The city agency of Bandar Lampung acts as the implementer regarding the implementation and development of mangrove forests on urban coasts so that the existence of mangrove forests can increase in size and experience protection and sustainable management.
4. Communities including non-governmental NGOs play a role as actors in carrying out efforts to manage mangrove forests in Karang City. The local community and several NGOs such as Walhi, Mitra Bentala, and Tangan have an interest and role in planting and conserving mangrove forests. Non-government NGOs carry out activities in the form of outreach to the community about the importance of the existence of mangrove forests, so that people have awareness to protect mangrove forests in these locations. The roles of both are in accordance with the social forest program, forest management starts from upstream to downstream, starting with nurseries and land management, maintaining and supervising, this will be effective if done directly by the community.

3.2.4. Policy Development Options. The choice of policy development is based on inaccurate implementation of national policies related to mangrove forest management and protection. Starting from an analysis of the costs and benefits of urban mangrove forests, both directly and indirectly, the choice is made by prioritizing the benefits and costs of managing and protecting sustainable mangrove forests through policy options for developing mangrove forests into ecotourism and green open spaces [29,30,31].

Several research studies specifically conducted in Brazil, Zanzibar and Vietnam describe the development of urban coastal mangrove forests as a vehicle for eco-tourism and eco-eduturism for urban communities. The impact felt in these 3 countries is the creation of jobs and increased welfare for the surrounding community [32]. A study in the Pagar Alam area, South Sumatra, illustrates that forest rehabilitation supports the preservation of traditional houses, which is one of the cultural-based tourism. Forest rehabilitation has relevance to the cultivation of the types of wood used as raw materials for the manufacture of traditional houses [33].

Mangrove forests on the urban coast, if managed into ecotourism, will provide results from the benefits of environmental services owned by mangrove forests, creating jobs and increasing welfare for the surrounding community. The costs to be incurred are in the form of policy implementation costs, in the form of building facilities and facilities. The cost of facilities and facilities is incurred by the state because the management to be built involves a cross-institutional planning program which is a transitional cost from the value of unused resources, considering that policies for resource reallocation costs such as rehabilitation and replanting of mangroves that have been logged by the surrounding community are not implemented [34].

Studies conducted in several countries as best practices such as in Indonesia, Bangladesh, as well as a study of the future of urban mangroves conducted in Asia, the United States of America, illustrates overcoming the damage caused by allowing the conversion of mangrove forests that have damaged the environment to be done through changing the face through protection of the remaining mangrove forests through one of the green open space models [30, 34, 35]. Rehabilitation of mangrove forests has expanded the area and will certainly be able to increase the need for green open space in the Bandar Lampung city which currently has only reached 21%. Likewise, the management of mangrove ecotourism will support the availability of green open spaces and green lanes in urban coastal areas.

4. Conclusion

The choice of policy development to protect and manage mangrove forests in urban areas is carried out through ecotourism policies and making mangrove forests a green open space. Collaboration of stakeholders including the Department of Marine Affairs and Fisheries of Bandar Lampung City and the Department of Housing, Settlements of Bandar Lampung City and the Forestry Service of Lampung Province is an effort to create smart regulation.

Several levels for the effectiveness of policy development that are needed are (a) evidence-based data development and selection of appropriate analytical methods as a basis for policy analysis; (b) public education; (c) Increasing government commitment through policy integration; (d) Together with NGOs to carry out a movement to replant mangrove forests.

Acknowledgments
The author would like to express his gratitude and appreciation to the University of Lampung for financial support through DIPA FISIP UNILA for Fiscal Year 2021. I also convey my thanks and appreciation to the Head of the Department of Marine Affairs and Fisheries, Bandar Lampung City; Head of Housing, Settlements, City of Bandar Lampung; Lampung Provincial Forestry Service, Government and Community of Karang City Village, Bandar Lampung, Lampung, Indonesia.

References

[1] Anon 2017 Forest Conservation Strategies to Mitigate the Impact of Climate Change on Human Security in East Africa: a Case Study of Mount Kenya Forest

[2] Anh N N and Hai-Hoa N 2017 Evaluating policy effectiveness on coastal mangrove management: case study in Kien Thuy and Do Son districts, Hai Phong city *J. For. Sci. Technol.* **2** pp 43–54

[3] Barbier E B 2016 The protective service of mangrove ecosystems: A review of valuation methods *Mar. Pollut. Bull.* **109** pp 676–81

[4] Basyuni M, Bimantara Y, Siagian M, Wati R, Slamet B, Sulistiyono N, Nuryawan A and Leidonad R 2018 Developing community-based mangrove management through eco-tourism in North Sumatra, Indonesia *IOP Conf. Ser. Earth Environ. Sci.* **126**

[5] Bhomia R K, Kauffman J B and McFadden T N 2016 Ecosystem carbon stocks of mangrove forests along the Pacific and Caribbean coasts of Honduras *Wetl. Ecol. Manag.* **24** pp 187–201

[6] Bryson J, Humphrey H H and Affairs P 2016 What to do when stakeholders matter : A guide to stakeholder identification and analysis techniques What To Do When Stakeholders Matter : A Guide to Stakeholder Identification and Analysis Techniques By University of Minnesota Visiting Professor for 2002-03 Academic Year Graduate School of Business University of Strathclyde Scotland A paper presented at the London School of Economics and Political Science 10 February 2003 .

[7] Carvalho B E de, Marques R C and Netto O C 2017 Regulatory Impact Assessment (RIA): from the State of Art until Conceptual and Framework Proposal Model *J. Contemp. Manag.* **6** 111–26

[8] Datta D, Chattopadhyay R N and Guha P 2012 Community based mangrove management: A review on status and sustainability *J. Environ. Manage.* **107** pp 84–95

[9] Duadji N and Tresiana N 2021 Cantrang and environment protection: Policy analysis of handling the risk of implementation of the Ban in Lampung Bay *IOP Conf. Ser. Earth Environ. Sci.* **739**

[10] Dwi Sulistyana M I C, Yuwono S B and Rusita R 2017 Kenyamanan Hutan Kota Linara Berbasis Kerapatan Vegetasi, Iklim Mikro Dan Persepsi Masyarakat Di Kota Metro *J. Sylva Lestari* **5** 78
[11] Febryano I G, Damai A A, Hardian D, Winarno G D and Tresiana N 2021 Cantrang : A Dilemma in Policy Implementation (Case in Lampung Bay , Indonesia) Sieć trałowa : Dylemat we wdrażaniu polityki (przypadek zatoki Lampung w Indonezji) **16** pp 133–42
[12] Febryano I G, Harum O M A, Wulandari C, Hidayat W, Banuwa I S, Prasetia H, Iswandaru D, Novriyanti N, Duadji N, Tresiana N, Zulfiani D, Ichsan A C and Salampessy M L 2021 Raw material of Besemah traditional house construction in Indonesia *Folia For. Pol. Ser. A* **63** 74–80
[13] Febryano I G, Suharjito D, Darusman D, Kusuma C and Hidayat A 2014 The Roles and Sustainability of Local Institutions of Mangrove Management in Pahawang Island *J.Manaj.Hutan Trop.* **20** pp 69–76
[14] Ferreira A C and Lacerda L D 2016 Degradation and conservation of Brazilian mangroves, status and perspectives *Ocean Coast. Manag.* **125** pp 38–46
[15] Giri C, Zhu Z, Tieszen L L, Singh A, Gillette S and Kelmelis J A 2008 Mangrove forest distributions and dynamics (19752005) of the tsunami-affected region of Asia *J. Biogeogr.* **35** 519–28
[16] Hartati F, Qurniati R, Febryano I G and Duryat D 2021 Nilai Ekonomi Ekowisata Mangrove Di Desa Margasari, Kecamatan Labuhan Maringgai, Kabupaten Lampung Timur *J. Belantara* **4** 1–10
[17] Hollweck T 2016 Robert K. Yin. (2014). Case Study Research Design and Methods (5th ed.). Thousand Oaks, CA: Sage. 282 pages. *Can. J. Progr. Eval.* **1** 108–10
[18] Ishatono I and Raharjo S T 2016 Sustainable Development Goals (SDGs) dan Pengentasan Kemiskinan *Share Soc. Work J.* **6** 159
[19] K. Haris O and Syahbudin 2019 Legal Environmental Action Aspects, Based on Environmental Conservation, According to Law 32/2009 Concerning Protection and Management of Environment **363** 127–32
[20] Khan M F A, Rahman M S and Giessen L 2020 Mangrove forest policy and management: Prevailing policy issues, actors' public claims and informal interests in the Sundarbans of Bangladesh *Ocean Coast. Manag.* **186** 105090
[21] Maharani M K, Febryano I G, Tresiana N and Banuwa I S 2021 Perubahan Luasan Lahan Mangrove Sebagai Ruang Lampung Change in The Mangrove Areas As Open Green Space In The Coastal Area Of Bandar Lampung City **4** pp 18–24
[22] O'Malley M 2019 Securing the future of AgTech *Netw. Secur.* 2019 12–4
[23] O'Malley M 2019 Securing the future of AgTech *Netw. Secur.* 2019 12–4
[24] Pakedai K T and Raya K K 2014 Arahan Kebijakan Pengelolaan Hutan Mangrove: Kasus Pesisir Kecamatan Teluk Pakedai, Kabupaten Kuburaya, Provinsi Kalimantan Barat *J. Geogr. Media Inf. Pengemb. dan Profesi Kegeografian* **11** pp 43–57
[25] Pham T D and Yoshino K 2016 Impacts of mangrove management systems on mangrove changes in the Northern Coast of Vietnam *Tropics* **24** pp 141–51
[26] Puspitojati T and Samsoedin I 2015 Kajian Pengembangan Ruang Terbuka Hijau Di Kota Bandung *J. Anal. Kebijak. Kehutan.* **12** pp 55–66
[27] Quinn C H, Stringer L C, Berman R J, Le H T V, Msuya F E, Pezzuti J C B and Orchard S E 2017 Unpacking changes in mangrove social-ecological systems: Lessons from Brazil, Zanzibar, and Vietnam *Resources* **6**
[28] Reed M S and Curzon R 2015 Stakeholder mapping for the governance of biosecurity: a literature review *J. Integr. Environ. Sci.* **12** pp 15–38
[29] Romañach S S, DeAngelis D L, Koh H L, Li Y, Teh S Y, Raja Barizan R S and Zhai L 2018 Conservation and restoration of mangroves: Global status, perspectives, and prognosis *Ocean Coast. Manag.* **154** pp 72–82

[30] Roy A K D 2016 Local community attitudes towards mangrove forest conservation: Lessons from Bangladesh *Mar. Policy* **74** 186–94
[31] Suhendra.Adi 2017 Matra pembaruan *J. Matra Pembaruan* **1** pp 75–84
[32] Tresiana N and Duadji N 2021 Environment and polemic of cantrang ban in Lampung bay: The importance of stakeholder mapping *IOP Conf. Ser. Earth Environ. Sci.* **739**
[33] Tresiana N and Duadji N 2016 Multi Stakeholders Governance Body Model In Achieving the Excellence Public Policy *MIMBAR, J. Sos. dan Pembang.* **32** 401
[34] Wisler A K 2009 'Of, by, and for are not merely prepositions': teaching and learning Conflict Resolution for a democratic, global citizenry *Intercult. Educ.* **20** 127–33
[35] Yilmaz O, Mansuroglu S and Yilmaz R 2013 Swot analysis of ecotourism as a tool for sustainable development: A case research in north-west black sea coastal zone of Turkey *J. Environ. Prot. Ecol.* **14** 786–98

Purification of selenium-containing leachate in wastewater tanks at a tunnel construction site

K Okazaki[1], **T Kurahashi**[1] **and M Sakakibara**[2,3]

[1] Civil Engineering Research Institute for cold region, Public Works Research Institute, 1-34 Hiragishi 1-3, Toyohira-ku, Sapporo, Hokkaido, 062-8602, Japan
[2] Graduate School of Science and Engineering, Ehime University, 2-5 Bunkyo cho, Matsuyama, Ehime, 790-8577, Japan
[3] SRIREP Project, Research Institute for Humanity and Nature, 457-4 Motoyama, Kamigamo, Kita-ku, Kyoto, 603-8047, Japan

Corresponding Author: 90185@ceri.go.jp

Abstract. The leachate containing heavy metals which generated from excavated soil and rock in the civil engineering works appropriate measures are required. We conducted indoor and outdoor experiments to purify heavy metal-containing leachate using a water tank, and proposed a method for treating heavy metals using the purification function of plants at civil engineering works. The plant used was *Eleocharis acicularis*, which belongs to the Cyperaceae family. The results of the indoor tank experiment on selenium indicated a decrease in the selenium concentration in the leachate from 0.093 to 0.008 mg/L in 14 days. The selenium content per kg of *Eleocharis acicularis* based on dry weight increased from 0.010 to 0.205 mg, indicating that the plant absorbed selenium. The outdoor tank experiment revealed that the selenium concentration in the leachate decreased from 0.090 to 0.007 mg/L in 21 weeks, and the daily absorption of selenium was 0.0005 mg/L. The lead concentration decreased from 0.004 to 0.001 mg/L in one week, and the arsenic concentration decreased from 0.003 to less than 0.001 mg/L in three weeks. Calculation of the leachate purification cost using *Eleocharis acicularis* under the current outdoor experiment conditions resulted in a 56% treatment cost reduction compared with the regular treatment cost.

1. Introduction

In a tunnel, cutting of earth and other civil engineering works may generate excavated soil and rock containing arsenic (As), lead (Pb), selenium (Se) and other heavy metals [1]. Temporary storage of such excavated soil and rock may generate leachate containing heavy metals and other substances at levels exceeding the environmental quality or standard effluent values due to rainfall or other reasons. In view of such potential above-standard contents of heavy metals and other substances in leachate from excavated soil, appropriate measures are required in terms of protection of the surrounding environment and project costs. In recent years, the heavy metal-absorbing effect of phytoremediation, an environmental remediation technology using *Eleocharis acicularis* and other plants, has been attracting attention. This *Eleocharis acicularis* absorbs and accumulates heavy metals has been shown to be effective in mine effluent treatment [2-4]. *Eleocharis acicularis* is a perennial that grows in clumps at ponds, marshes, reservoirs, and paddy fields, and it is distributed throughout all in Japan. *Eleocharis acicularis* propagates with runners that branch off from the root area. The grown height is about 15 to 20 cm. The authors have so far confirmed a 75% to 89% reducing effect on the total amount of arsenic contained in the effluent as well as a 40% to 60% reduction in the effluent itself by purification experiments using an artificial channel [5-7].

In this study, an indoor experiment to confirm the selenium absorption effect of *Eleocharis acicularis* was conducted in a small water tank using leachate from temporarily stored soil at a tunnel construction site to elucidate the heavy metal absorption/accumulation effect of *Eleocharis acicularis* and to establish a treatment technique for efficient absorption/accumulation. Another experiment at the tunnel site involved adding *Eleocharis acicularis* to a tank storing leachate containing selenium, lead and arsenic, and analysis of the concentration of heavy metals in the leachate and the amount absorbed by *Eleocharis acicularis* provided the daily absorption capacity of the plant, which is necessary for purification design. This paper also presents the cost reduction effect of leachate purification using *Eleocharis acicularis*, which was calculated by comparing the purification costs using the plant with regular treatment costs.

2. Materials and Method

2.1. Purification experiment using an indoor water tank

Eleocharis acicularis of 0.4 kg in wet weight was placed on a plastic sheet floating in a small water tank (W20 x D40 x H30 cm) as shown in Figure 1. Then 10 litres of leachate from tunnel construction site of Hokkaido Prefecture Japan. (selenium concentration: 0.093 mg/L) from tunnel excavation soil was poured into the tank to submerge the plant. Water was collected from the tank and the pH, EC and water depth were measured at the beginning of the experiment and on the 1st, 3rd, 5th, 10th and 14th days (end of the experiment). Approximately 10 g of *Eleocharis acicularis* was also collected at the beginning and end of the experiment to analyse the selenium concentration. At the same time, water from a blank case (on the left in Figure 1) of leachate without *Eleocharis acicularis* was collected and analysed.

2.2. Purification experiment using an outdoor water tank

In an open water tank (W1.8 x L7.0 m) placed in a roofed space, 33.6 kg wet weight of *Eleocharis acicularis* (wet weight of *Eleocharis acicularis* per 1 m^2 of area is 4 kg) was put in plastic containers (with holes in the bottom) floated with Styrofoam. In the tank, 17.5 m^3 of leachate containing 0.090 mg/L of selenium was stored (Figure 2) in tunnel construction site of Yamanashi Prefecture Japan. Water was weekly collected from the tank from the beginning of the experiment, and concentration of heavy metals (lead, arsenic, selenium), pH, electric conductivity and turbidity were measured.

At the time of each water sampling, a pump (lift capacity 3.5 L/min) was installed at the bottom of the water tank, and the water was stirred for 15 minutes before sampling. The target reduction value of

Figure 1. Indoor tank experiment. Left: blank tank. Right: tank with *Eleocharis acicularis*.

Figure 2. Outdoor tank experiment.

the selenium concentration at this site was 0.01 mg/L, which is the environmental quality standard value in Japan. The difference between the initial concentration of selenium and the target value was divided by the number of days required from the beginning of the experiment to the time when the concentration fell below the target value, to calculate the daily amount of selenium absorbed by *Eleocharis acicularis*. The cost reduction rate was also analysed by comparing the leachate purification cost using the plant with the regular treatment cost.

2.3. Analysis method

The concentrations of selenium, lead and arsenic in the leachate were analysed using the Inductively coupled plasma atomic emission spectroscopy (ICP-AES, Hitachi High technology Co., Ltd. Japan. P-4010), after adding and storing with nitric acid (Hokkaido Wako Junyaku Co., Ltd. Japan. Concentration 60.0-63.0 %) to make the selenium concentration 1% after filtering. As pre-treatment, *Eleocharis acicularis* was washed thoroughly with ultrapure water, dried for two days at 40°C in a dryer and reduced to a fine powder. The powder specimen, to which 30% hydrogen peroxide (Hokkaido Wako Junyaku Co., Ltd. Japan. Concentration 30.0-35.5 %), 61% nitric acid and 38% hydrofluoric acid (Hokkaido Wako Junyaku Co., Ltd. Japan. Concentration 46.0-48.0 %) were added, was then evaporated to dryness at 98°C. After adding 61% nitric acid and evaporating again at 98°C, the specimen was dissolved with 30% nitric acid. The concentrations of selenium, lead and arsenic in the solution were analysed by the ICP-AES method. Here, the heavy metal content of *Eleocharis acicularis* is a value corresponding to the dry weight at the time of analysis.

3. Results and Discussion

3.1. Indoor experiment results

Figure 3 shows the results of the indoor experiment. The selenium concentration of the leachate in the tank with *Eleocharis acicularis* decreased to less than one tenth, from 0.093 to 0.008 mg/L, in 14 days, whereas the concentration in the tank without the plant increased by 20% to 0.112 mg/L. This increase in concentration was thought to be due to the decrease in the initial water volume of 10 litres to 7.5 litres through evaporation in both tanks. The pH level that was measured concurrently changed from 6.3 to 7.4 in the tank with *Eleocharis acicularis* and from 7.2 to 7.9 in the tank without the plant. The EC level changed from 124 to 162 mS/m in the tank with *Eleocharis acicularis* and from 124 to 151 mS/m in the tank without the plant. The cause of the increase in pH and EC was also considered to be due to evaporation. The selenium content of *Eleocharis acicularis* increased from the initial value of 1.0 mg/kg to 22.1 mg/kg in 14 days. The converted weight of selenium per kg of dry weight of *Eleocharis acicularis* changed from 0.010 mg to 0.205 mg. After these results confirmed that the plant absorbs selenium, an outdoor experiment was conducted. During the indoor tank experiment, no leaf withering or new leaf growth was confirmed.

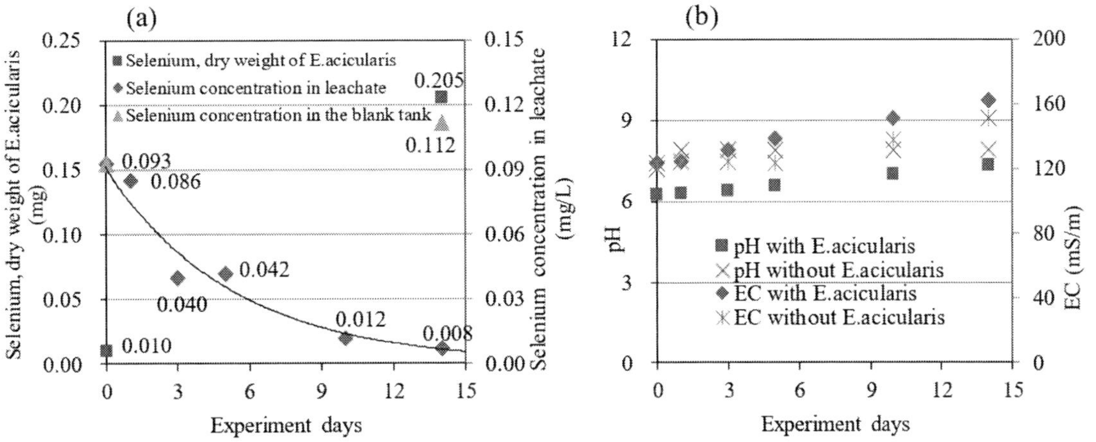

Figure 3. Indoor tank experiment results. (a) Selenium, dry weight of *Eleocharis acicularis*, selenium concentration in leachate, selenium concentration in the blank tank. (b) pH and EC levels of leachate.

Table 1. Outdoor tank experiment results.

Measuring data week passed	Pb (mg/L)	As (mg/L)	Se (mg/L)	pH	EC (mS/m)	Turbidity
initial value	0.004	0.003	0.090	7.2	601	170
1 week	less than 0.001	0.002	0.084	6.8	604	23
2 week	less than 0.001	0.002	0.081	6.6	608	11
3 week	less than 0.001	less than 0.001	0.074	6.8	612	7
4 week	less than 0.001	less than 0.001	0.074	6.9	617	5
5 week	less than 0.001	less than 0.001	0.070	6.9	622	7
7 week	less than 0.001	less than 0.001	0.062	6.9	630	9
9 week	less than 0.001	less than 0.001	0.051	7.0	636	4
10 week	less than 0.001	less than 0.001	0.046	7.1	640	10
11 week	less than 0.001	less than 0.001	0.042	7.2	642	11
12 week	less than 0.001	less than 0.001	0.036	7.1	644	6
13 week	less than 0.001	less than 0.001	0.032	7.2	647	11
15 week	less than 0.001	less than 0.001	0.024	7.0	654	12
17 week	less than 0.001	less than 0.001	0.016	7.1	658	7
19 week	less than 0.001	less than 0.001	0.011	7.2	663	8
21week	less than 0.001	less than 0.001	0.007	7.4	665	6
[illegible]	[illegible]	[illegible]	[illegible]	[illegible]	[illegible]	[illegible]

3.2. *Outdoor experiment results*

Table 1 shows the results of the outdoor experiment. The initial selenium concentration of 0.090 mg/L decreased to 0.007 mg/L after 21 weeks, and fell below the target value of 0.01 mg/L set at this site. The initial concentration of lead (0.004 mg/L), which was below the target (environmental quality standard) value of 0.01 mg/L, decreased to less than 0.001 mg/L in one week. The initial concentration of arsenic (0.003 mg/L), which was also below the target (environmental quality standard) value of 0.01 mg/L, decreased to less than 0.001 mg/L in three weeks. The pH and EC levels changed from 7.2 to 7.4 and 601 to 665 mS/m, respectively. Turbidity decreased from the initial value of 170 to 23 in one week. This reason thought to be that the particles in the water settled naturally. Meanwhile, the selenium content of *Eleocharis acicularis* increased from the initial value of 0.1 mg/kg to 8.9 mg/kg. The converted weight of selenium per kg of dry weight of *Eleocharis acicularis* changed from 0.005 mg to 0.414 mg. As demonstrated by the experiment, the selenium concentration in leachate was successfully reduced by using *Eleocharis acicularis*. Next, the difference between the initial concentration and the reduction target value of 0.01mg/L was divided by the number of days required from the beginning of the

Table 2. Heavy metals concentration of water and *Eleocharis acicularis*.

Heavy metals	21 weeks passed data			25 weeks passed data [8]		
	Consentration of water (mg/L)	Consentration of Eleocharis.acicularis (mg/kg-DW)	Daily content by Eleocharis acicularis (mg/kg-DW)	Consentration of water (mg/L)	Consentration of Eleocharis.acicularis (mg/kg-DW)	Daily content by Eleocharis acicularis (mg/kg-DW)
Fe	0.049	2,440	16.60	0.144	7,960	45.00
Cu	-	-	-	0.004	215.0	1.60
Zn	0.006	61.7	0.420	0.106	2,380	17.40
As	0.002	0.2	0.001	0.059	399.0	2.90
Pb	-	-	-	0.080	1,401	10.20

experiment to the time when the concentration fell below the target value, to calculate the amount of selenium absorbed by *Eleocharis acicularis*. The reduction target value of 0.01 mg/L was first subtracted from the initial concentration of 0.090 mg/L, and the concentration difference of 0.080 mg/L was found. This concentration difference was divided by the number of days required until the concentration fell below the target value, 147 (21 weeks x 7 days), and the daily amount of selenium absorbed by

Eleocharis acicularis was found to be 0.080 mg/L/147 days = 0.0005 mg/L. This outdoor experiment in a tank revealed the daily amount of selenium absorbed by *Eleocharis acicularis*. However, the value was calculated under the condition whereby 8.4 m^2 (33.6 kg wet weight) of the plant was placed in 17.5 m^3 of leachate. Table 2 shows the results of the heavy metals concentration of water and *Eleocharis acicularis*. Nurfitri et al [8] shown the content of heavy metals in the roots and leaves of *Eleocharis acicularis* in a 25-week mine wastewater purification experiment using *Eleocharis acicularis*. Our experimental period was 21 weeks. In addition, our experiment used the water tank, and the experiment of Nurfitri et al [8] was the result of under the condition that water flows. Comparing both, although the water concentration was different, *Eleocharis acicularis* contained copper and lead in addition to iron, zinc and arsenic. Our experiment also contains selenium in the aquarium. The daily selenium content of *Eleocharis acicularis* was calculated by dividing the selenium content of *Eleocharis acicularis* by the experimental period, and the result was 0.114 mg/kg.

3.3. Treatment cost reduction rete with the use of Eleocharis acicularis
Based on the results of the outdoor tank experiment in the previous section, the leachate purification cost by using *Eleocharis acicularis* was calculated and compared with the regular treatment cost to find the reduction rate, assuming the following site conditions.

3.3.1 Site conditions
i) The volume of the leachate was 17.5 m^3
ii) The concentrations of the heavy metals (lead, arsenic, selenium) in the leachate were maintained at the environmental quality standard value of 0.01 mg/L.
iii) Multiple tanks were prepared assuming large amounts of generated leachate due to heavy rainfall.
iv) Regular unit price of leachate treatment: USD367/m^3 (JPY40,000/m^3)

3.3.2 Costs and comparison
i) Regular treatment cost =USD367/m^3 (JPY40,000/m^3) × 17.5 m^3 =USD6,422(JPY700,000)
ii) Purification cost using *Eleocharis acicularis*
 · Cost of *Eleocharis acicularis* =USD110 m^3 (JPY12,000/m^2) ×8.4 m^3 =USD925(JPY100,800)
 · Analysis cost =USD110/set (JPY12,000/set) x 17 number of analysis =USD1,872(JPY204,000)
Total: USD 2,796(JPY304,800)

3.3.3 Cost difference and reduction rate
i) Difference =regular treatment cost -purification cost using *Eleocharis acicularis*
 =USD6,422(JPY700,000) -USD2,796(JPY304,800) =USD3,626(JPY395,200)
ii) Reduction rate = purification cost using *Eleocharis acicularis* / regular treatment cost
 = USD2,796(JPY304,800) / USD6,422(JPY700,000) =0.435
An on-site purification experiment using *Eleocharis acicularis* was conducted based on the above results, and leachate was successfully treated at approximately 44 % of the regular treatment cost.

4. Conclusions
This experiment revealed the following findings.

- The results of an indoor purification experiment using a water tank confirmed the absorption of selenium by *Eleocharis acicularis*.
- The results of an outdoor tank experiment showed that the daily amount of selenium absorbed by *Eleocharis acicularis* was 0.0005 mg/L.
- Calculation of the treatment cost of leachate containing heavy metals under the conditions of the current outdoor experiment revealed that leachate can be purified at 44 % of the regular treatment cost.

Since a sufficient effect may not be obtained with leachate containing mud due to the adhesion of mud to the stems and leaves of the plants, the authors intend to conduct studies on purification methods using plant fibber in the future.

Acknowledgment: The authors wish to express their gratitude to the staff of the Hokkaido Regional Development Bureau of the Ministry of Land, Infrastructure, Transport and Tourism, Japan for their cooperation with this study.

References

[1] Committee for review of the manual for handling rock and soil containing naturally occurring heavy metals at construction sites 2010 Manual for handling rock and soil containing naturally occurring heavy metals at construction sites (In Japanese) pp 72-84

[2] Ha N T H Sakakibara M Sano S Hori R S and Sera K 2009 The potential of *Eleocharis acicularis* for phytoremediation Case study at an abandoned mine site *CLEAN - Soil, Air, Water* **37** pp 203-208

[3] Ha N T H Sakakibara M and Sano S 2009 Phytoremediation of Sb, As, Cu and Zn from Contaminated Water by the Aquatic Macrophyte *Eleocharis acicularis CLEAN - Soil, Air, Water* **37** pp 720-725

[4] Sakakibara M Harada A Sano S and Hori R S 2009 Heavy Metal Tolerance and Accumulation in *Eleocharis acicularis* a Heavy Metal Hyper Accumulating Aquatic Plant Species *Geo-pollution Science, Medical Geology and Urban Geology* **5** pp 1-8

[5] Okazaki K Yamazaki S Kurahashi T and Sakakibara M 2016 An Experiment for Purifying Drainage Water by Using a Plant that Absorbs Heavy Metals (In Japanese with English abstract) 53rd Environmental Engineering Forum Japan Society of Civil Engineers *Journal of the Japan Society of Civil Engineers G (Environmental Engineering Group)* **72** 7 pp 179-185

[6] Okazaki K Yamazaki S Kurahashi T and Sakakibara M 2017 An Artificial Channel Purification Experiment for Arsenic rich Drainage Water from the Abandoned Mine by using *Eleocharis acicularis* International Conference on Transdisciplinary Approach Research 2017 *Earth and Environmental Science* **536** (2020) 012003 doi:10.1088/1755-1315/536/1/012003

[7] Okazaki K Kurahashi T Yamazaki S and Sakakibara M 2018 Temperature dependence for purification of leachate containing heavy metals by phytoremediation using the artificial channel Transdisciplinary Research on Environmental Problems in Southeast Asia 2018 *Earth and Environmental Science* **589** (2020) 012019 doi:10.1088/1755-1315/589/1/012019

[8] Nurfitri AG Masayuki S and Koichiro S 2017 Phytoremediation of Heavy Metal-Polluted Mine Drainage by *Eleocharis acicularis Environmental Science, An Indian Journal* **13** 1 pp 1-11

Bioethanol production from seaweed solid waste biomass of agar processing

I Winarni[1], Uju[2], J Santoso[2] and T Wibowo[3]

[1] Forest Product Research and Development Center, Jl. Gunung Batu No. 5 Bogor, 16610, Indonesia
[2] Faculty of Fisheries and Marine Sciences, Bogor Agricultural University, Jl. Agatis No.1, Dramaga, Bogor, 16680, Indonesia
[3] Ministry of Finance, Jl. Dr. Wahidin Raya No.1, Jakarta, 10710, Indonesia

Corresponding Author: inawinarni@yahoo.com

Abstract. This paper describes about the potential application of renewable energy sources from biomass (seaweed waste) to replace fossil fuels combustion which produced CO_2 emission and heavy metal pollution. The objective of the research is to utilize the seaweed solid waste from agar production industry for bioethanol feedstock. The enzymatic method was carried out for glucose production with the addition 1% of surfactant Tween 20 during the saccharification process of three kinds of substrate concentration (20, 25, 30% dw) and 10 FPU/g substrate for the cellulose concentration and used *S. cerevisae* as yeast in the fermentation process for bioethanol production. The result showed that the highest reducing sugar value (122.4 mg/mL) and 2.86% for the bioethanol content were from 30% dw treatment of substrate concentration of solid agar waste. Therefore, the waste from agar industry is a potential raw material to be used as biofuel and could reduce the heavy metal problems.

1. Introduction

Currently, the increase of energy demand of fossil fuel continues to occur. This phenomenon is influenced by the increase in economic needs and population. The utilization of energy continues to increase, but it is not balanced with processing raw material reserves so that energy availability in Indonesia is decreasing. The largest consumption of fossil fuel in the transportation sector as a solution and gasoline reaches 88% [1]. Therefore, renewable energy is needed as an alternative to energy needs. On the other hand, the potential application of renewable energy sources from biomass (seaweed waste) to replace fossil fuels combustion which produced CO_2 emission and heavy metal pollution. Therefore, bioethanol, as one alternative energy from biomass raw materials can be developed because it is environmentally friendly and widely available in nature. On the other hand, bioethanol also has a high octane number and can reduce toxic emissions such as formaldehyde, benzene and 1,3-butadiene [2].

Bioethanol as a renewable biomass fuels could be derived from sugarcane, corn, switch grass, algae, etc. as raw materials resources [3]. But, nowadays, lignocellulosic or cellulose materials are considered to be one of the most promising renewable resources as an alternative because not competed with the food purposes [4]. Meanwhile, the use of cellulose as a raw material for making bioethanol can be sourced from agricultural waste, plantation waste, wood and forestry waste, and organic waste [4-6]. The utilization development of waste as raw material for bioethanol can reduce the capacity of waste, besides that waste is easy to obtain. One of the wastes that can be used as raw material for bioethanol is seaweed solid waste from industry of agar processing.

Cellulose are processed for bioethanol production through three major operations: hydrolysis, fermentation, and distillation [6]. Hydrolysis is carried out enzymatically and the addition of Tween 20 as a surfactant. Tween 20 surfactant functions as an enzyme stabilizer so that it can help convert cellulose to glucose [7]. Fermentation is a process of converting reducing sugars from the results of hydrolysis in the form of glucose into bioethanol. The last stage is distillation, to produce pure bioethanol from the fermentation results. In this study, bioethanol content from seaweed solid waste was measured using an alcohol meter and gas chromatography. Glucose levels can affect the acquisition of bioethanol levels during the fermentation process; the higher the glucose levels, the higher the bioethanol levels [8].

Seaweed is a marine plant that is at the bottom of the waters. Seaweed is divided into three classes, namely red seaweed (*Rhodophyceae*), brown seaweed (*Phaeophyceae*), and green seaweed (*Chlorophyceae*). In general, seaweed contain large amounts of complex sulphated carbohydrates (polysaccharides) that are often extracted for use mainly in the hydrocolloid industry. Each seaweed group uniquely consists of different types of complex polysaccharides serving as storage and supporting structural tissue such as the cell wall [9]. Red seaweed or *Rhadophyceae* has a very high use in the Indonesian industrial sector, especially in the manufacture of agar production. *Gracilaria* sp. is one of the seaweeds that is commonly used as a raw material for making agar. The agar production that is carried out continuously in line with consumer demand causes an increase in the abundance of manufactured waste. Solid waste in agar is a by-product of processing seaweed into agar. Meanwhile, solid agar waste contains cellulose, lignin, hemicellulose, pectin, and other organic materials. Triwisari (2010) reported that cellulose content in agar waste is around 27.8–39.45% [10]. The resulting waste reaches 65-70% of the raw materials used. According to the report of seaweed industries, if the cellulose content of dry seaweed was about 10-15%; the available cellulose in Indonesia was 6,691-10,036.5 tons of cellulose/year [11]. The amount of the cellulose content from seaweed solid waste were high enough and very potential to be developed as biofuel.

Dried seaweed for agar production should treated with hot alkali solution to reduce sulphate content and to improve gel strength [12-15]. The agar yield was 20-38% [12,16-17], in this case approximately 60-75% of seaweed biomass will produce solid waste biomass of agar processing (SWBA). Some advantages of the SWBA feedstock are the feedstock not compete with food source and it can create the unwanted wastes into value added product of the seaweed industry [18]. Furthermore, SWBA was easy to be hydrolyzed due to low lignin content [14]. Moreover, the seaweed cellulose is different from terrestrial plants. Seaweed is mainly composed of cellulose Iα (triclinic crystalline form) [19-20] which have weak hydrogen bonding, whereas terrestrial plant is dominated by cellulose Iβ (monoclinic crystalline form) as a result cellulose of seaweed will easily be hydrolyzed [21]. The fact solid waste biomass was easier to be hydrolyzed than terrestrial biomass such bagasse reported by previous articles [22-23]. In addition the structural of solid SWBS more swelling and amorphous than bagasse, the structure was presumably solid seaweed waste biomass was easily to be hydrolyzed [24-25]. Therefore, agar waste could be used as a raw material for bioethanol production and could reduce the abundance of solid waste production from the industries Sudhakar *et al.* reported that red seaweed solid waste as raw materials in the manufactured of bioethanol by enzymatic hydrolysis could obtained bioethanol content about at 4.0% [26].

The purpose of this preliminary research is to make bioethanol from seaweed solid waste from the agar processing as a by-product that could reduce the environmental pollution due to the waste disposal, and it is also expected to increase the economic value of seaweed solid waste.

2. Methodology

2.1. Materials and preparation

The material used was solid waste of processing agar from agar production factory namely CV Agar Sari Jaya, Malang, East Java, Indonesia with a semi-wet sample condition. Sample preparation was initiated by washing using clean water until a neutral pH. Then, the sample were air-dried until dry pulp were obtained.

Cellulase from Novozyme 188 (Sigma–Aldrich, Co., Louis, USA) was used in all enzymatic saccharification in this research. Cellulase activity was 88.09 FPU/mL according to a method in a NREL technical report, NREL/TP-510-42628 [27] and Saccharomyces cerevisiae as yeast was used in fermentation part. The surfactant we used in the enzymatic saccharification process has a chemical name polyoxyethylen e sorbitan monolaurate or also known as Tween 20; with molecular formula $C_{58}H_{114}O_{26}$ and molecular weight of 1227.51 g / mol. Surfactant chemical structures were shown in Figure 1.

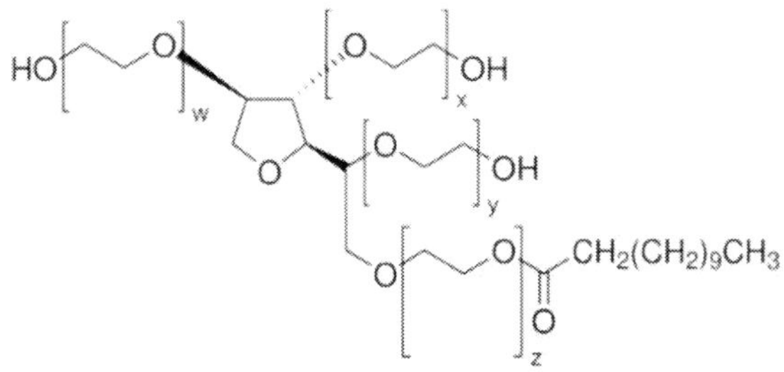

Figure 1. Chemical structure of Tween 20

2.2. Enzymatic saccharification of seaweed solid waste

One percent (1%v) of Tween (Tween® 20, Merck 822184) was dissolved in 100 mL of 0.05 M citric acid ($C_6H_8O_7$) buffer at buffer at pH 4.8. Cellulase enzyme (10 FPU/g substrate) from Novozyme was added to the solution, and the mixture was stirred for 1 h in room temperature. Finally, several concentrations (20, 25, 30% dw) of substrate were added to the solution, and the suspension was shaken at 50° C for 48 h. After saccharification, as much as 5 mL of saccharification product from each experiment were filtered. The reducing sugar concentration in the filtrate was measured by DNS (3,5-dinitrosalicylic acid) method. The rest of saccharification product of each experiment was fermented by adding yeast (Pakmaya®) (1% dw substrate). The fermentation process was performed in room temperature for 96 h. The resulted fermentation broth was distilled twice with water then distilled until the distillate volume reached 100 mL Ethanol concentration in distillate was determined using gas chromatography GC (Agilent 6890 N) equipped with flame ionization detector (FID) and computer integrator software. The temperature of FID detector and the injection port was set at 220°C and 250°C, respectively. Helium with the flow rate 3°C/minute was used as a carrier gas. A capillary column (Carbowax 20 M) was used. Oven temperature was initially set at 60°C and then increase to 80°C by programming at 20°C.

3. Result and discussion

3.1. Chemical analysis of seaweed solid waste

The raw material of waste seaweed dry pulp was obtained from CV Agar Sari Jaya Malang were analyzed of the moisture content, ash content and cellulose. Characterization was carried out to determine the constituent components of dry powder gelatinous pulp. The results of the observation of the raw material testing for the sample can be seen in Table 1.

Table 1. Raw materials characterization seaweed solid waste

Fraction	Content	Previous study	
		Nurhayati and Kusumawati [28]	Fauziah [29]
Water (%)	4.65±0.55	12.15	16.22
Ash (%)	47.44±0.09	7.07	42.15
Cellulose (%)	14.79±0.35	72.02	-

From Table. 1, it can be seen that, the results of the water, ash and cellulose contents of the raw material of seaweed solid waste were 4.65%, of 47.44%, and 14.79%, respectively. According to another research, the water content showed lower value compared to the previous research [28-29]. The water content in the sample was influenced by the type of seaweed waste and the level of dryness of the samples. The ash content indicates the total mineral contained in a material. The ash content of the seaweed solid waste showed high enough value compared to the previous research which were 7.07% and 42.15%, respectively [14-15]. The ash content is quite high because the waste is a concentrate of inorganic materials left behind over from agar production. Solid waste contains minerals, such as Na, Mn, Ca, Mg, K, Fe, P, Zn, and Cu. The mineral content of seaweed processing waste is also influenced by the process given during the processing [10]. Meanwhile, the agar processing at CV Agar Sari Jaya-Malang using H_2SO_4, NaOH, NaOCl and Perlite; in the bleaching process they also using chemicals such as CaO or NaOCl, various acids and alkaline solutions in the extraction process, so that it could be the reason why the ash content in solid seaweed waste still high [23] .

Cellulose is the main component of plant cell walls. The content of cellulose obtained was lower than the previous research [28]. One of the factors for cellulose content is determined by the type of plant used as a source of cellulose. High-fiber plants contain higher cellulose than low-fiber plants. Israel et al. reported that the cellulose content of corn stalks reached 60% and banana stem waste was around 37.5% [30]. Other result showed that cellulose content of seaweed waste could reached 26.92% [23].

3.2. Crystallinity Analysis of seaweed waste pulp

XRD is a tool used to characterize the crystal structure, the crystal size of a solid material. When analyzed using XRD, all materials containing certain crystals will reveal specific peaks [31]. Based on the results of the XRD diffractogram, seaweed waste pulp has a crystallinity index of 25.97%. It is necessary to know the degree of crystallinity to determine the level of regularity in the placement of atoms in a particular unit cell. It can be seen from the resulting crystallinity index value, it can be ascertained that this sample was belong to the amorphous structure, making it easier to hydrolyze the cellulose into glucose monomers. Sulfida reported that, the XRD diffractogram showed of pineapple leaf cellulose including type I cellulose [20]. The highest degree of crystallinity was obtained by pineapple leaf cellulose with a delignification time of 90 minutes which was equal to 65.98%.

3.3. Reducing sugar and bioethanol analysis of solid waste seaweed of agar processing

3.3.1. Reducing sugar analysis. The content of cellulose in seaweed solid waste as raw material for bioethanol production, it is necessary to process cellulose into monosaccharides, namely reducing sugars such as glucose, through saccharification or enzymatic hydrolysis. Cellulase enzymes are used in the hydrolysis process. The mechanism of cutting the chain by the cellulase enzyme to produce a short glucose chain occurs in a complex manner. In the early stages, there is a random termination of cellulose bonds on the internal side of the cellulose fibers by endo-1,4-β-D-glucanase so that it makes it easier for the cellobiohydrolase enzyme to work to break the cellulose bonds into shorter chains. Furthermore, exo-β-1,4-gluconase or cellobiohydrolase attacks the outer part of cellulose to produce cellobiose as its main structure.

Completed saccharification of cellulose will produce glucose. The addition of Tween 20 surfactant to the saccharification process was carried out to facilitate the breakdown of cellulose into glucose. Tween 20 surfactant has a function as an enzyme stabilizer and maintains denaturation, Tween 20 can also affect the structure of the substrate used in the saccharification process and can affect the interaction of the enzyme with the substrate and works specifically to maintain the absorption of enzymes in lignin and substrate [32].

The result of reducing sugar value of seaweed waste can be seen in Table 2. Based on Table 2, the high reducing sugar content in the 30 % (dw) substrate is equal to 122,40 (mg/mL). In general, the amount of substrate affects reducing sugar levels. The greater the amount of substrate, the higher the reducing sugar content produced [33]. However, the reducing sugar content in the amount of substrate 25 g is smaller than the amount of substrate 20% (dw). This can be caused by the saccharification process that occurs incompletely due to the suboptimal work of the enzyme. The non-optimal work of the enzyme can be caused by obstacles in the interaction between the enzyme and cellulose on the substrate. The conversion process of cellulose into reducing sugars can still occur during the fermentation process, as long as there is sufficient cellulase enzyme.

Tween 20 surfactant has a function as a stabilizer to increase reducing sugar levels. According to the previous research, the reducing sugar levels in sengon wood against the control (without the Tween 20 surfactant) used an enzyme concentration of 15 FPU/g of substrate with the same test treatment, indicating that the control results were lower than using the Tween 20 surfactant; and 15% of sengon wood with the addition of Tween 20 surfactant produced a reducing sugar content of 149.7 mg/mL, whereas without using Tween 20 surfactant, it was obtained a reducing sugar content of 82.9 mg/ mL [34]. Therefore, it can be concluded that without the addition of the Tween 20 surfactant will result in lower reducing sugar levels. The reducing sugar that has been obtained can be used by *Saccharomyces cerevisiae* to produce bioethanol at the fermentation stage.

Table 2. Reducing sugar analysis of seaweed solid waste of agar processing

Samples (% dw)	Reducing sugar value (mg/mL)
20	92.16±4.07
25	82.56±2.71
30	122.40±25.12

3.3.2. Bioethanol content of seaweed solid waste. Bioethanol obtained after the fermentation stage was then purified by distillation. Distillation is carried out to separate bioethanol from fermentation broth, which is mostly water [35] with a ratio of boiling points. Bioethanol has a boiling point of 78 °C.

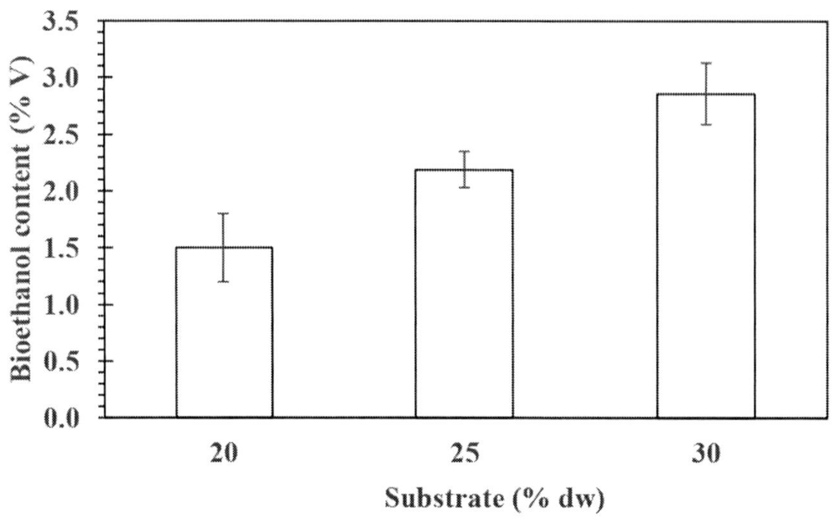

Figure 2. Bioethanol content of seaweed solid waste

Based on Figure 2, the highest bioethanol content was produced (2.86%v) when we loading substrate of 30%dw in the enzymatic saccharification process with 10 FPU/g substrate of cellulose enzyme. This is directly proportional to the high levels of reducing sugars that have been obtained (Table 2.). In general, reducing sugar content will be directly proportional to the level of bioethanol produced. Higher the reducing sugar content produce higher the bioethanol content. Meanwhile, 25%dw of substrate with low reducing levels resulted in high bioethanol content compared to 20%dw of substrate. This could be caused during the fermentation process occurs, not only *S. cerevisiae* works, cellulase enzymes still work to break down cellulose into glucose, so that *S. cerevisiae* could still be broken down into bioethanol during the fermentation process [7]. Therefore, the amount of substrate 25%dw could produce a higher bioethanol content than the substrate 20%(dw).

Previous research reported that bioethanol content from *Saccharina japonica* with the Simultaneous Saccharification and Fermentation (SSF) method using a concentration of 10% *Pichia angophorae* and *S. cerevisiae* at 30 °C for 64 h resulting in an bioethanol content of 1.00% (9.8 mL / L) [36]. Another research reported that the highest bioethanol content in simultaneous saccharification and fermentation (SSF) was obtained on 8 days fermentation at 35 °C and pH 4.8 with bioethanol content of 1.05% [23]. With shorter time fermentation and low cellulase dosage concentration in the enzymatic saccharification stage, this result produced higher bioethanol content than previous reports.

4. Conclusions

The highest production of reducing sugar and bioethanol content of seaweed solid waste were 122.40 (mg/mL) and 2.86%v, respectively; and could achieved from treatment with addition 30% of substrate; 10 FPU/g substrate of enzyme and 1%v Tween 20. The crystallinity index of seaweed solid waste was around 25.97%. It could be concluded that, seaweed solid waste is a potential raw material for bioethanol production based on its reducing sugar and bioethanol content and crystallinity index value.

5. References

[1] Edomah N 2018 *Reference Module in Earth Systems and Environmental Sciences, Elsivier*
[2] Byadgi SA and Kalburgi P 2016 *Procedia Environ. Sci.* **35** 555
[3] Sarkar N Ghosh SK Bannerjee S and Aikat K 2012 *Renew. Energy* **37** 19
[4] Alkasrawi M Eriksson T Borjesson J wingren A Galbe M Tjerneld F and Zacchi G 2003 *Enzyme Microb. Technol.* **33** 71

[5] Winarni I Oikawa C Yamada T Igarashi K Koda K and Uraki Y 2013 *BioResources* **8** 2195
[6] Vohra M Manwar J Manmode Padgilwar RS and Patil S 2014 *J. Environ. Chem. Eng.* **2** 573
[7] Winarni I and Bardant TB 2017 *J Forest Product Research* **35** 231
[8] Wang J Chae M Beyene D Sauvageau D and Bressler DC 2021 *Bioresour. Technol.* **330**
[9] Offei F Mensah M Thygesen A and Kemausuor F 2018 *Fermentation* **4** 01
[10] Triwisari DA 2010 Bogor Agricultural University
[11] Anonymous 2019 Perpres Nomor 33
[12] Li H Yu X Jin Y Zhang W and Liu Y 2008 *Bioresour. Technol.* **99** 3301
[13] McHugh D 2003 *FAO Fisheries Technical Paper*
[14] Meinita MD Marhaeni B Hong Y and Jeong G 2017 *J. Appl. Phycol.*l. **29** 3201
[15] Villanueva RD Sousa MM Gonçalves MP Nilsson M and Hilliou L 2009 *J. Appl. Phycol.* **220** 211
[16] Morinho-Soriano E and Bourret E 2005 *Bioresour. Technol.* **96** 379
[17] Shukla R Kumar M Chakraborty S Gupta R Kumar S Sahoo D and Kuhad, R.C 2016 *Bioresour. Technol.* **220** 584
[18] Goh C Tan K Lee K and Bhatia S 2010 *Bioresour. Technol.* **101** 4834
[19] Siddhanta aK Prasad, K Meena R Prasad G Mehta GK Chhatbar MU Oza MD Kumar S and Sanandiya ND 2009 *Bioresour. Technol.* **100** 6669
[20] Sulfida D 2020 Bogor Agricultural University
[21] Daroch M Geng S and Wang G 2013 *Appl. Energy* **102** 1371
[22] Roldán IUM Mitsuhara AT Munhoz Desajacomo JP de Oliveira LE Gelli VC Monti R Silva do Sacramento LV and Masarin F 2017 *Biomass and Bioenergy* **107** 233243
[23] Hakim A Chasanah E Uju and Santoso J 2017 *Squalen Bull. of Mar. and Fish. Postharvest and Biotech.* **12** 41
[24] Uju Goto M and Kamiya N 2016 *Bioresour. Technol.* **214**, 487
[25] Ulises A Roldán, IUM Tiemi A Pedro J Desajacomo M Ezequiel L Oliveira D Cress V Monti R Vitor L and Masarin F 2017 *Biomass and Bioenergy* **107** 233
[26] Sudhakar M Jegaatesan A Poonam C Perumal K and Arunkumar K 2017 *Renew. Energy* **105**
[27] Ghose A 1987 *Pure Appl. Chem.* **59** 257
[28] Nurhayati and Kusumawati R 2014 *J. Pascapanen dan Bioteknol. Perikan* **9** 097
[29] Faujiah F 2012 Bogor Agricultural University
[30] Israel AU Obot IB and Asuquo JE 2008 *J. Chem.* **5** 81
[31] Darwis A 2017 Scince article
[32] Lestari WW Winarni ID and Rahmawati F 2017 *IOP Conference Series: Materials Science and Engineering* vol 172
[33] Sari L 2017 Bogor Agricultural University
[34] Winarni I and Bardant TB 2017 *J Forest Product Research* **35** 231
[35] Anindyawati T 2009 *Ber. selulosa* **44** 49
[36] Jang JS Cho YK and Jeong G 2012 **35** 11

IOP Conf. Series: Earth and Environmental Science 1027 (2022) 012030 doi:10.1088/1755-1315/1027/1/012030

Feasibility of rooftop solar PV program for 9 tourism villages towards green village development in Bali

DK Halim[1] **and AANS Wahyuni** [1]

[1] Politeknik Internasional Bali, Pantai Nyanyi, Beraban, Kediri, Tabanan, Bali, 82121, Indonesia

Corresponding Author: dk.halim@pib.ac.id

Abstract. Electricity in Bali majorly is supplied by power plants in Java which use fossil-fuelled. With petroleum and coal reserves to run out by 2025, Bali Government has issued Governor Ordinance No. 45/2019 concerning clean energy in encouraging Balinese to use rooftop solar photovoltaic (PV). As backboned tourism industry, Bali has drastically been declining due to COVID-19 causing most hotels and their supporting industries closed, the government then has changed the orientation from urban tourism to rural tourism. This paper proposes rooftop solar PV power plant program in the tourism village representing all 8 regencies and 1 municipality (Sudaji in Buleleng, Catur and others in Bangli, Tenganan in Karangasem, Kerta in Gianyar, Blimbingsari in Jembrana, Paksebali in Klungkung, Bongan in Tabanan, Bongkasa in Badung, and Sanur Kauh in Denpasar). Recent studies show huge potential for solar energy in Bali and the program is aligned with Sustainable Development Goals (Bali SDGs). The study elaborates problems in implementing the program since solar PV is still new and traditions could hinder the people in the tourism village to utilize it. Behaviour approach must be explored to make the program can be successfully done.

1. Introduction

The current rising global temperatures are the impact of Greenhouse Gas (GHG) emissions. GHG is caused by increased emissions of gases produced by fossil fuels such as carbon dioxide (CO_2), methane (CH_4), dinitro oxide (N_2O), and chlorofluorocarbons (CFCs) so that the energy resulted is also trapped in the atmosphere. The increasing global temperatures will be faster due to human activity, causing an oversupply of GHG emissions released into the atmosphere. Solar PV to reduce GHG emissions is in line with Nationally Determined Contribution (NDC) climate change mitigation efforts to build resilience in safeguarding food, water and energy resources. Indonesia shares its development towards low carbon and climate security as commitment to contribute to global efforts to achieve sustainable development goals. It is necessary to strengthen the climate resilience as stated in the NDC as a strong basis, particularly in renewable energy issues.

The potential of renewable energy in Indonesia comes from the sun, water and waterfall flows, wind, geothermal, bioenergy, movement, temperature differences in the sea layers, and hydrogen. Based on the results of technical studies for Bali's Energy Planning, the greatest potential of energy generation comes from the sun, which can produce energy power of 1,254 MW, as shown in Table 1. With the presence of solar energy, which is the largest producer of energy compared to other sources, the utilization of solar needs to be developed into a power that can be utilized. Solar energy can be converted into electrical energy that can be utilized for daily life in the long term. The need for electrical energy every year is increasing, where currently, the electricity spread to the entire archipelago is supplied by oil and coal power plants. Relying on national oil reserves, which run out in 2019, requires Indonesia to become a net importer of oil, while other coal reserves will run out in 2025 according to Government Ordinance No. 79/2014 on National Energy Policy. This critical condition, anticipated by the Government, resulted in the issuance of Ordinance No. 79/2014, revision of Presidential Regulation No. 5/2006 on National Energy Policy. Generating renewal of solar energy is highlighted to be a considerable potential as a transition to clean energy.

Table 1. Renewable energy potentials in Bali [1]

No.	Type	*Potency MW (Megawatt)*	Obstacle
1	Solar	1254.00	Land & Economic Scale
2	Wind	1019.00	Technology & Economic Scale
3	OTEC (Ocean Thermal Energy Conversion)	320.00	Immature Technology
4	Geothermal	262.00	Socio-cultural values
5	Hydro	208.00	Water Crisis in Bali
6	Biomass (incl. Waste to Energy)	146.90	Tipping Fee
7	Biogas	44.70	Economic Scale
8	Microhydro & Minihydro	15.00	Ineffective usage
	Total	3269.60	

The electrical energy in Bali Island is majorly supplied by the National Electricity Company (PLN) from fossil fuel power plants in Java Island. With petroleum inventories running low and coal inventories to run out by 2025, PLN-Bali has prepared a projection planning to anticipate the energy crisis as well as modeled the Energy Mix. This projection is important in assisting Bali's transition to renewable energy sources before all fossil fuels run out. According to the Global Solar Atlas, Bali is well-positioned to generate 1286 kWp on average, as shown in Figure 1, with its peripheral areas in the south, northwest and northeast providing the best value. The value kWp (kilo Watt peak) refers to kWh (kilo Watt per hour) is the unit used for solar PV because the solar flux is not stable or intermittent, sometimes disturbed by the cloud & rain. While PLN-Bali's 2020 target for photovoltaic output was 159 MW, the CORE (Centre of Renewable Energy, Udayana University) study suggested this might be compromised due to climatic variability and will increase up to 182 MW in 2022 for Hotel/Tourism, Government/Publics, Business/Commercials, Housing, and others. Nevertheless, the CORE predicted the number achieved was only 108 MW due to the intermittent nature of solar power [2].

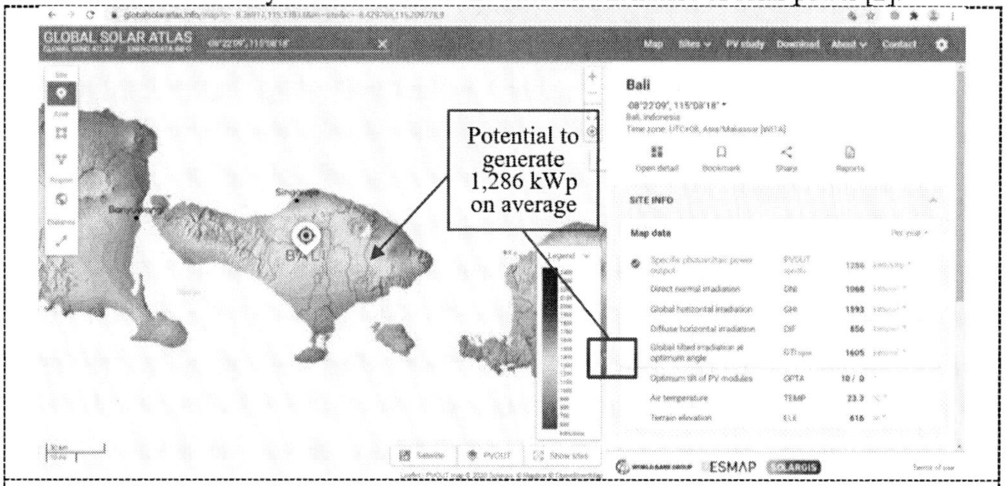

Figure 1. Solar energy potential in Bali island

IOP Conf. Series: Earth and Environmental Science 1027 (2022) 012030 doi:10.1088/1755-1315/1027/1/012030

2. The Program

Bali province has 8 local regencies (Badung, Tabanan, Bangli, Karangasem, Klungkung, Jembrana, Buleleng, Gianyar) and 1 municipality (Denpasar). The need for large electrical energy encourages PLN to construct a coal-fired power plant in Bali through its continuing development which unable to meet the needs of 1200 MW or more. Recent studies show that there is significant potential to produce renewable energy in Bali. A joint study by CORE and Greenpeace had set a Road Map of Solar Power Plants throughout the island [2]. The Rooftop PV program aligns with NDC for energy security and strategic issues with SDGs of Bali Province in the Medium-term Development Plan Year 2018-2023.

Barriers to holding of the Rooftop PV program from the market aspect usually are the lack of public education on the importance of the use of Rooftop PV for household needs and the availability of tools and land/space to place solar panels. In Bali, the barriers even more complicated due to Socio-cultural aspect such as tradition and local politics because of leadership dualism between Traditional Village management (Banjar Adat) and Government-structured Village management (Banjar Dinas). Although legal aspects already empower the program as the utilization of energy by installing Solar Power Plants on the roof (PLTS) such as stipulated in the Regulation of the Minister of Energy & Mineral Resources of the Republic of Indonesia No. 16/2019 concerning the 2nd Amendment to the Regulation No. 49/2018 concerning the Use of Rooftop Solar Power System by Public Consumers of PLN, Governor Ordinance No. 45/2019 concerning Bali Clean Energy, and other regulations governing a number of cost for holding solar panels. These regulations provide the importance of renewable energy to be realized that will result in the increasing number of uses of solar panels, especially in tourism which is the largest industry on the island of Bali. Currently, the development of the tourism industry on the island of Bali is drastically decreasing due to Covid-19, which caused hundreds of hotels and their supporting industries closed as the visiting tourists has declined from 6,275,210 in 2019 to 1,069,473 in 2020 and only 43 in 2021 until July [3]. This forced the Government to change the tourism orientation from urban tourism to rural tourism as it allows social distancing. Thus, it will greatly influence investment in the uptake of domestic renewable energy infrastructure. This program is a pilot project of the Solar PV rooftop power plant (PLTS) implemented in the tourism village.

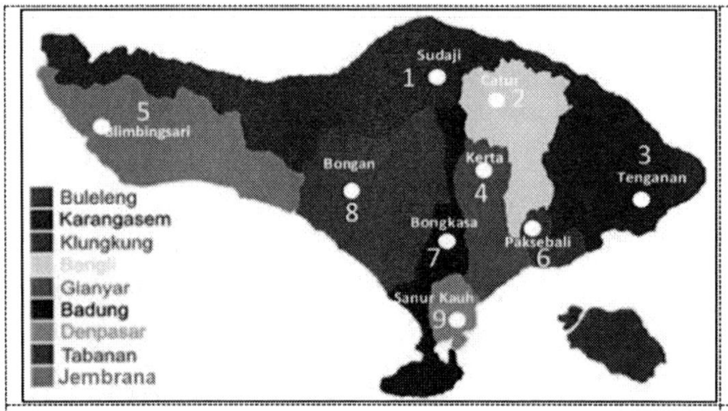

Figure 2. Nine targeted tourism villages in the program

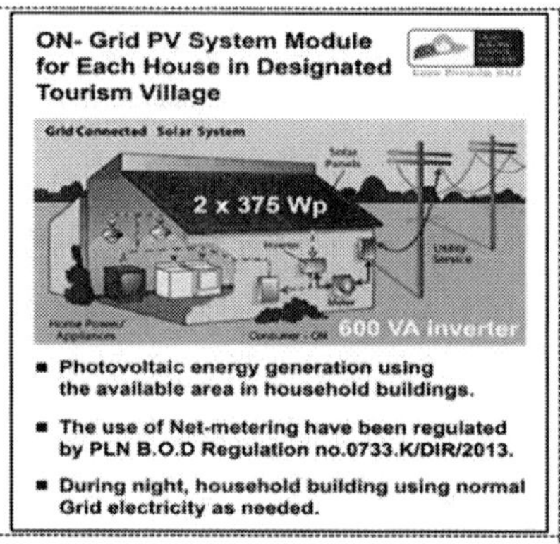

Figure 3. Solar PV package

As shown in Figure 2, the solar PVs would be distributed for nine tourism villages as follow: (1) Sudaji in Buleleng, (2) Catur in Bangli, (3) Tenganan in Karangasem, (4) Kerta in Gianyar, (5) Blimbingsari in Jembrana, (6) Paksebali in Klungkung, (7) Bongan in Tabanan, 8) Bongkasa in Badung, and (9) Sanur Kauh in Denpasar. The realization of the application of solar PV can only be done through cooperation with partners, including financial partners, village stakeholders, governments, and organizations related. The partners are tasked to prepare supporting components, to form special bodies, to monitor and care when activities have been realized. Applying renewable energy to tourist villages on the island of Bali will encourage economic recovery, such as encouraging small and medium enterprises, encouraging digital technology development, all of which require reliable and sustainable energy.

2.1. Sustainable Tourism Development Potentials.
As shown in Figure 3 above, the program would give a package of two (2) units of solar PV panel @ 375 Wp/panel (1038 mm x 1755 mm) and one inverter to an individual who wants to take the offers, so that the total electricity will be max. 750 Wp for one house. The cost of installation is about USD 1500 and would be given incentives by the Government but must be borne by the community, which can be cash or in a form of soft loans given by appointed Cooperative Bodies. The tenor can be arranged up to 5 years with a special low-interest scheme as an incentive. With associated indicators, and baseline and target values, this project covers all areas as follows:
2.1.1. Economic co-benefits. The creation of jobs by using local people as manpower source for solar PV installation and maintenance and help the community to save money regarding cut the cost of electricity in enhancing community welfare.
2.1.2. Social co-benefits. To improve community resilience for people can give a common concern to maintain energy source, to intensify village security because of good lighting, facilitate and accelerate communication among rural communities in obtaining information from internet and other media.
2.1.3. Environmental co-benefits. Solar PV does not produce pollution and in line with governments program to reduce GHG emissions by 29-41% to reduce microclimate and global warming widely.

3. The Resistance

Some barriers create resistance among the community to implement this program, and those are the lack of public education, the availability of tools and land/space to place solar panels, money to buy, and local politics because of leadership dualism between Traditional Village management (Banjar Adat) and Government-structured Village management (Banjar Dinas). Some theories can be applied to solve the problems as described below.

According to the Theory of Reasoned Action (TRA), behavior that changes is the results of behavioral intention, and behavioral intention is influenced by social norms and individual attitudes toward behavior [4]. Subjective norms describe individual beliefs about normal and acceptable behavior in society, while individual attitudes toward behavior are based on individual beliefs about that behavior. The theory states that the best predictions of a person's behavior are based on that person's interests. Behavioral interest is based on 2 main factors: 1) individual belief that resulted from his/her behaviors, 2) individual perception of the closed people to him/her regarding his/her behavior [5].

Theory of Planned Behavior (TPB) is the extension of the Theory of Reasoned Action (TRA), and TPB has been widely accepted as a tool to analyze the difference between attitudes and intentions and intentions and behaviors. In this regard, attempts to use TPB as an approach to persuade people, or embrace the Solar PV program, can overcome the resistance, and provide a means of understanding the gap between attitudes and behavior [6].

4. Methodology

The survey was used for this research as data collected through a questionnaire by taking samples from the population. This research was a descriptive study to analyze the socio-cultural dynamics of the tourism village communities by gathering facts without employing any hypotheses testing. Data collection was done online by collecting primary data thru Google form. The approach was to describe the results of the data obtained in the form of tables and diagrams. Secondary data such as documents, and Forum Group Discussion (FGD) that involving 9 respondents whom representing each tourism village, were used to describe the tourism villages' condition

The number of respondents was 324 living in 9 tourism villages mentioned above that randomly and voluntarily participated, consisting of 37 of Tenganan people, 22 of Sudaji people, 66 of Bongan people, 44 of Paksebali people, 17 people of Blimbingsari, 40 of Kerta people, 4 Bongkasa people, 68 of Sanur Kauh people, and 26 people of Catur and Other tourism villages surrounding as shown in table 2. The number of respondents in this research required representative sample size to represent the population [7]. It can be calculated that about 380 samples are ideally needed to represent population above 50.000 up to indefinite numbers with normal distribution assumption. Therefore, 324 samples were considered sufficient for this research.

Table 2. Samples of 9 tourism villages

No	Tourism Village	Regency / Municipality	Population	Sample	Percentage (%)
1	Catur & Others	Bangli	2,079	26	8.02
2	Tenganan	Karangasem	4,627	37	11.42
3	Sudaji	Buleleng	9,852	22	6.79
4	Bongan	Tabanan	8,058	66	20.37
5	Paksebali	Klungkung	5,531	44	13.58
6	Blimbingsari	Jembrana	1,300	17	5.25
7	Kerta	Gianyar	4,940	40	12.35
8	Bongkasa	Badung	6,107	4	1.23
9	Sanur kauh	Denpasar	15,167	68	20.99
Total			57,661	324	100.00

In planning a program, the step taken after determining the main problem was identifying the cause of the problem. This process could be done by several methods such as Problem Tree, Flow Chart Diagram, Fish Bone Diagram, organizational elements, and others. A problem tree was employed in this research to find solutions by mapping the anatomy of cause and effect around the problem in a similar way to a Mind Map but in a more structured way. The problem tree has three parts; trunk, roots, and branches. The trunk represents the main problem; the root is the cause of the core problem, while the branches represent the impact. The use of this problem tree was related to project planning. It happened because the causal component in the problem tree would affect the design of possible interventions. Figure 4 below is an example of a problem tree chart regarding renewable energy like Solar PV.

A problem tree provides an overview of all the known causes and effects of a problem becoming a problem. This overview was important in planning a community integration or behavior change project because it established the context in which a project occurs. Understanding context helps reveal the complexities of life, and this is essential in planning a successful change project [8].

A problem tree includes writing the causes in negative form. Inverting the problem tree by replacing negative statements with positive ones creates a solution. This method provides a means of reviewing the existing understanding of problem causes and how to solve it. It will reveal the branches that will lead us to the main problem. This very valuable existing intervention may not overcome the existing factors. The objectives to use the problem tree are: 1) to identify the cause of the problem, 2) to show the main problems that exist to villagers and decision-makers related to tourism village, 3) to facilitate decision-making or the next step in the planning [9].

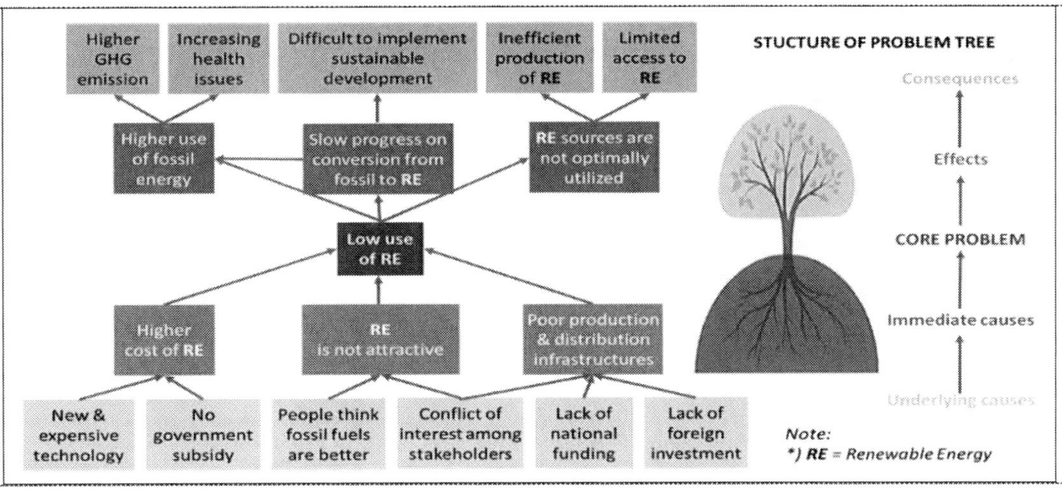

Figure 4. Example of problem tree for renewable energy [10]

5. Results

The profile of respondent could be described as follows: 233 males (71.9%) and 91 females (28.1%) participated based on gender, with 27 people (8.3%) graduated from elementary school (SD), 4 people (1.3%) from junior high school (SMP), 127 people (39.2%) from high school (SMA/ SMK), 164 people (50.6%) from college, 2 (0.6%) from postgraduate (master/doctoral) (Figure 5). Of role/position in the tourism village, 34 people (10.5%) are village government officers, 12 people (3.7%) are community organization activist, 47 people (14.5%) tourism village managers, 231 people (71.3%) villagers. Based on Age are 72 people (22.2%) aged 17-24 y.o., 89 people (27.5%) aged 25-34 y.o., 66 people (20.4%) aged 35-45 y.o., and 97 people (29,9%) aged over 45. Based on Occupation, 26 (8%) are civil servants, 124 (38,3%) private employees, 58 (17.9%) self-employed, 13 (4.05%) farmers, 38 (11,7%) unemployed, and 52 (16%) are others. Of the 324 respondents, 311 people (96%) recognized that their village is a Tourism Destination, 13 people (4%) did not know that their area is Tourism Destination. The low percentage of farmers and largest percentage of private employee as respondents explained that many villagers have been moving to the city and turn their job into tourism industry that is dominated by private sector. As Bali economy backbone is tourism, so that the island suffers the most (minus 12,3% of economic downturn) when Covid-19 is come.

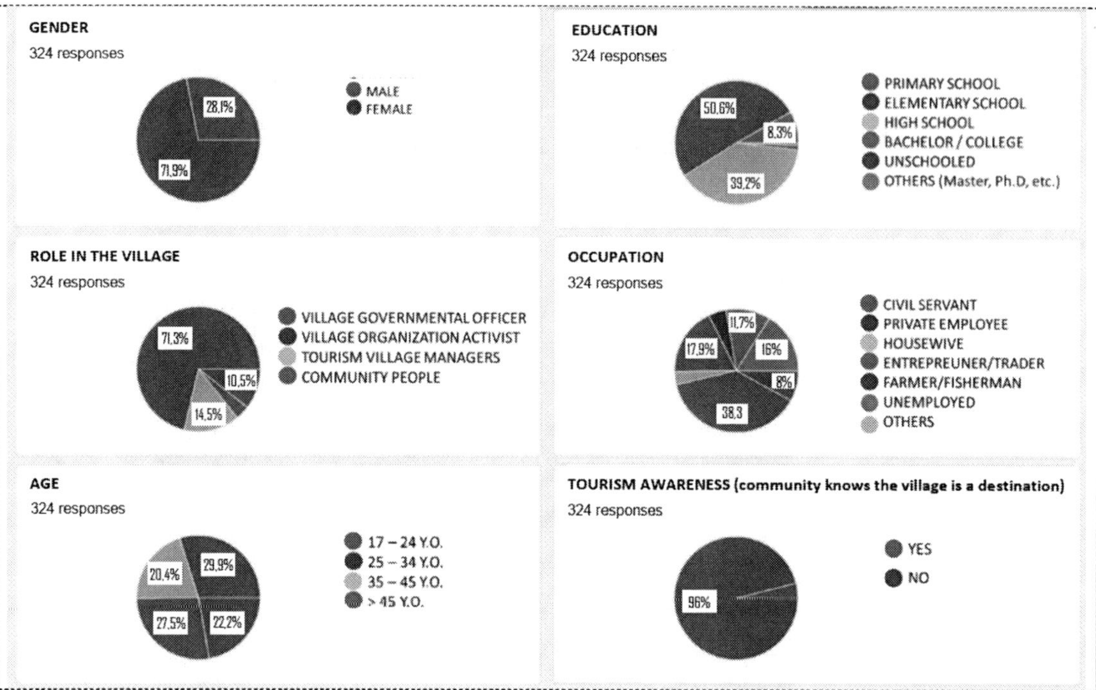

Figure 5. The profile of respondents.

5.1. Community Participation.

Regarding participation in tourism development activities, 250 villagers (77.2%) participate, and 74 (22.8%) do not participate. 180 (55.6%) said there are barriers to participation, and 144 (44.4%) said no barriers (Figure 6). This might because of the power and politics among village leaders, along with poor public communication, have hindered community participation on tourism development in the village. On homestay business, 53 villagers (16.4%) own homestay, and 271 (83.6%) did not own homestay. On willingness to make their house for a homestay, 196 villagers (60.5%) will do, and 128 (39.5%) do not want to. Furthermore, 219 villagers (67.6%) want to change their rooms according to the standard of a healthy homestay, and 105 (32.4%) do not want to change it. On the Rooftop Solar PV, 278 villagers (85.8%) want to install it, and 46 (14.2%) do not want to.

Figure 6. Community participation on tourism development in the village

5.2. Dynamics of Solar PV Program.

The program for tourism villages was related to socio-cultural dynamics of the community as the results show that 154 villagers (47.5%) use 450-900 KVA for their homes, 147 (45.4%) use 1,300-2,200 KVA, and 23 (7.1%) use 3,500-6,600 KVA (Figure 7). They were also asked if the monthly electricity cost could be cut by >50%, and the panel durability was guaranteed up to 25 years, 297 villagers (91.7%) were willing to invest about USD 350-700, and 27 (8.3%) were willing to invest more than USD 700. If they were given soft loan, the instalments duration expected by 76 villagers (23.5%) were within 1-2 years; 119 (36.7%) were within 3-5 years; 95 (29.3%) were within 5-10 years, and 34 (10.5%) > 10 years. The instalment value expected by 238 villagers (73.5%) is <USD 35/month; 61 (18.8%) within USD 35-70/month; 25 (7.7%) >USD 70/month. On financial institution preferred to provide soft loan/credit, 176 villagers (54.3%) chosen bank, 69 (21.3%) chosen cooperative, 49 (15.1%) chosen vendor, and 30 (9.3%) chosen non-bank/others.

IOP Conf. Series: Earth and Environmental Science 1027 (2022) 012030 doi:10.1088/1755-1315/1027/1/012030

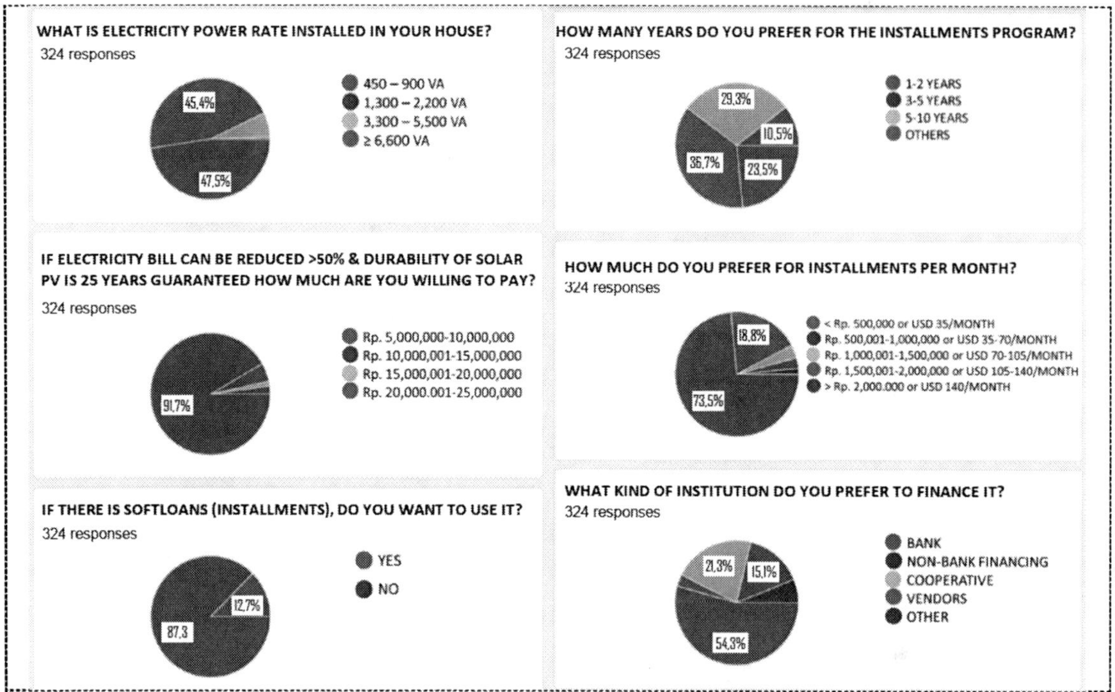

Figure 7. The dynamics of solar PV program.

5.3. *The Obstacles.*

The survey also shows that 257 villagers (79.9%) would be happy if their houses were installed with solar PV so they could save electricity costs, 63 (19.4%) were not sure, and only 2 (0.7%) was not happy. However, this research revealed a paradox as 115 villagers (35.5%) perceive that would be obstacles, 121 (37.3%) thought oppositely, and 88 (27.2%) were not sure (Figure 8). The obstacles were varied; 104 villagers (32.1%) thought there would be costly, 120 (37%) thought the community do not understand the energy crisis in Bali and do not care, 62 (19.1%) thought about maintenance and technology, 38 (11.8%) thought of politics among leaders, customs, and others.

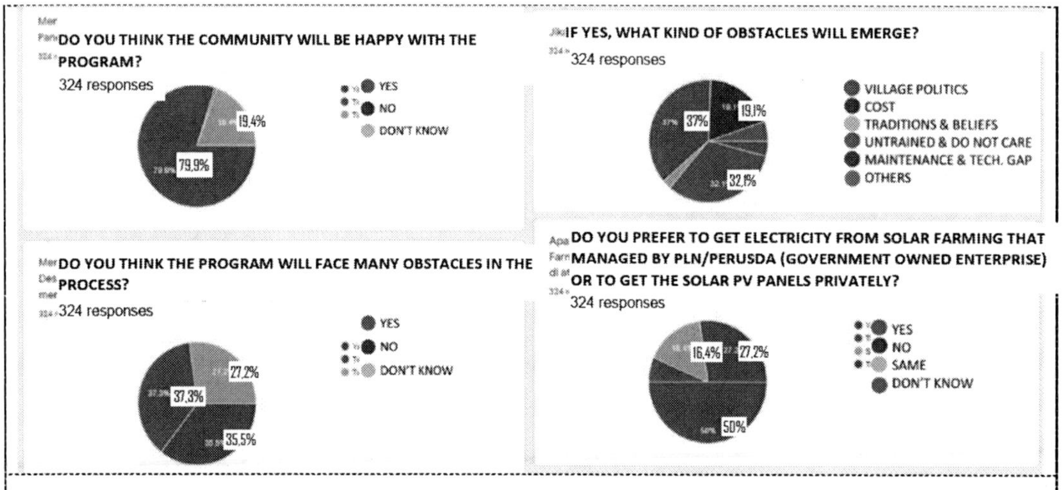

Figure 8. The obstacles.

Respondents also preferred to have electricity from solar farming managed by PLN or local government-owned company (Perusda) rather than to have it on their house roof with 162 villagers (50%), but only 21 (6.4%) preferred to have it privately, while 53 (16.4%) were okay with either way, and 88 (27.2%) are not sure.

6. Analysis and Discussion

Based on a survey of 8 tourist villages, namely Sudaji, Kerta, Blimbingsari, Catur, Bongkasa, Paksebali, Tenganan, Bongan in Bali regarding the planning of the Rooftop Solar Power Plant (PLTSA) program, the electricity generated currently installed in the Tourism Village community was around 450 - 2,200 VA. The cost of installing solar PV (solar panels/solar power) was around Rp. 5,000,001-10,000,000 (USD 350-700), and most people wanted to take advantage of soft loans/credit facilities with instalments scheme below Rp. 500,000/month (USD 35/month)

As problem tree analysis shown in Figure 4, the core problem was electricity would be hampered in 2025, as the coal resource will be run out in Indonesia while the oil reserves have been depleted since 2019, if Bali did not anticipate this issue because electricity for Bali has majorly supplied from the Java that is powered by fossil fuels. The villager would be the most suffered group as their purchasing power was the lowest to afford, so the Provincial Government of Bali has initiated this Solar PV program. There were also two other immediate causes that trigger the core problem: high-cost investment to build Renewable Energy Power Plant in Bali in providing about 1200-1500 Megawatt (as calculated by PLN-Bali, the state-owned electricity company), and low demand of solar PV. The underlying causes were the production cost which was still high, and low willingness to pay as there were no incentive given to those who install solar PV at their homes. Besides, for Tourism Village, this is usually located in remote or rural area that creates geographical challenge in logistics and after sales issues, eventually created only small market.

The unsolved core problem would create at least three effects, 1) the use of diesel generator set, which produces 2 consequences such as limited supply and GHG emission and air pollution, 2) low productive activity, which creates insufficient basic services and poor tourism facilities, and 3) lesser local products, along with low productive activity will result in low economic growth. These would affect villagers' income and lower the community climate resilience, for eventually tourism village would be undeveloped.

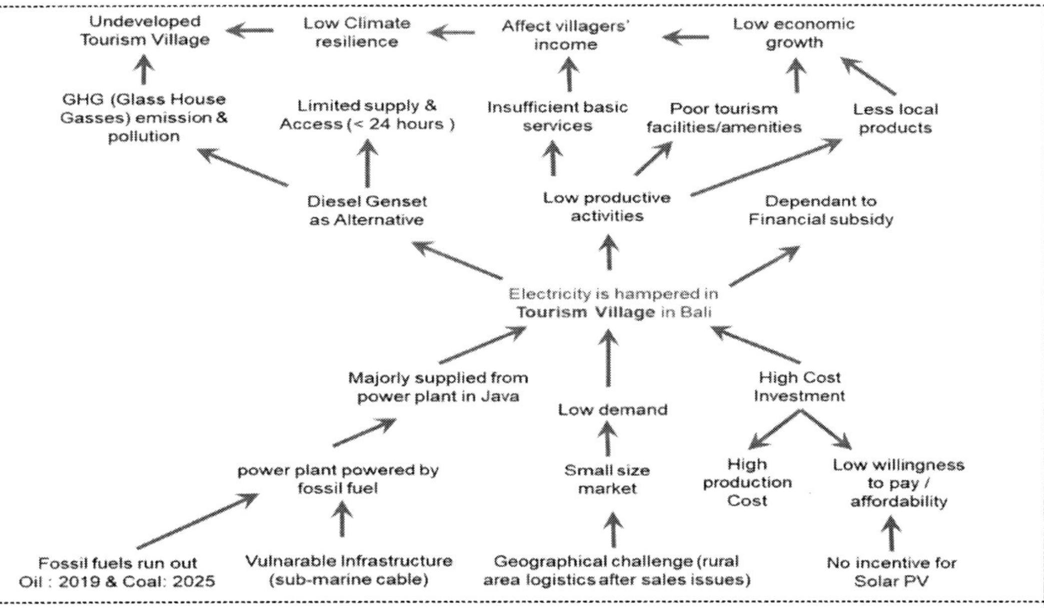

Figure 9. The problem tree of the rooftop solar PV program in tourism village

Regarding the Theory of Reasoned Action (TRA) [5], describing behavior that changes based on the results of behavioral intentions, which is influenced by social norms and individual attitudes towards behavior [4], the Governor Regulation No. 45/2019 concerning Bali Clean Energy that regulates all buildings to install Rooftop Solar PV by 2025 of at least 20% of the subscribed power to PLN would create a new social norm and would affect the behavior of the villagers, particularly because most of them were happy if their houses were installed with solar PV panels. The happiness would be the best prediction for a person's behavior based on that person's interests. As behavioral interest was based on 2 main factors, namely individual belief in the results of the behavior carried out and individual perceptions of the views of those closest to the individual on the behavior carried out, the views and the testimonies of influential figures in the targeted villages regarding the planning of the program who use solar PV for electricity for their personal and household purpose presumably will encourage villagers to follow the step. Particularly because God blesses the whole region with abundant in sunlight as Balinese is a religious community. These figures' efforts at the village level could answer Bali's challenges in realizing energy independence. Bali can harvest sunlight to become an unlimited source of electricity starting from the household scale. This belief will strongly support the Rooftop Solar PV program, as the survey results also reveal that people are happy to save electricity costs.

Another theory known as Theory of Planned Behavior (TPB), has been widely accepted as a tool for analyzing the differences between attitude and intention as well as intention and behavior and can overcome some of the limitations of previous theory as well as to provide a means of understanding the observed gaps between attitude and behavior [6]. In the program, there were also differences between attitude and intention that become an obstacle. The obstacle was that people do not fully understand the energy crisis in Bali and do not care about the situation, as they might think it is not their responsibility. There were also a mental block barrier on the maintenance costs and solar technology knowledge issues. The community also preferred a Solar Farming power plant managed by PLN/Perusda (local government owned company) rather than rooftop solar PV installed in their house to avoid further problems.

7. Conclusions

The survey of 324 respondents, based on variables gender, education, tourism awareness, role in the village, age, and occupation of the respondents, can be concluded as follows:

1) Of the development of tourism villages, 250 people (77.2%) have participated, although they encountered obstacles that is conveyed by 180 people (55.6%). 271 villagers (83.6%) do not have homestays, but most of them 196 (60.5%) also want to rent out their house for tourists as well as complying with the standard of healthy homestay by 219 villagers (67.6%), and about 278 villagers (85.8%) want to have a green homestay by installing solar PV.
2) With electricity subscribed 450-2,200 VA, solar PV installation cost is expected to be US$ 350-700 by utilizing soft loans or credit of 3-5 years instalment with USD 35/month preferred from the Bank.
3) The program is very welcome by the community because it can cut electricity costs as revealed from the Forum Group Discussion (FGD) among 9 representatives of respective tourism village. However, there are indeed obstacles caused by the Social aspects, such as socio-cultural dynamics of the community related to the fact that the community is not so familiar with solar panels, lack of energy sense of crisis, lack of solar PV knowledge as well as technological phobic, power and politics among village leaders, along with poor public communication, that hinders community participation of tourism development in the village. Economic aspects related to procurement as most components are still imported, installation and maintenance cost, costly advance solar PV technology, access to get the finance particularly in the Covid-19 era nowadays where the Bali economy has been declining to -12% because tourism, as the backbone of Bali is suffering the most. Environmental aspects related to reducing air pollution and GHG emissions, are positive from the villagers' point of view.

It is recommended that public education on solar PV program for the communities in tourism villages through workshops and seminars should be carried out in a more intense way, and the Bali Government must facilitate soft loans with long tenor installments, as well as provide incentives for those who install the solar PV, such as subsidy/special low price, and deduction on land and property tax.

References

[1] Halim DK and Setiawan IB 2020 *Bali: Towards A Green Island* (Proceedings 56th ISOCARP World Planning Congress) pp 1446-1457
[2] CORE 2019 *Peta Jalan Pengembangan PLTS Atap: Menuju Bali Mandiri Energi.* (Denpasar : Udayana University & Greenpeace Indonesia)
[3] Badan Pusat Statistik Provinsi Bali 2021. https://bali.bps.go.id/
[4] Eagle L *et al* 2013 *Social Marketing* (London: Pearson Prentice Hall)
[5] Fishbein M and Ajzen I 1975 *Belief, Attitude, Intention, and Behavior: An Introduction to Theory and Research* (Reading, MA: Addison-Wesley)
[6] Park H and Blenkinsopp J 2009 Whistleblowing as Planned Behaviour – A Survey of South Korean Police Officer *Journal of Business Ethics* **85** pp 545-556
[7] Krejcie RV and Morgan DW 1970 Determining Sample Size for Research Activities. *Educational and Psychological Measurement* **30** pp 607-610
[8] Asmoko H 2014 *Memahami Analisis Pohon Masalah* http://www.bppk.depkeu.go.id/
[9] Dillon LB 2014 *Problem Tree Analysis* http://www.sswm.info/
[10] E.Co Ltd 2021 *The project logic and Problem tree analysis Section B.2 of GCF concept note*

Heavy metals concentration in green macroalgae *Spirogyra* sp. of Way Ratai River Pesawaran Regency Lampung

F Alhafizoh [1], E L Widiastuti [1] *, Tugiyono [1], G N Susanto [1] and D F Mumtazah [1]

[1] Faculty of Mathematics and Natural Sciences, University of Lampung, Indonesia

Corresponding Author: endang.linirin@fmipa.unila.ac.id (E L Widiastuti)

Abstract. The Way Ratai River was affected by heavy metals due to gold and tailings processing activities. Certain metals in high concentrations will be dangerous if found in the environment, including the organisms in them, such as *Spirogyra* sp. Sampling was carried out at 4 observation stations around the Way Ratai River, and this research was conducted in November 2020 – February 2021. Analysis of metal content was performed by using the ICP-OES. The results showed that the heavy metal content in *Spirogyra* sp. namely, Ag, Cd, Cr, Cu, Fe, Mn, Ni, Pb, Zn of station A, B, C were above the predetermined quality standards, but Co content was undetectable. In river water, the content of heavy metals Ag in station A, Cu in all stations, Cd and Fe at in station B-D, Pb in station C, Mn and Zn in station D have exceeded the river water quality standards PP RI No. 22/2021 and ANZECC & ARMCANS 2000. Based on calculating bioconcentration factor (BCF) of heavy metals in *Spirogyra* sp, accumulation of heavy metals was in high to very high category with the highest BCF value of 159381.643 found for heavy metal Mn at station A.

1. Introduction

Ratai Bay is a bay located in Pesawaran Regency and is one of the areas that can be affected by heavy metals due to gold mining activities. The potential for gold in the area has attracted various parties to exploit it. Mining is carried out scattered in several places, either by a company or by local communities on a small scale [1]. Based on field observations, several gold processing places can be seen at several points, one of which is around the banks of the Way Ratai river.

Mining areas, especially gold mines, produce waste that contains toxic compounds such as heavy metals. The presence of heavy metals has potential to cause dangerous environmental impacts on the health of the community around the mining site. Gold mining tailings contain one or more toxic, hazardous materials such as arsenic (As), cadmium (Cd), lead (Pb), mercury (Hg), cyanide (CN) and others [2].

Heavy metals are generally toxic to living things, but some of them are needed in small amounts. Heavy metals are grouped into two types, namely essential metals in the form of heavy metals in certain amounts needed by the body such as zinc (Zn), copper (Cu), iron (Fe), cobalt (Co), and manganese (Mn), and non-essential metals whose presence is not needed by the body and are toxic such as Hg, Cd, Pb, Cr, and others [3].

The factors that cause heavy metals to be classified into pollutants are because heavy metals cannot be degraded through biodegradation, such as those of organic pollutants, but they can accumulate into the environment. The increase in heavy metal levels in the river and seawater will cause an increase in heavy metal levels in the body of aquatic biota [4]. As in the macroalgae group, green algae that live in

IOP Conf. Series: Earth and Environmental Science 1027 (2022) 012031 doi:10.1088/1755-1315/1027/1/012031

water areas by sticking to the substrate of plant roots, rocks, wood, and objects in the water, so it tends to receive more pollutants than other aquatic biotas. The presence of algae has an important role both in stagnant waters and flowing waters, namely as an important food source for invertebrates and some fish, as a biological parameter that provides information for evaluating the quality of water, and can be a vital accumulator of heavy metals. Since macro algae can absorb heavy metals through a biological system known as biosorption.

With the existence of high human activities around the Way Ratai river flows, such as daily community activities and gold mining waste disposal in the Way Ratai river flow which empties into Ratai Bay, it is assumed that the waters of the Ratai Bay have been polluted. In order to determine whether there is any pollution from gold mining activities to the Ratai Bay ecosystem, it is necessary to determine the concentration of heavy metals in the surrounding ecosystem, such as in small rivers that empty into Ratai Bay, one of which is the Way Ratai river. Therefore, this study was conducted to determine heavy metals content in river water of Way Ratai river by looking at the green macroalgae, such as *Spirogyra* sp.

2. Materials and methods

2.1. Study area

This study was conducted from November 2020 to February 2021. Sampling was carried out at 4 observation stations along the Way Ratai river flow from Bunut Seberang Village, Way Urang Village, Khepong Jaya Village, and Sanggi Pematang Awi Village. Sample preparations were carried out at the Biomolecular Laboratory, Department of Biology, Faculty of Mathematics and Natural Sciences, University of Lampung. Analysis of heavy metal content of samples was carried out at the Integrated Laboratory Unit and Technology Innovation Center (LTSIT), University of Lampung, Lampung Province, Indonesia. For more details, the research locations and observation stations can be seen in Figure 1 and Table 1.

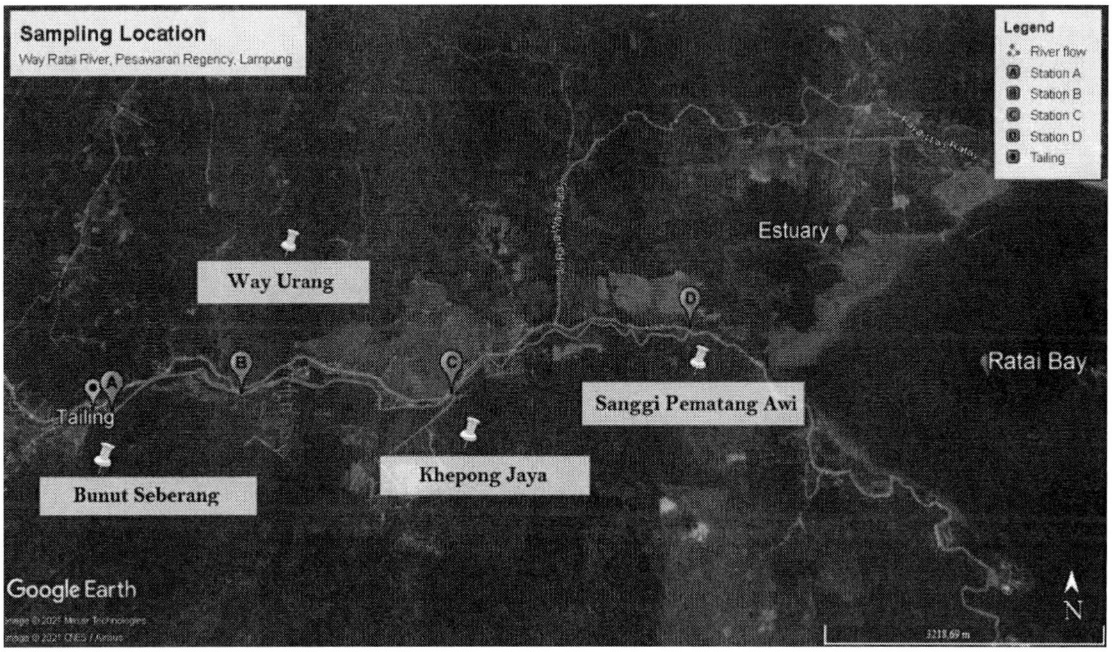

Figure 1. Sketch the sampling location

Table 1. Sampling coordinates from 4 stations (study sites)

Station	Location	South Latitude	East Longitude
A	Bunut Seberang Village River	5 ° 36'39.5 "S	105 ° 05'31.5 "E
B	Way Urang Village River	5 ° 36'24.0 "S	105 ° 06'46.0 "E
C	Khepong Jaya Village River	5 ° 36'30.4 "S	105 ° 06'26.0 "E
D	Sanggi Pematang Awi Village River	5 ° 36'01.0 "S	105 ° 09'29.0 "E

2.2. Materials
The tools used in this research are GPS (Global Positioning System), cool box, plastic zipper bag, plastic bottle, HDPE bottle, stationery, label paper, mortar and pestle, digestion flask, Erlenmeyer, analytical balance, oven, hot plate, measuring cup, metal spatula, volumetric pipette, and Agilent Varian 715-ES Inductively Coupled Plasma Optical Emission Spectrometry (ICP-OES) from Rocklin, California, USA. Collected samples were river water and green macroalgae *Spirogyra* sp., 30% HCl solution from Avantor Performance Materials Inc, Thailand, 65% HNO_3 solution from EMD Miliopore Corp. Germany, 30% H_2O_2 solution from Avantor Performance Materials Inc. USA, distilled water, and filter paper Whatman No. 41.

2.3. Research procedure

2.3.1. Research design. The sampling method used was purposive sampling based on the distance between the location of gold mining activities and the input of waste, and the ease of sampling (Figure 1). Water and *Spirogyra* sp. were randomly collected for each station with a distance between stations closed to 500 m.

2.3.2. Sample collection and preparation. A sampling of green macroalgae Spirogyra sp. was taken from the surface or bottom of the water, adjusted to the algae condition in the field. Then washed with river water at the sampling location and then put into a sample bag (plastic zipper). The water sampling technique followed Indonesian National Standard number 6989.57/2008 concerning sampling methods for surface water [5]. Water samples were taken by using plastic bottle directly on the river water closed to the surface and they were put in dark sample bottle containers

Algae samples were dried in an oven at 105°C for 24 hours and cooled. The dried samples were then pounded until it became powder and weighed as much as 1 g and then put into Erlenmeyer. 2 mL of HCl, concentrated HNO_3, and H_2O_2 were added, respectively, and then closed and left for 24 hours. Furthermore, the solution obtained was heated on a hot plate at 95°C for 1 hour. After that, the overall volume was increased to 50 mL with distilled water [6]. The concentration of heavy metals in dissolved algae was determined by ICP-OES.

As much as a 10 mL water sample was put into the digestion flask. Then given 0.2 mL of HNO_3 and 0.1 mL of HCl and homogenized. The sample was heated using a hot plate with a temperature of 95°C in 30 minutes. The sample was then cooled and transferred to a 10 mL measuring flask. In order to remove sediment, water sample was filtered by using filter paper in advance. The sample was added with distilled water until the sample volume became 10 mL and transferred to HDPE bottles, and ready to be analysed [7].

2.3.3. Heavy metal analysis using ICP-OES. Samples that have been prepared were analysed using ICP-OES to see the concentration of heavy metals Ag, Cd, Co, Cr, Cu, Fe, Mn, Zn, Pb, and Ni

2.4. Data analysis
The heavy metal concentration data obtained were then analysed descriptively, then they were compared according to the environmental quality standards that had been set. The quality standards used were

based on Australian and New Zealand Guidelines for Fresh and Marine Water Quality year 2000 [8] and refers to the standard water quality standards established by the Indonesian Government Regulation No. 22 the Year 2021 on the Implementation of the Protection and Management of the Environment, Appendix VI National Water Quality Standards [9]. The quality standard for algae was based on the regulations of the National Standardization Agency, regarding the maximum limit of heavy metal contamination in food [10] and national environmental quality standard as set by Pakistan EPA 1993 [11].

2.5. *Calculation of bioaccumulation of heavy metals in algae Spirogyra sp.*

The study of the ability of algae to accumulate heavy metals in the waters of the Way Ratai river was analysed using a bioconcentration factor (BCF). The bioconcentration factor was calculated using a formula [12] as follows:

$$\mathrm{BCF} = \frac{C\ biota}{C\ ambient\ media}$$

Description: c biota: heavy metal concentration in algae; c ambient media: concentration of heavy metals in water.

3. Results and discussion

3.1. *Heavy metals in Spirogyra sp.*

The value of heavy metal content in algae *Spirogyra* sp. can be seen in Table 2 below:

Table 2. Average heavy metal content in *Spirogyra* sp.

Station	Metals (mg/kg)									
	Ag	Cd	Cu	Pb	Fe	Mn	Co	Cr	Zn	Ni
A	10.585	2.073	14.506	41.082	6853.495	2231.343	<0.50*	14.488	197.725	5.695
B	9.791	5.023	19.610	56.474	14615.570	4158.216	<0.50*	11.825	365.735	7.517
C	7.093	4.283	13.264	29.119	14708.750	3520.102	<0.50*	11.641	188.035	6.393
Quality Standard	1.0[b]	0.1[a]	1.0[b]	0.3[a]	8.0[b]	1.5[b]	0.1[b]	1.0[b]	5.0[b]	1.0[b]

A: Bunut Seberang Village river station.
B: Way Urang Village river station.
C: Khepong Jaya Village river station.
(*): The measured value is below the detection limit, the detection limit value is stated.
[a]National Standardization Agency, regarding the maximum limit of heavy metal contamination in food (SNI 7387: 2009).
[b]National Environmental Quality Standard as set by Pakistan EPA 1993

The result of heavy metal concentrations in *Spirogyra* sp algae was compared with quality standards based on the Regulation of the National Standardization Agency, concerning Maximum Limits of Heavy Metal Pollution in Food and the national environmental quality standard as set by Pakistan EPA 1993. Because the research has not found a quality standard that regulated heavy metal concentrations for *Spirogyra* sp. in river water. So that the quality standard is used by looking at the biomagnification process that occurs, namely the tendency of a pollutant to be concentrated and move from one trophic level to the next trophic level in the food pyramid [13]. So that heavy metals will accumulate more in the next level consumers, namely aquatic biota such as fish that consume algae contaminated with heavy metals.

Based on the analysis of heavy metal content in green macroalgae *Spirogyra* sp. in the Way Ratai River, Pesawaran Regency, obtained the result that exceeds the quality standards, namely Ag, Cd, Cu, Pb, Fe, Mn, Cr, Zn, and Ni metals found at 3 flow stations in A, B, and C. While at stations D, namely the Way Ratai River, Sanggi Pematang Awi Village, samples of green algae *Spirogyra* sp. could not be found. Because when taking algae samples, station D which is the estuary area is in a state of high tide. Therefore, it was unable to determine the heavy metals content in *Spirogyra* sp at this station (D). The heavy metal content for Co metal was not detected in ICP-OES, where the detection limit or the level of accuracy of this tool was <0.500 mg/kg.

The results of the metal analysis showed that Fe metal at station A-C had a higher average content than other heavy metals. The average content of Fe metal ranged from 6853.495 – 14708.750 mg/kg, with the highest content at station C and the lowest at station A. The second-highest average heavy metal content was found in Mn metal, indicated by the highest Mn metal content at station B, which was 4158.216 mg/kg and the lowest average was 2231.343 mg/kg at station A.

Overall, based on the analysis results obtained, it can be seen that the value of heavy metal content at station B is higher than stations A and C. The difference in the value of heavy metal content between stations B and C is not too big, but the most visible difference is in the value of heavy metal content at station A. The difference between the three stations was presumably caused by where the *Spirogyra* sp. was found in the water. *Spirogyra* sp, collected in station A was floating on the rock surface and not fully submerged in water, unlike those found at stations B and C. At stations B and C *Spirogyra* sp. was found attaching to the rock and submerged in water so that the metal accumulation process would be predicted to be higher. Heavy metal contamination on *Spirogyra* sp. of Way Ratai River presumably occurs due to community activities around the river, such as tailing process from gold mining residue which found at several points around the riverbanks and dumping activity. This tailing activity of gold mining could be found in the riverbank of station A up to station D. The improper disposal of tailings at some gold processing locations will cause environmental pollution. This activity causes the river water to become cloudy, due to sludge sedimentation from the tailings processing process which contains heavy metals. Furthermore, it settles on the riverbed or clings to rocks so that it prevents algae from using light for the photosynthesis process. Furthermore, daily community activities such as washing and domestic waste dump around the banks of the Way Ratai River also could trigger heavy metal pollution.

Biologically, metals affect enzymatic reactions so that they indirectly affect the processes of photosynthesis, respiration, and reproduction so that they will inhibit the growth of algae [14]. Furthermore, if the algae contaminated by heavy metals is consumed by other aquatic biotas such as fish, the heavy metals will accumulate in the water biota network. Furthermore, if consumed by humans who occupy the highest trophic level in the food chain in a food pyramid, the human body will accumulate the highest heavy metals.

Commonly, the heavy metal accumulation process occurs in two biochemical pathways, namely the adsorption of metal ions into the cell surface and the absorption of metals into cells referred to as bioaccumulation. The adsorption process (passive uptake) is fast, back and forth, and occurs in both dead and living cells [13]. This process might occur in the accumulation of heavy metals in *Spirogyra* sp.

Metal ions enter through the cell walls of algae which contain a wide variety of polysaccharides and proteins that have several active sites. These active sites are capable of binding to metal ions through chemical and physical mechanisms, namely ion exchange, complex formation, and adsorption involving functional groups such as carboxylate groups, amines, thiolates, hydroxides, imidazoles, sulfhydryls, phosphodiester, and phosphate groups [15,16]. Furthermore, in the absorption or bioaccumulation mechanism (active uptake), algae will move metal ions bound to the cell wall to the deeper cell organelles. Bioaccumulation occurs as a result of exposure to metal ion materials continuously into the cell, the accumulation of ions will remain in the cell because the rate of release of metal ions is much lower than the rate of absorption into the cell [13].

3.2. Heavy metals in Way Ratai river water, Pesawaran Regency

Analysis of heavy metal content was also carried out on river water samples, the results of the analysis were presented in Table 3 below.

Table 3. Average heavy metals in river water.

Station	Metal (mg/L)									
	Ag	Cd	Cu	Pb	Fe	Mn	Co	Cr	Zn	Ni
A	0.103	0.006	0.052	Nd	0.252	0.014	0.069	0.013	0.009	0.048
B	Nd	0.014	0.057	Nd	0.447	0.030	0.060	0.016	0.034	0.031
C	Nd	0.029	0.041	0.038	0.472	0.027	0.023	0.011	0.035	0.020
D	Nd	0.013	0.041	Nd	0.902	0.112	0.042	0.012	0.151	Nd
River water quality standards	0.003[a]	0.01[b]	0.02[b]	0.03[b]	0.3[c]	0.1[c]	0.2[b]	0.05[b]	0.05[b]	0.05[b]

A, B, C, D: Observation station.
Nd: Not detected
[a] Australian and New Zealand Guidelines for Fresh and Marine Water Quality (ANZECC / ARMCANZ) year 2000.
[b] Indonesian Government Regulation No. 22 year 2021 on the Implementation of the Protection and Management of the Environmen, Appendix VI National Water Quality Standards class 3.
[c] Indonesian Government Regulation No. 22 year 2021 on the Implementation of the Protection and Management of the Environment, Appendix VI National Water Quality Standards class 1.

Based on the results of research on the content of heavy metals in Way Ratai River water at four observation stations, it can be seen that the Ag metal at station A exceeds the predetermined quality standard, which is based on ANZECC/ARMCANZ 2000 (Australian and New Zealand Guidelines for Fresh and Marine Water Quality). Meanwhile, the content of Ag metal at station B-D was not detected by ICP-OES with an Ag metal detection limit of 0.003 mg/L.

The metal content of Cd at stations B, C, and D, Cu metal at all observation stations, Pb at station C, and Zn at station D exceeds the Class 3 water quality standard, which is water intended for freshwater fish cultivation, animal husbandry, and water for irrigation plantations regulated in Government Regulation of the Republic of Indonesia No. 22 of 2021. The content of Fe metal at stations B, C, and D, Mn metal at station D based on PP RI No. 22 of 2021 are above the standard value for Class 1 quality standards, namely water designated as drinking water, where the normal content limit is 0.3 mg/L for Fe and 0.1 mg/L for Mn metal. While the metal content of Fe and Mn is in the Class 3 water group, there is no limiting factor for the designation of that class.

The content of heavy metals Co and Cr in the water in the Way Ratai river for all stations has a value below the quality standard based on PP RI No. 22 of 2021, as well as for Mn, Zn, and Ni at stations A-C. Once again, 5 of the heavy metal concentration in station D which closed to the estuary exceed the predetermined standard values, namely for Cd, Cu, Fe, Mn and Zn. Yet, this still has to be considered carefully, considering that these waters have been contaminated by heavy metals Co, Cr, Mn, Zn and Ni even though they are still below the threshold, since the local activity in riverbank of the Way Ratai is still going on.

The results of the metal analysis showed that the heavy metal Fe in all sampling stations had the highest content compared to other metals in the range of 0.252 - 0.902 mg/L. The highest Fe content is at station D which is the estuary area. The high content of Fe metal in the estuary is thought to be caused by domestic waste disposal, agricultural activities, industrial waste deposits, iron water reservoirs, and corrosion from water pipes containing ferrous metal [17]. The accumulation of the oxidation process that occurs as a result of the gold mining process and the processing of tailings waste from gold mining also contributes to the high concentration of Fe carried by the river flow to the estuary.

Overall, the heavy metal content in water shows a lower level of metal content compared to the heavy metal content in the algae *Spirogyra* sp. One of the factors that could explain the low analysis results of heavy metal content in river water is presumably because several observation stations are located in areas with turbulence, such as at stations A, B, and C compared to those in station D with much laminar flow. This difference in water current then might cause the levels of Fe, Mn, and Zn at station D to be higher than those at other stations. River currents cause the presence of heavy metals in the water to become unstable[18].

The dynamic of water current causes pollutants such as heavy metals in the water to be evenly distributed in the river water column or undergo dilution. In addition, the deposition process of some heavy metals can also accumulate into the body of these aquatic organisms[17].

Heavy metal contamination in the Way Ratai River can occur due to the processing of tailings from gold mining residue found at several points around the riverbanks. Heavy metal contamination can also result from the disposal of domestic waste or household waste and agricultural activities. Heavy metal contamination in the air will lead to the accumulation of aquatic organisms and ultimately be harmful to humans who consume these aquatic organisms [19]. Moreover, if the metal content exceeds the specified threshold value because the metal can enter the body of aquatic organisms through the air and the food consumed so that the accumulation of food in the aquatic body will be toxic to the life of the air organism. This also applies to humans who consume waters polluted by heavy metals such as will harm their bodies [18].

3.3. Bioconcentration factor (BCF) analysis

In addition to testing the heavy metal content in the green algae of *Spirogyra* sp and river water in the Way Ratai river, Pesawaran Regency, a calculation of the value of the BCF was also carried out to determine the ability of algae to accumulate heavy metals in water. This BCF value can be obtained from the ratio between the metal concentration absorbed into the algae network and the metal concentration in river water. The values of the BCF are presented in Table 4.

Table 4. Value of metal bioconcentration factor (BCF) in *Spirogyra* sp.

Metal	Station		
Parameters	A	B	C
Ag	102	-	-
Cd	345	359	147
Cu	279	344	323
Pb	-	-	766
Fe	27196	32697	31162
Mn	159381	138607	130374
Co	-	-	-
Cr	1114	739	1058
Zn	21969	10757	5372
Ni	118	242	319

Table 5. Categories of heavy metal bioconcentration factor (BCF) [20].

BCF Range	BCF Category
>1000	Very high
100-1000	High
30-100	Moderate
<30	Low

The presence of heavy metals in water causes an accumulation process in the body of aquatic organisms, where this accumulation can occur through direct absorption of heavy metals in water and the food chain. Bioaccumulation of chemicals in water is an important criterion for evaluating the ecology and level of pollution of an environment. In addition to it, determining of the level pollution in water could be done by measuring the value of the bioconcentration of the biota that lives in it [21].

Based on the results, BCF values above 1000 were obtained for heavy metals Mn, Fe, and Zn, so they were categorized as very high accumulative properties. Meanwhile, for Ag, Cd, Cu, Pb and Ni metals on green algae *Spirogyra* sp. in the high accumulative category because the BCF values are in the range of 100-1000. In this study, the bioconcentration factor for metal Co could not be determined because the metal content in the organism (algae) was not detected. Based on these categories, it can be seen that the accumulation ability of *Spirogyra* sp in accumulating heavy metals is in the high to the very high accumulative category.

The high value of the bioconcentration factor (BCF) proves that the algae *Spirogyra* sp tends to absorb and accumulate heavy metals, so it can be used as a bioindicator of the aquatic environment. The results showed *Spirogyra* sp. in the Way Ratai River, is easier to accumulate Mn than other metals. The sequence of heavy metal accumulation in green algae *Spirogyra* sp. is Mn> Fe> Zn> Cr> Pb> Cu> Cd> Ni> Ag, while the heavy metal Co cannot be determined.

If bioaccumulation continues, biomagnification can occur, which involves the food chain as a link. Biomagnification tends to increase the concentration of pollutants along with the increase in trophic levels in the food chain. So that producers accumulate the lowest toxic materials and the last consumers accumulate the most [13].

4. Conclusion

The heavy metal content of Ag, Cd, Cu, Pb, Fe, Mn, Cr, Zn, and Ni in the green algae *Spirogyra* sp. at each station is above the quality standard that has been set. The accumulation ability of the *Spirogyra* sp. in water is in the high to the very high accumulative category. Heavy metal content in *Spirogyra* sp. and water that exceeded the quality standard indicates that the Way Ratai River waters are contaminated with heavy metals.

5. Acknowledgement

This study was supported by The Research Institute for Humanity and Nature (RIHN: a constituent member of NIHU), Project No. 14200102.

References

[1] Bppesawaran 2018 Mining Potential of Pesawaran Regency *https://potensi.pesawarankab.go.id*
[2] Mirdat, Patadungan Y S and Isrun 2013 Status of Heavy Metal Mercury (Hg) in Soil in the Gold Mine Processing Area in Poboya Village, Palu City *E-journal Agrotekbis* **1** 127–34
[3] Vouk V 1986 *Handbook on the toxicology of metals* ed G F N B A F M N L T Friberg (New York)
[4] Yusuf Y 2011 Analysis of Lead (Pb) Levels in Crossed Carp Cultivated in Floating Cages Cirata Reservoir West Java *Sci. Res. Appl. Chem.* **1** 98
[5] National Standardization Agency 2008 *SNI 6989.57:2008 Water and Wastewater – Part 57: Surface Water Sampling Methods*
[6] Yozukmaz A, Yabanli M and Sel F 2018 Heavy Metal Bioaccumulation in Enteromorpha intestinalis , (L .) Nees , a Macrophytic Algae : The Example of Kadin Creek (Western Anatolia) **61** pp 1–12
[7] USEPA (US Environmental Protection Agency) 1994 *Determination of macro and trace elements in rare earth magnesium cast iron by inductively coupled plasma atomic emission spectrometry* (Washington, DC.)
[8] NZECC and ARMCANZ 2000 Australian and New Zealand guidelines for fresh and marine water quality The Guidelines Australian and New Zealand Environment and Conservation Council Agriculture and Resource Management Council of Australia and New Zealand *Natl. Water Qual. Manag. Strateg.* **1** 314
[9] Government of Indonesia 2021 *Government Regulation Number 22 of 2021 concerning Guidelines for Environmental Protection and Management* (Indonesia)
[10] National Standardization Agency 2009 *SNI 7387:2009. Maximum Limit of Heavy Metal Contamination in Food*
[11] Gazette of Pakistan 1993 National Environmental Quality Standards Goverment of Pakistan, Islamabad, Pakistan 2–3

[12] Potipat J, Tangkrock-olan N and Helander H F 2015 Bioconcentration Factor (BCF) and Depuration of Heavy Metals of Oysters (Saccostrea cucullata) and Mussels (Perna viridis) in the River Basins of Coastal Area of Chanthaburi Province, Gulf of Thailand *Environ. Asia* **8** 118–28
[13] Purnamawati F S, Soeprobowati T R and Izzati M Potential of Chlorella vulgaris Beijerinck in Laboratory Scale Cd and Pb Heavy Metal Remediation *Bioma* **16** 102–13
[14] Conti M E and Cecchetti G 2003 A biomonitoring study: Trace metals in algae and molluscs from Tyrrhenian coastal areas *Environ. Res.* **93** pp 99–112
[15] Wang J and Chen C 2006 Biosorption of heavy metals by Saccharomyces cerevisiae: A review *Biotechnol. Adv.* **24** 427–51
[16] Yilmaz M, Tay T, Kivanc M and Turk H 2010 Removal of copper(II) ions from aqueous solution by a lactic acid bacterium *Brazilian J. Chem. Eng.* **27** 309–14
[17] Supriyantini E and Endrawati H 2015 Content of Heavy Metal Iron (Fe) in Water, Sediment, and Green Shellfish (Perna viridis) in Tanjung Emas Waters, Semarang *Trop. Mar. J.* **18** 133–40
[18] Wardani D A K, Dewi N K and Utami N R 2014 Accumulation of Heavy Metal Lead (Pb) in Green Mussel Meat (Perna viridis) at the Estuary of the West Flood Canal, Semarang *J. Life Sci.* **3** 1–8
[19] Rochyatun E, Kaisupy M T and Rozak A 2006 Distribution of Heavy Metals in Water and Sediment in the Waters of the Cisadane River Estuary *Makara Sains* **10** pp 35–40
[20] Ghosh M and Singh S P 2005 A Reveiw on Phytoremediation of Heavy Metals and Utilization of Its Byproducts *Appl. Ecol. Enviromental Res.* **3** pp 1–18
[21] Ivanciuc T, Ivanciuc O and Klein D J 2006 Modeling the Bioconcentration Factors and Bioaccumulation Factors of Polychlorinated Biphenyls with Posetic Quantitative Super-structure / Activity Relationships (QSSAR) *Mol. Divers.* **10** 133–45

IOP Conf. Series: Earth and Environmental Science 1027 (2022) 012032 doi:10.1088/1755-1315/1027/1/012032

Changes in Land cover using the NDVI (*Normalized Difference Vegetation Index*) Method in Kedamaian Subdistrict, Bandar Lampung City as Urban City

Ali Rahmat[1], Azan Noer Ramadhan[2], Winih Sekaringtyas Ramadhani[2], Indah Listiana[2], Helvi Yanfika[2], Raden Ajeng Diana Widyastuti[2], Abdul Mutolib[3]

[1] Research Center for Limnology, Badan Riset dan Inovasi Nasional, Indonesia.
[2] Faculty of Agriculture, University of Lampung, Indonesia.
[3] Postgraduate, Siliwangi University, Indonesia.

Corresponding Author: alirahmat911@gmail.com

Abstract. Land cover change occurs naturally, but in urban areas, Land cover change usually occurs more rapidly than in rural areas. Therefore, it is necessary to analyse Land cover changes using spatial data to determine Land cover changes that occurred in urban areas such as Bandar Lampung City in 2000 and 2020. This study uses Landsat 7 and 8 imagery processed by the NVDI (Normalized Difference Vegetation Index) analysis method with an arithmetic formula to determine the vegetation index NDVI = (NIR-RED) / (NIR + RED). After obtaining the land cover map from the results of the NDVI analysis, a ground check was carried out at 30 location points to determine the actual condition of the field and calibrate the data from NDVI analysis and actual condition. In this study, it was found that the results of Land cover from 2000 to 2020 experienced changes, namely in open land/settlements increased by 74% or 217 ha, shrub land decreased by 27% or 32 ha, low vegetation land decreased by 4.8% or 7 ha, and vegetation land is decreasing by 72.5% or 100 ha. Based on the results of ground checks, changes in Land cover are caused by the construction of settlements, housing, and industrial estates due to an increase in population and urban growth.

1. Introduction

Urban areas experience changes in Land cover due to the development process that occurs. Land cover change is an unavoidable phenomenon in relation to the transformation process in allocating natural resources from one use to another [1]. The regional hierarchy is determined by the number of residents, the number, and the type of public service facilities. In general, it can be said that the higher the population and the greater the number and types of facilities in an area, the higher the hierarchy owned by the region's local community. Therefore, it is necessary to make communication efforts from various parties (community, local government, businessmen) to reach an agreement regarding the rules of Land cover change that minimize environmental damage[2].

Land cover change analysis by utilizing temporal-spatial data is instrumental, especially determining the locations where Land cover changes occur [3]. For the purposes of regional development, local governments tend to expand the region towards the countryside. After the area developed, many migrants joined and settled in the area with the consequence that new activities emerged that demanded the provision of land for non-agricultural activities [4]. Land cover change is usually related to regional development, and it can even be said that Land cover change is a consequence of regional development.

IOP Conf. Series: Earth and Environmental Science 1027 (2022) 012032 doi:10.1088/1755-1315/1027/1/012032

The level of regional development shows the number of service facilities, namely the existing infrastructure facilities and infrastructure. The more existing infrastructure facilities tend to be an attraction for development activities.

Based on the above explanation, remote sensing is one solution for monitoring a green space area. It can be used to find out information about the green condition in the specific location, both species and vegetation density, which is very complex using data from satellite imagery. The vegetation index is used to describe the intensity of plants in an area in the image. The vegetation index is a mathematical combination between the red band and the NIR band (Near-Infrared Radiation), which has long been used to indicate the presence and condition of vegetation [5]. The calculation of NDVI (Normalized Difference Vegetation Index) is based on the principle that green plants are very effective at absorbing radiation in the visible light spectrum (PAR or *Photosynthetically Active Radiation*), while green plants reflect near-infrared radiation [6].

This study aims to determine the distribution of Land cover and Land cover changes in 2000 and 2020 in the Kedamaian subdistrict, Bandar Lampung City, Indonesia. Some research has already done using NDVI to analyze the Land cover change [7, 8, 9]. In this case, it is necessary to use Landsat satellite imagery which will be processed to determine the value of the vegetation index, area, and vegetation density class using the NDVI method in the Kedamaian subdistrict. So that the results of this research are expected to provide information about changes in Land cover in the Kedamaian subdistrict, Bandar Lampung City.

2. Research Methodology

Observations were made in the District of Kedamaian, Bandar Lampung City. The research was conducted from October – December 2020. Data analysis was carried out at the Soil Survey and Land Evaluation Laboratory, Department of Soil Science, Faculty of Agriculture, University of Lampung.

The materials used consist of Landsat 8 imagery data for 2020, Landsat 7 image data for 2000, data on administrative boundaries for the Bandar Lampung region in 2020, and field survey data for 2020. The supporting tools used include a computer set with ArcGIS 10.3 software for spatial analysis, Microsoft Excel for data processing, and Android devices with the Avenza Map application.

The implementation phase broadly consists of the preparation phase, data collection, data analysis, field checking, interpretation of results, and preparation of reports. At the preparatory stage, regional topic selection, literature study, and determination of the method used for data analysis were carried out. At the data collection stage, data were collected consisting of several sources, namely earth explorer, the center for planning and development of the Bandar Lampung region, including spatial data in the form of a map of the city's administrative boundaries of Bandar Lampung. The analytical methods used in the data analysis stage include spatial analysis, NVDI (Normalized Difference Vegetation Index) analysis, and comparative Land cover analysis. At the field checking stage at 30 location points, the Avenza Map application on Android devices is used to determine the field checkpoints' condition. After the field checking stage, the result interpretation stage is carried out.

The results of the NDVI were then used in the Land cover calculation to determine the high percentage of Land cover. The analysis stages consisted of 2 stages, Land cover classification and Land cover change analysis. The analysis is based on Geographic Information System (GIS), following the research needs in spatial terms. The method used is the Normalized Difference Vegetation Index (NDVI) landscape-scale method.

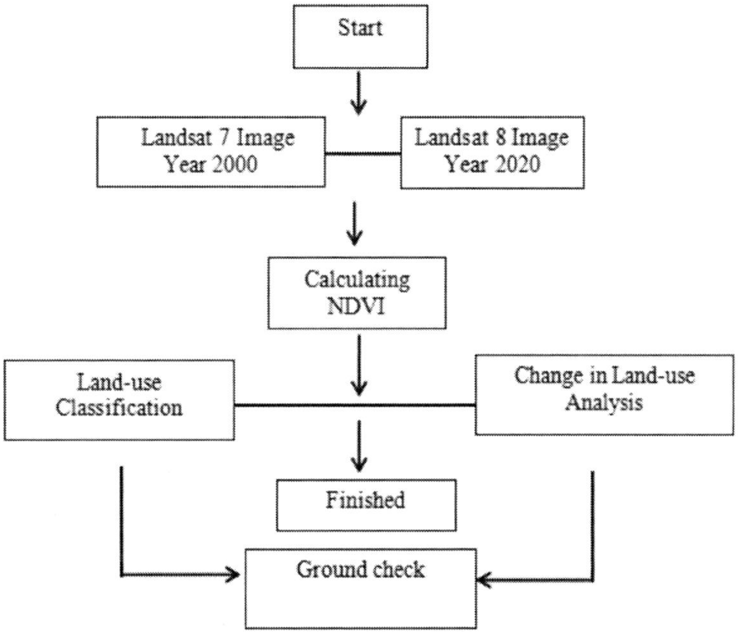

Figure 1. Research's Flowchart

3. Results and Discussions

Land cover in the Kedamaian subdistrict is spatially presented on a Land cover map in the Kedamaian subdistrict. The results of the interpretation of Land cover in the Kedamaian subdistrict from Citra Landsat grouped Land cover into four types of Land cover, namely, open land/settlement, shrubs, low vegetation, and medium vegetation.

Table 1. Land cover of Kedamaian subdistrict in 2000.

Land cover	Symbol	Index	Area	
			Ha	%
Cloud		-0,13 – -0,08	11	1,6
Open land/settlement		-0,08 – 0,01	294	41,5
Shrub land		-0,01 – 0,11	120	16,9
Low vegetation		0,11 – 0,32	146	20,6
Medium vegetation		0,32 – 0,52	138	19,5
Total			709	100

The Landsat 7 image analysis results in 2000 using the NDVI method showed that the use of open land was 294 ha, shrubland was 120 ha, low vegetation was 146, and medium vegetation was 138. There was cloud cover in some areas of 11 ha. The most significant Land cover is on open land/settlement. This figure 2 is still relatively low in urban areas. This happened because, in 2000, the Kedamaian subdistrict had not yet experienced division. Kedamaian subdistrict used to cover the subdistricts of Tanjung Karang Timur and Sukarame. The high use of open/residential land is due to the use of Land coverd as a residential area considering that Bandar Lampung City is the capital city of Lampung Province.

In shrub Land cover, low and moderate vegetation in 2000 was relatively high in urban areas. This is because, in 2000, population growth was still low. In 2000 the population in Bandar Lampung City was 743,109 people [10]; when compared to the population in 2020 of 1,051,500 people, the population is still relatively low, so that the use of open land/settlements is still relatively stable.

IOP Conf. Series: Earth and Environmental Science 1027 (2022) 012032 doi:10.1088/1755-1315/1027/1/012032

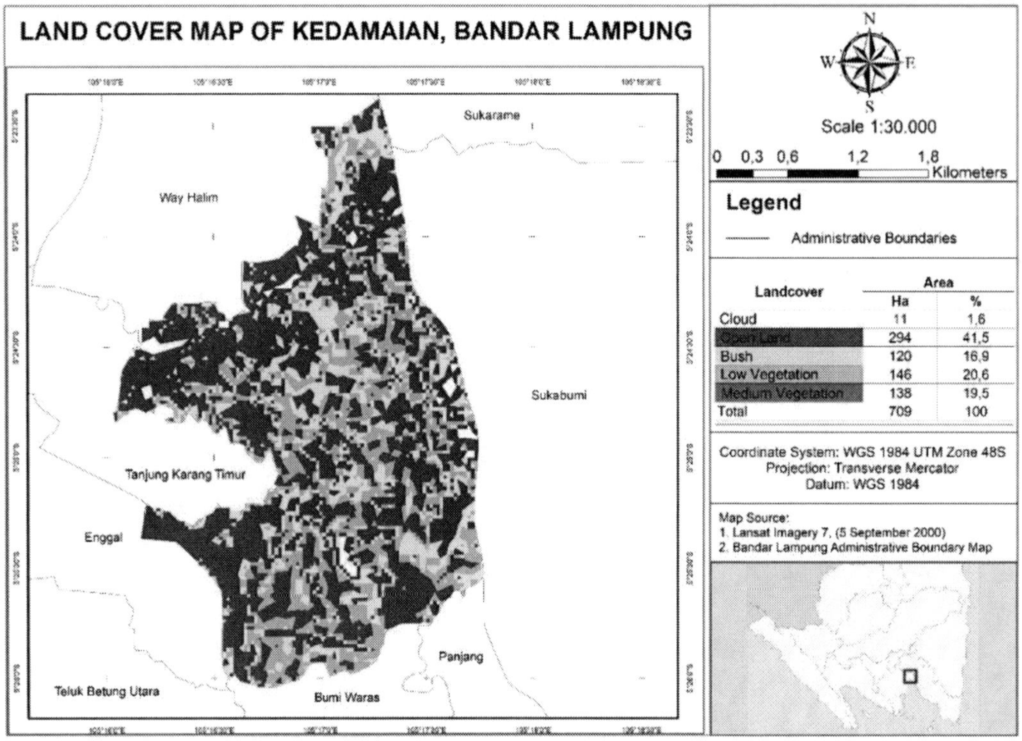

Figure 2. Land cover map of Kedamaian subdistrict in 2000.

Table 2. Land cover area of Kedamaian subdistrict in in 2020

Land cover	Symbol	Index	Area	
			Ha	%
Cloud		0,024 – 0,03	0	0
Open land/settlement		0,03 – 0,25	469	63,9
Shrub land		0,25 – 0,35	88	11,99
Low vegetation		0,35 – 0,45	139	18,93
Medium vegetation		0,45 – 0,58	38	5,18
Total			734	100

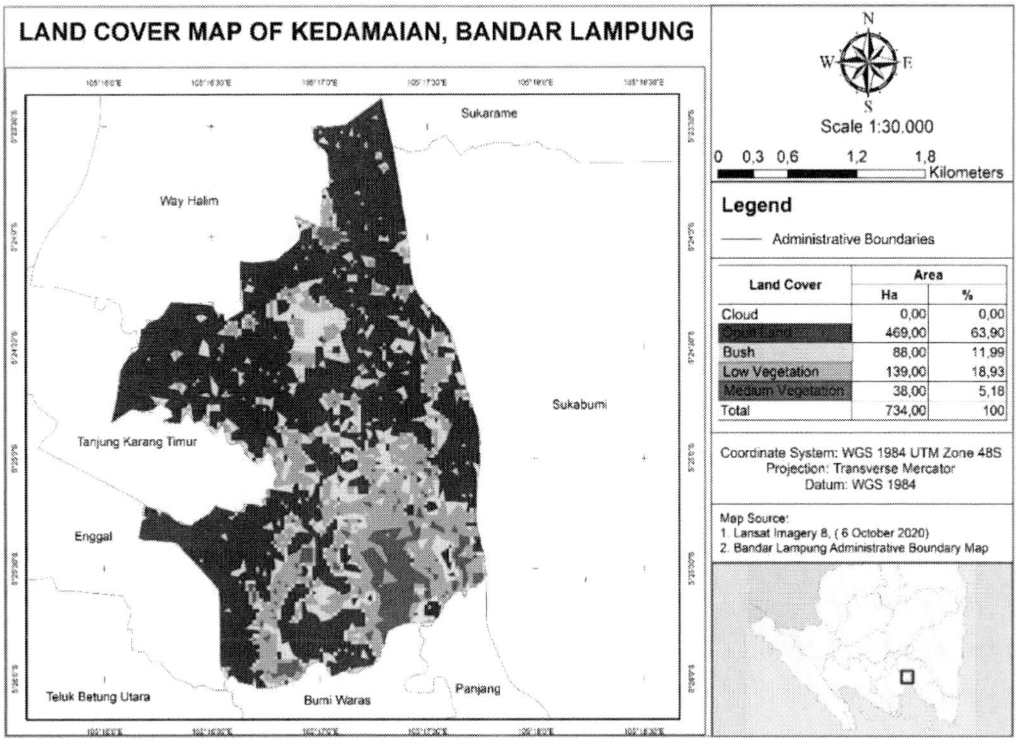

Figure 3. Land cover map of Kedamaian subdistrict in 2020.

Land cover in the Kedamaian subdistrict has undergone significant changes over the past 20 years. The Land cover map of the Kedamaian subdistrict in 2000 and 2020 is presented in Table 3. Land cover that has increased in the area is open land/settlement, and Land covers that have decreased in the area are shrubs, low vegetation, and medium vegetation.

The type of Land cover that has increased in area in Kedamaian District is the use of open land/settlements. In the 2000-2020 period, there was an increase of 217 ha, spread throughout the Kedamaian District. In the type of Land cover that experienced a decrease in area in the period 2000 - 2020, namely the use of shrubland which decreased -32 ha, low vegetation by -101 ha, medium vegetation by -100 ha.

Table 3. Land cover area 2000-2020 and its changes

Land cover	Year 2000 (ha)	Year 2020 (ha)	Change 2000 – 2020 (ha)	Change 2000 – 2020 (%)
Open land/settlement	294	469	217	74
Shrub land	120	88	-32	-27
Low vegetation	146	139	-7	-4,8
Medium vegetation	138	38	-100	-72,5

The analysis of Landsat 8 imagery in 2020 using the NDVI method obtained 469 ha of open Land cover, 88 ha of shrubs, 139 ha of low vegetation, and 38 ha of medium vegetation. The most significant Land cover is in open land/settlement by 469 ha or an increase of 217 ha compared to 2000. The increase in Land cover as open land/settlements is due to the increasing number of residents, thereby increasing residential development, opening up work areas and industrial areas. This implies that the conversion of agricultural land will be accompanied by changes in the community's economic, cultural, and political orientation, which are also generally irreversible. The development of an area is generally always followed by population growth [11]. This significant increase occurred due to the increasing number of residents in the Kedamaian District. In 2019 the Kedamaian subdistrict had a population of 55,533 people or an increase of 19.24% compared to 2010 [12].

In shrub land cover, the area decreased by 32 ha and medium vegetation by 100 ha. Based on the results of field checks at several points on the 2000 Land cover map, several residential buildings, housing, open land, offices, and industrial areas were found at the field check points in 2020. This indicates that the conversion of shrub land and vegetation land to residential area and urban development.

The decrease in the area of Land cover for shrubs, low vegetation, and moderate vegetation is due to the increase in urban growth so that the conversion of land functions as residential areas and industrial areas is getting higher. This is due to the rapid increase in Land cover change or urbanization resulting in the conversion of green land or agriculture into buildings. The process of Land cover by humans from time to time continues to change along with the development of civilization, and human needs will be higher for land needs. Changes in Land cover occur due to population growth and the development of life demands; the need for housing, which requires space as a container, is increasing.

According to Hartini et al. [11], the increase in land use as open land/settlements is due to the increasing number of residents, thereby increasing residential development, opening up jobs and industrial areas (see figure 4). This implies that the conversion of agricultural land will be accompanied by changes in the community's economic, cultural, and political orientation, which are also generally irreversible. The development of an area is generally always followed by population growth.

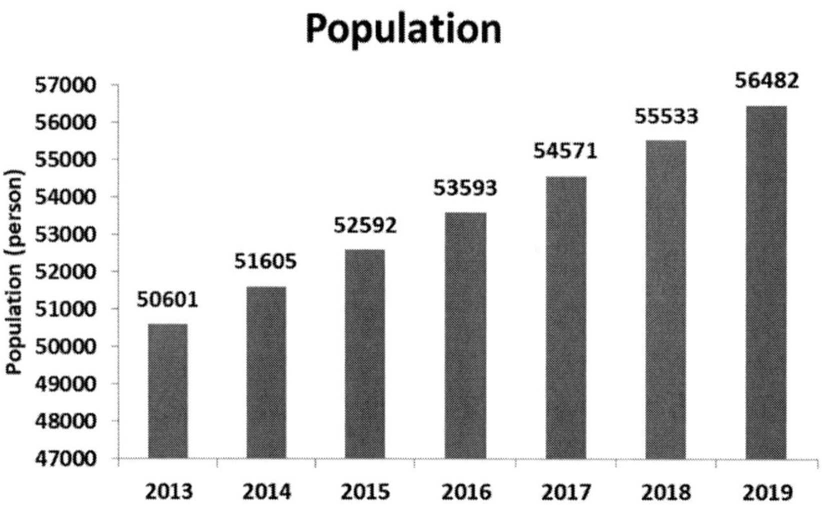

Figure 4. Graph of Population Increase.

Based on the results of field checks (figure 5) at several points, residential buildings, housing, open land, offices, and industrial areas were found at the field checking points. According to Tian & Li [13], changes in non-agricultural Land cover are characterized by the growth of industrial estates. Industrial growth is one of the things that causes urbanization to accelerate further the growth and development of cities [14]. Migration of workers to industrial areas will speed up the urbanization process, which is automatically followed by changes in the surrounding Land cover.

Open land/settlement	
 Coordinate : -5.415194°, 105.293516° Condition : Industrial Area	 Coordinate : -5.413508°, 105.288794° Condition : Housing Area
Low vegetation	
 Coordinate : -5.428539°, 105.278369° Condition : Graveyard	 Coordinate : -5.409622°, 105.292008° Condition : Corn Field
Medium vegetation	
 Coordinate : -5.422436°, 105.290686° Condition : Hill	 Coordinate : -5.417189°, 105.290479° Condition : Teak Forest

Source: Analysis, 2020

Figure 5. Field condition as a results of ground check based on vegetation density

Figure 6. Google earth image of residential area 1 2020 [15].

Figure 7. Google earth image of residential area 2 2020 [15].

Figure 8. Google earth image of industrial estate area 2020 [15].

Rapid population growth and the increasing demands of community needs for land often result in conflicts of interest over Land cover and a mismatch between the Land cover and the designation plan. Limited land in urban areas also causes the city to develop physically towards the outskirts of the city. Regarding the Land cover, suburban areas are areas that experience many changes in Land cover, especially changes in agricultural Land cover to non-agriculture due to the influence of the development of nearby cities [16].

This population growth has implications for increasing the need for quality and quantity of facilities and infrastructure so that the increase in open land/settlements follows urban growth. Urban growth is a multidimensional spatial and population process in which cities and urban settlements are considered the center of population focus because of specific economic and social facilities, which are vital components in meeting human needs [17]. Urban growth is accelerated by increasing urban population, followed by changes in the rural economy and increasingly urban lifestyles as the workforce increases in the non-agricultural sector.

4. Conclusion

The conclusions of this research are: The Kedamaian subdistrict experienced a very large increase in Land cover as open/settlement land from 2000 – 2020. This can be seen from the analysis of Landsat images using the NDVI method, which increased the value of the increase in open/settlement Land cover increasing to 74% or 217 ha. The decrease occurred in shrub Land cover by 27% or 32 ha, low vegetation by 4.8% or 54 ha, and medium vegetation by 72.5% or 100 ha. This increase in the use of open land/settlements is due to an increase in population and urban growth as residential and industrial areas.

Authorships

Ali Rahmat and Azan Noer Ramadhan are main authors of this article.

References

[1] Sugiharto. 2007 *Pembangunan dan Pengembangan Wilayah*. Universitas Sumatera Utara Press. Medan.

[2] Munibah K, Sitorus SRP, Rustiadi E, Gandasasmita K, and Hartrisari 2010 Dampak perubahan penggunaan lahan terhadap erosi di DAS Cidanau, Banten. *Jurnal Tanah dan Iklim* **32** pp 55-69

[3] As-Syakur AR, Suarna IW, Adnyana IWS, Rusna IW 2010 Studi perubahan penggunaan lahan di DAS Badung. *Jurnal Bumi Lestari* **10** 2 pp 200-207

[4] Ashari. 2003 Tinjauan tentang alih fungsi lahan sawah ke non sawah dan dampaknya di Pulau Jawa. *Forum Penelitian Agro Ekonomi FAE* **21** 2 pp 83-95

[5] Lillesand T. M., and Kiefer R.W., 2000 *Remote Sensing and Image Interpretation*. Second Edition, John Wiley & Sons, New York.

[6] Ryan, L. 1997 *Creating a Normalized Difference Vegetation Index (NDVI) image Using MultiSpec*. University of New Hampshire.

[7] Muhaimin, A.R., Ramadhani, W.S., Rahmat, A. 2021 Analisis Perubahan Penggunaan Lahan di Kecamatan Tanjung Karang Timur, Kota Bandar Lampung dengan Menggunakan Metode NDVI. *Open Science and Technology* **1** 1 pp 1-7

[8] Ramadhani, W.S., Pratama, D.L., Rahmat, A., Istiawati N.F. 2021 Analisis Perubahan Penggunaan Lahan di Kecamatan Kedondong, Kabupaten Pesawaran dengan Pemanfaatan Citra Landsat. *Open Science and Technology* **1** 1 pp 58-69

[9] Pangestu, R., Santika, N., Ramadhani, W S., Rahmat, A., Mutolib, A., Nurhaji, S. 2021 Land Use Change Analysis Using NDVI Approachment in Terbanggi Besar, Central Lampung at 2000 and 2020 *Applied Research in Science and Technology* **1** 1 pp 46-57

[10] Badan Pusat Statistik Bandar Lampung. 2000 *Bandar lampung Dalam Angka Tahun 2000*. Badan Pusat Statistik. Bandar Lampung

[11] Hartini, S., Harintaka, dan Istarno 2008 Analisis konversi ruang terbuka hijau menjadi penggunaan perumahan di Kecamatan Tembalang Kota Semarang. *Jurnal Media Teknik* pp 470-478

[12] Badan Pusat Statistik Bandar Lampung 2020 *Bandar lampung Dalam Angka Tahun 2020*. Badan Pusat Statistik. Bandar Lampung.

[13] Tian, L., Ge, B., & Li, Y 2017 Impacts of state-led and bottom-up urbanization on Land cover change in the peri-urban areas of Shanghai: Planned growth or uncontrolled sprawl? *Cities* **60** Part B pp 476– 486

[14] Pawan. 2016 Urbanization and Its Causes and Effects: A Review. *International Journal of Research and Scientific Innovation* **3** 9 pp 110–112

[15] Google Inc. 2020 *Google Maps: Kedamaian Subdistrict Location Map, Bandar Lampung City*. In *https://maps.google.com/*

[16] Rahayu, S. 2009 Kajian konversi lahan pertanian di daerah pinggiran kota yogyakarta bagian selatan (studi kasus di sebagian daerah kecamatan umbulharjo). *Jurnal Pembangunan Wilayah dan Kota* **5** pp 365-372

[17] Dadras, M., Shafri, H. Z. M., Ahmad, N., Pradhan, B., & Safarpour, S. 2015 Spatio-temporal analysis of urban growth from remote sensing data in Bandar Abbas city, Iran. Egyptian. *Journal of Remote Sensing and Space Science* **18** 1 pp 35–52

The Relationship Between The Health of Mangrove Forests and The Level of Community Welfare

Irlan Rahmat Maulana[1], Rahmat Safe'i[2], Indra Gumay Febryano[1], Hari Kaskoyo[1], Ali Rahmat[3]

[1] Department of Forestry, Faculty of Agriculture, University of Lampung, Indonesia.
[2] Master of Forestry Study Program, Department of Forestry, Faculty of Agriculture, University of Lampung, Indonesia.
[3] Research Center for Limnology, Indonesian Institute of Sciences, Indonesia.

Corresponding Author: rahmat.safei@fp.unila.ac.id

Abstract. An assessment of the forest health condition needs to be carried out to ensure that the condition of mangrove forest health is maintained. Margasari Village is where the average community works as fishermen and utilizes the mangrove forest as ecotourism. Therefore, the condition of the mangrove forest needs to be in good condition. This research was conducted in Margasari Village, Labuhan Maringgai District, East Lampung Regency. This study aims to obtain the value of mangrove forest health status and determine the relationship between mangrove forest health indicators and community welfare. Research methods to determine the value of forest health status using Forest Health Monitoring (FHM). SPSS 20 statistical spearman rank correlation test was used to determine the relationship between mangrove forest health and community welfare, with 44 respondents using village communities who operate mangrove forests. The results showed that the average value of mangrove forest health status in Margasari Village was 5.40 (moderate category). Indicators of the level of community welfare related to the health level of mangrove forests in Margasari Village are health and nutrition, education, and indicators of housing and environment.

1. Introduction

Mangrove forests are forests that grow in river estuaries, tidal areas, or the edge of the sea and are highly productive ecosystems with the same primary production level as humid tropical forests that are evergreen throughout the year [1]. Mangrove forest in Margasari Village is a mangrove forest ecosystem with potential both physically, economically, and ecologically [2]. The mangrove forest in Margasari village has an area of 817.59 ha [3]. The site has increased because it only had an area of 700 ha [4]. The functions of mangrove forests can be categorized into three, namely biological/ecological functions, physical functions, and socio-economic functions [5]. One of the benefits of mangrove forests in Margasari Village is to increase people's standard of living. This can be seen from two conditions, namely the condition of the mangrove ecosystem as a whole (pond land, agricultural land, salt ponds, ecotourism) and the condition of the ecosystem components as a primary biotic component (flora and fauna) [6]. Based on this, the existence of mangrove forests in Margasari Village is vital for the life around them, especially for the welfare of coastal communities, so there is a need for forest health management. The community in Margasari Village is very dependent on the existence of mangrove forests as an increase in community welfare. Furthermore, the various benefits of mangrove forests impact the community's economy, especially in improving the welfare of rural communities, so that mangrove forests need to be maintained to maintain the community's economy [4].

Forest health management is an effort to integrate knowledge about the ecosystem, dynamics, and genetics of disturbing organisms with economic considerations, so that the risk of damage is below the threshold of loss through monitoring forest health [7]. Forest health monitoring is a method of monitoring, assessing, and reporting on the current status, changes, and long-term trends of forest ecosystem health [8]. A healthy forest is formed, where biotic and abiotic factors are not limiting factors in achieving forest management objectives [9]. Forest health monitoring is carried out to anticipate concerns that climate change could cause new types of damage that were never previously discovered [10]. In addition, forest health monitoring is carried out to protect a mangrove forest conditions from damage caused by biotic and abiotic factors. According to [11], the function of mangrove forests on the biological environment is considerable, considering that these forests are primary ecosystems that are crucial for supporting life in coastal and ocean areas. According to [12], forest health monitoring aims to find out the current forest condition, changes, and trends that may occur. This problem makes forest health monitoring essential to be applied in various types of woods, one of which is mangrove forests in Margasari Village, which uses mangrove forests to fulfill needs.

The relationship between mangrove forests and community welfare cannot be separated so that the community around the mangrove forest needs to maintain and use the mangrove forest optimally to preserve the forest so that it remains healthy and can improve the welfare of the surrounding community. In addition, research on the relationship between forest health and community welfare in Margasari Village has not been carried out. This research is expected to optimize forest use for community welfare while maintaining good forest conditions. This research was conducted to determine the relationship between the level of community welfare through the health condition of mangrove forests in Margasari Village, Labuhan Maringgai District, East Lampung Regency.

2. Research Methodology

2.1. Time and study site

This research was conducted in August-September 2020 in the mangrove forests of Margasari Village, Labuhan Maringgai Subdistrict, East Lampung District.

2.2. Materials

The tools used consisted of a forest health tally sheet, colored mica plastic, tacks, paralon (1.5 inches), compass, permanent marker, density scale card, and tree canopy transparency, Global Positioning System (GPS), binoculars, meter, forest health book, welfare level questionnaire, and digital camera.

2.3. Methods

2.3.1 Making clusters-plot of Forest Health Monitoring (FHM). The determination of the number of cluster-plots to be made is based on a Regulation of the Director General of Forestry Planning and Environmental Management No P.1/PKTL/IPSDH/PLA.1/1/2017 regarding Technical Guidelines for Forest Inventory in Protected Forest Management Units (KPHL) and Production Forest Management Unit (KPHP). The number of sampling plots is carried out proportionally to the number of sampling plots considering the size of the dominant stratum (strata B and C) and the dominant type of mangrove in Margasari Village (the types are *Avicennia marina* and *Rhizophora mucronata*), so there are 4 cluster-plots. The shape of the FHM cluster-plot can be seen in Figure 1.

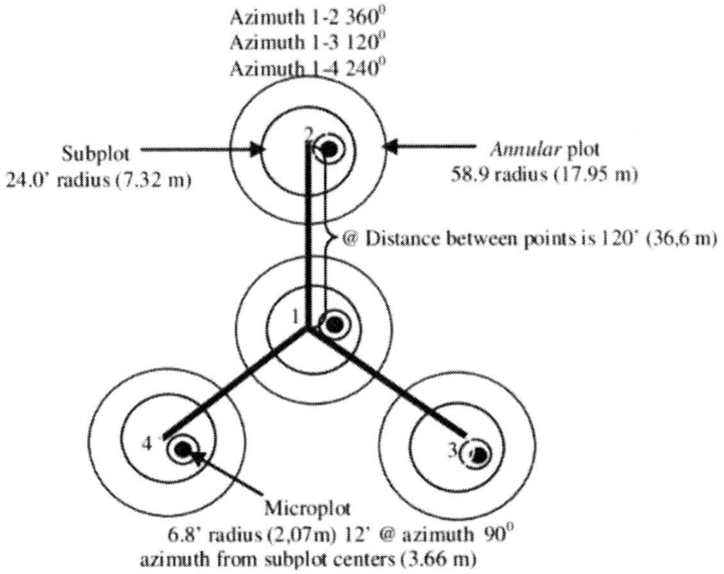

Figure 1. FHM cluster-plot design (USDA Forest Service, 1999).

Field data collection was obtained by measuring the parameters of the ecological indicators of forest health using the FHM method. Ecological indicators of forest health used include productivity through measuring stem diameter, vitality through measuring tree damage and crown conditions, and site quality through measuring the Cation Exchange Capacity (CEC) [13][14].

Three indicators of forest health measurement are used, namely productivity, vitality, and site quality. The data obtained were used to determine the health status of the forest as measured by FHM. The mangrove forest health assessment is obtained from the final value of the forest health condition, which results from the multiplication of the weighted value and the score value of each forest health indicator. The formula for the final value of forest health [15], namely:

$$NKH = NT \times NS$$

Where, NKH is the final value of forest health condition, NT is the parameter weighted value of each forest health indicator, NS is the parameter score value of each forest health indicator.

2.3.2. Data Collection of Community Welfare Indicator. The community welfare data was collected by direct interviews with respondents who are determined based on the purposive sampling method through a questionnaire. Interviews were conducted to obtain data on indicators of community welfare to determine their relationship with forest health. The parameters used to determine the levels are according to [16], which states that knowing the level of community welfare is through indicators of health and nutrition, education, employment, housing, and the environment. The selected variables are directly related to the condition of the community in utilizing the mangrove forest and affect the health of the mangrove forest. The selection of Margasari Village because this village operates the mangrove forest they have by increasing their household, the use of welfare is feared to threaten the condition of the existing mangrove forest. Therefore, village selection was carried out to determine whether the relationship between community welfare and mangrove forest health was optimal or not.

The number of respondents is determined by taking 15% of the population in Margasari Village. The size of the sample depends on the level of accuracy or error tolerance desired by the researcher. However, if the error tolerance level in the study is 15%, the maximum error rate taken is 5% (0.05).

The more significant the error rate, the smaller the number of samples, and conversely, the smaller the error rate, the greater the number of pieces obtained. With a selection of 44 respondents, it requires a smaller error rate [17]. According to [18], the population of Margasari Village is 2,106 households. Based on this population, the number of samples to be taken in Margasari Village is 44 respondents, especially in the community who carry out management and utilization around the mangrove forest. The determination of the number of samples is calculated using the Slovin formula, namely:

$$n = \frac{N}{N\,(e^2) + 1}$$

Explanation:
n : Number of respondents
N : The total number of family leaders (kepala keluarga/KK) in Margasari Village
e : Precision 15%

2.3.3. Determining the Relationship Between Forest Health Indicators and Community Welfare Indicators. The method of analyzing data on the relationship between forest health and community welfare around the mangrove forest in Margasari Village used in this study was inferential analysis using Rank Spearman analysis using the SPSS Statistical 20. According to [17], Spearman Rank correlation is used to find a relationship or test the significance of the hypothesis if each variable is ordinal, and the data sources between variables do not have to be the same, so use this instrument to determine the relationship between two existing variables. that is, the difference between the health of mangrove forests and the level of community welfare. Variable Y is obtained from welfare indicators (health and nutrition, education, employment, and level of housing and environment). Then variable X is the health of the forest which is obtained from internal factors of forest health (productivity, tree damage / CLI Cluster Plot Index, damage to the canopy / Voliage Crown Ratio (VCR), and soil CEC). That data is obtained from the calculation of ecological indicators on each cluster-plot. Then interpret the strength of the relationship of each variable can be determined through the following values [19]:

Table 1. Guidelines for the strength of the relationship for each variable

Coefficient	Strength of the relationship
0.00 – 0.25	Very weak correlation
0.26 – 0.50	Sufficient correlation
0.51 – 0.75	Strong correlation
0.76 – 0.99	Very strong correlation
1.00	Perfect Correlation

3. Results and Discussions

3.1. General Condition of the Location

Margasari Village is located in Labuhan Maringgai subdistrict, East Lampung District, Lampung Province. This village has an area of ± 1,702 hectares [6]. The total population in Margasari Village is 8,784 people with 2,106 family heads consisting of 12 hamlets and 48 RT [18]. Margasari Village is a coastal or coastal village typology. The village, which is located at an altitude of 1.5 meters above sea level, has a flat landscape and is one of the coastal villages. Margasari Village has a sandy soil texture, with the color of the soil mainly being gray [18]. Margasari Village has natural potential in the form of mangrove forests which are currently being developed into an ecotourism area.

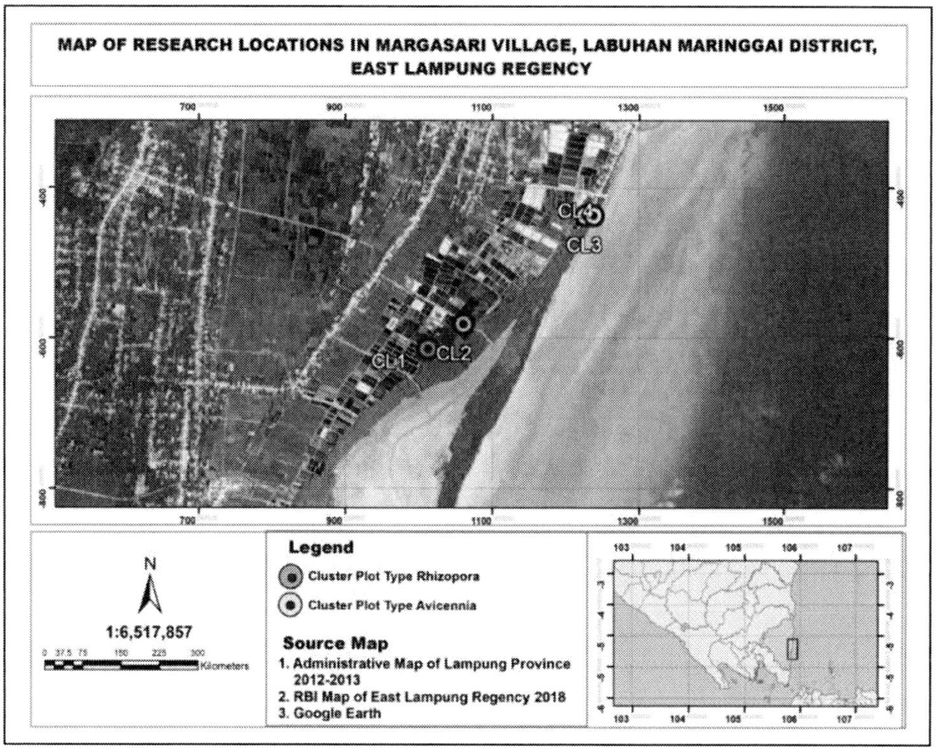

Figure 2. Map of the research location and the location for the formation of the FHM plot cluster.

The mangrove forest in Margasari village has an area of 817.59 ha [3]. This area has increased because previously, it only had an area of 700 ha [4]. The tourism management by the community is managed by a combination of fishermen groups, the Margajaya group, pamswakarsa, and a group of women farmers who love maritime. The community plays an active role in ecotourism management due to their awareness that mangrove forests play an essential role in the lives of the surrounding community [20].

The types of mangroves in this village include white fires (*Avicennia marina*), oil mangroves (*Rhizophora apiculata*), jeruju (*Acanthus ilicifolius*), and red pidada (*Sonneratia caseolaris*). The mangrove forest in this village is dominated by white fire mangroves (*Avicennia marina*) and oil mangroves (*Rhizophora apiculata*) [21]. This forest health assessment was conducted in an area dominated by both types of mangroves in the village. Most of the residents work as fishermen. There are also shrimp paste processing groups, fish processing groups, farmer groups, and mangrove groups in this village [22]. The Cinta Bahari Women's Group also carries out the utilization and processing of mangrove forest products [23]. This group uses mangrove forest products to make various forms of food and drink while still paying attention to its sustainability aspects.

3.2. Respondents Characteristics

Age affects the physical ability of people in an area to do work. Productive age is the age with an age range between 15-64 years. Ages under 15 years and over 64 years are categorized as unproductive and no longer productive [24]. The average age of community respondents around the mangrove forest in Margasari Village is in the productive category (Table 2).

Table 2. Age distribution of respondents in villages around the mangrove forest based on the economically productive age group in Margasari Village

Age Group (years)	Number (People)	Percentage (%)
0-14	0	0.00
15-64	42	95.45
> 64	2	4.55
Total	44	100

Source: Processed from field data

The majority of community respondents in Margasari Village are in the 15-64 year age group whose jobs depend on the existence of mangrove forests. This age group shows that people whose jobs are related to mangrove forests are people of productive age to carry out their work. The majority of the productive age allows people to optimize their work in improving the welfare of the community's household. According to [16], the productive working age is 15-64 years old. The practical age level of the community can optimize their work for household welfare. The community can also understand the importance of optimally utilizing the forest [25].

Education is one aspect that can be used as an indicator in measuring the quality of human resources. A higher level of education will promote the rational thinking pattern of a person in the community around the forest. The knowledge that farmers have can affect the factors of sustainable forest management [26]. The distribution of respondents to the village community around the mangrove forest based on the level of education can be seen in Table 3.

Table 3. Distribution of community respondents around mangrove forests based on the level of education in Margasari Village

No	Education level	Amount (people)	Percentage (%)
1	No school	1	2.27
2	Elementry School	15	34.10
3	Junior High School	11	25.00
4	Senior High School	16	36.36
5	Undergraduate/Graduate	1	2.27
	Total	44	100

Source: Processed from field data

The majority of respondents' education level is Senior High School with a percentage of 36.36%, and the second highest level of education is Elementary School with a percentage of 34.10%. The high level of education shows that education is necessary for the communities around the mangrove forest. However, not all people have the ability to get higher education. Higher education can also better influence a job and adopt technology, innovation to get a higher income.

The number of family members is the total number of family members who live under one roof and are the family's responsibility. The distribution of respondents in the number of family members of the village community around the mangrove forest of Margasari Village can be seen in Table 4.

Table 4. Distribution of respondents the number of family members of the community around the mangrove forest in Margasari Village

Number of Family Members (people)	Amount (people)	Percentage (%)
1-2	16	36.37
3-4	22	50.00
5-6	6	13.63
Total	44	100

Source: Processed from field data

The majority of respondents' families around the mangrove forest of Margasari Village have between 3 and 4 family members. The number of family members affects the level of the daily cost of a family. The more members of a family will affect the more costs incurred to meet family needs.

3.3. Assessment of forest health condition based on ecological indicators

Ecological indicators of mangrove forest health used in this study include productivity through measurement of LBDs, vitality by measuring tree damage and canopy conditions, and site quality through measuring soil CEC. The results of measuring the health condition of the forest are shown in Table 5.

Table 5. Results of measurement of mangrove forest health value for each indicator

Cluster-plot FHM	LBDs (m^2/ha)	CLI	VCRc	Soil CEC (me/100g)
1	0,031	2,80	2,52	15,79
2	0,028	1,96	2,99	16,76
3	0,045	2,70	2,70	16,68
4	0,51	1,58	3,58	13,39

Source: Processed from field data

Ecological indicators of mangrove forest health used in this study include productivity through measurement of LBDs, vitality by measuring tree damage and canopy conditions, and site quality. The highest productivity value is found in cluster-plot 3 (0.045 m^2 / ha), while the lowest productivity value is in cluster-plot 2 (0.028 m^2 / ha) (Table 5). One of the factors that determine the size of the diameter is the age and distance of the plant. A wide spacing will make plants more adaptable to the environment [27, 28]. The mangrove forest in Margasari Village has quite a tight spacing, especially in plot clusters 1 and 2, to protect the coastline from being attacked by waves while reducing the risk of abrasion. The level of productivity needs to be considered in forest management, as stated by [15], stated that the high or low productivity in the forest indicates the success rate of forest management rough measuring soil CEC. The results of measuring the health condition of the forest are shown in Table 5.

The highest tree damage (CLI) occurred in clusters 1 (2.80), and the lowest occurred in cluster-4 plots. The level of tree damage depends on the location where the damage was found, the type of damage, and the severity of a tree. Damage that occurs to trees in the mangrove forest can occur due to biotic and abiotic factors. Several biotic factors can cause tree damage in mangrove forests, namely pathogens, pests, and fungi. In contrast, abiotic factors can cause tree damage in mangrove forests, such as human activities, natural disasters, and environmental conditions [29]. In addition, climate change causes increased forest destruction and tree mortality from direct and indirect causes [30].

The greatest canopy damage occurred in cluster plot 5 (3.58), while the lowest occurred in cluster plot 1 (2.52). The highest VCR value indicates that the canopy condition has a good crown density. The high density of a tree can be interpreted that the tree has a canopy cover with lush foliage so that the need for photosynthesis to support tree growth can be met [31].

The results showed that the highest CEC value was in cluster plot 2 (16.76). According to [31], in mangrove forests in East Lampung district, the CEC value of land is included in clay illite and Clay Klorite, having CEC values of 16.53-20.19 me / 100g with medium low capacity. Soils with high CEC can absorb and provide nutrients better than soils with low CEC. Because nutrient elements are present in the colloid absorption complex, these nutrient elements are not easily washed away by water [32].

3.4. The final value of mangrove forest health

Based on the multiplication result between the weighted value and the value of the ecological indicator parameter score for forest health in each FHM plot, the final value of mangrove forest health is obtained. Weighted values of tree growth (productivity), soil fertility, and tree damage conditions used to assess the health of mangrove forests refer to research conducted by Apriliyani et al. [31] shown in Table 6.

Table 6. Weighted values on indicators of mangrove forest health

Indicators	Weighted values
Productivity	0.00
Canopy condition	0.32
Tree damage	0.27
Soil fertility	0.36

Source : Apriyani *et al* [33]

The forest health value is obtained from the multiplication of the score and the weighted value. The weighted values used to assess forest health are shown in Table 6. The health condition category of mangrove forests is determined by calculating the forest health status threshold, which is divided into three criteria for the health condition of conservation forests, which include good, moderate, and bad conditions [33]. The threshold value of forest health status is based on the parameter values of forest health indicators (Table 7).

Table 7. Health threshold values for mangrove forests

The threshold value for mangrove forest health	Mangrove forest health condition criteria
6.17 – 7.08	Good
5.26 – 6.16	Moderate
4.33 – 5.25	Bad

Source: Processed from field data

Value of mangrove forest health condition in each cluster-plot with mangrove forest health condition category (Table 8).

Table 8. Value of mangrove forest health status in East Lampung Distric

FHM cluster-plots	The final value of mangrove forest health	Mangrove forest health condition category
1	5.99	Moderate
2	6.55	Good
3	7.08	Good
4	4.33	Bad
Average	5.99	Moderate

Source: Processed from field data

The average forest health condition in the Mangrove Forest of Margasari Village, Labuhan Maringgai subdistrict, is moderate, or it can be quite good (5.99). This medium category indicates that the level of forest health in Margasari Village still needs to be further improved to provide more optimal benefits and maintain its sustainability. The high level of forest health indicates that forest management has been carried out well. Each forest health indicator (productivity, vitality, biodiversity, and site quality) is in good condition [34]. Meanwhile, the low level of forest health indicates that the management of mangrove forests is not good. The mangrove forest in East Lampung Regency is a protected forest, so this forest area is naturally intended to regulate water management, prevent floods, and erosion and maintain soil fertility. So far, the government and the community have worked together to conserve mangrove forests. However, the surrounding community does not fully understand the importance of maintaining the condition of mangrove forests. Some people take advantage of mangrove forests but ignore their sustainability. Many people, especially tourists who litter around mangrove tourism, commit acts of vandalism against mangrove trees that cause damage to these trees. This understanding still needs to be improved to maintain the health condition of mangrove forests in order to stay good.

3.5. Determination of The Relationship Between The Level of Welfare Indicators and The Health Level of Mangrove Forests

The parameters used to determine the level of welfare are according to [16], which states that the determination of the level of community welfare is through indicators of health and nutrition, education, employment, housing, and the environment. The chosen variable is because it is directly related to the condition of the community in utilizing the mangrove forest and affects the health of the mangrove forest. Indicators of community welfare and forest health are related if the limit of the opportunity value is not less than the number 0 and the significant correlation is less than the value of 0.05, which means that the two variables have a significant relationship. The strength of the correlation is determined based on the value of the correlation coefficient based on the value of the interval in table 1. Based on the Spearman rank correlation analysis, the indicators of community welfare are related to the health level of mangrove forests. The related indicators are the level of health and nutrition, education, level of housing, and environment (Table 9).

Table 9. The relationship between the health of the mangrove forest and the indicators of the level of welfare of the people of Margasari Village around the mangrove forest

Forest Health	Welfare Level Indicator			
	Health and Nutrition	Education	Employment	Housing and the Environment
Correlation coefficient	0.904*	0.833*	0.211	0.822*
Opportunity Value	0.013	0.039	0.688	0.045

Information: *. Significant correlation at 0.05 level of significance (2-tailed)

Health and nutrition have a strong positive relationship with the health of mangrove forests, with a correlation value of 0.904 at a significance level of 0.05 (Table 9). This relationship shows that the higher the health level of mangrove forests, the higher the community's health and nutrition. A good forest health condition can indicate the level of stability of the ecological environment in a forest area to help maintain the ecological balance of the environment, including the surrounding community [35]. The state of good forest health shows that the presence of pollutants is relatively low. However, pollutants are the leading cause of soil quality decline, which can affect the health condition of mangrove forests, affecting the surrounding community's health. Most of the people's health and nutrition are in good condition. The distribution of the level of health and nutrition of the community around the mangrove forest is presented in Figure 3.

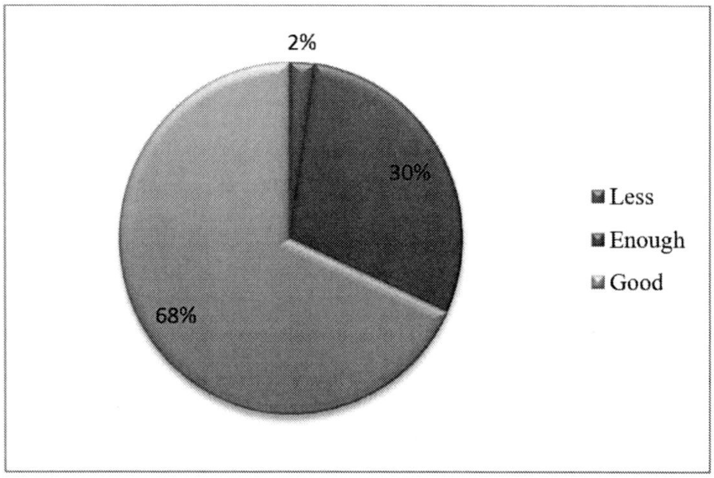

Figure 3. Distribution of community health and nutrition levels

The people around the mangrove forest in Margasari Village do not complain about the existence of mangrove forests as a cause of health problems experienced by the community. The existence of mangrove forests is felt to impact their health and nutrition, especially for life positively. Mangrove forests provide a food source because they are the home of marine animals consumed by the surrounding community and have high nutritional value. It can be an additional income to meet their daily needs. One of the forest health indicators that affect the health conditions of the people around the mangrove forest is the site quality indicator. According to [36], the condition of polluted soil will indirectly affect the health condition of the community, so that the community needs to prevent contamination of soil condition. Soil conditions, especially water quality, which is not affected by the surrounding polluting conditions, among others, due to changes in soil physical. The presence of these pollutants is the leading cause of the decline in soil quality which can affect the health condition of mangrove forests and public health [35].

Good site quality conditions indicate that the availability of nutrients for mangrove forests and for plant cultivation activities is sufficient; from this, the community is very concerned about soil quality conditions. Contaminated soil conditions will indirectly affect public health conditions, so people need to prevent contamination of soil conditions [36]. The people of Margasari Village apply concern for soil conditions, one of which is by not disposing of waste carelessly, which can damage soil nutrients. Various hazardous and toxic chemicals that contaminate the soil are sourced from seepage of garbage accumulations, wastewater treatment plants, and other sources. The presence of these pollutants is the leading cause of soil quality decline, which can affect the health condition of mangrove forests [35, 37].

The education indicator has a strong positive/unidirectional relationship with the health of the mangrove forest, with a correlation value of 0.833 at a significance level of 0.05 (Table 9). This relationship shows that the higher the level of education, the higher the level of forest health. Community education here is intended as a parameter of community adoption of the knowledge and insights possessed by the community. The community carries out management and utilization of mangrove forests as an effort to increase income and production of forest products. Most community education is in good condition. The distribution of the education level of the community around the mangrove forest is presented in Figure 4.

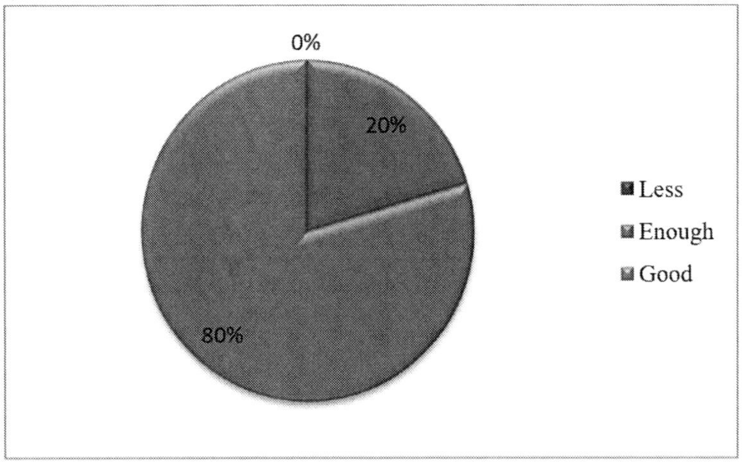

Figure 4. Distribution of community education levels

A high level of education can have a broader understanding of the benefits and ways of optimizing mangrove forests. According to [38], the education factor is one of the socio-economic factors that influence people's perceptions of forests and their functions, forest and land rehabilitation programs, and the level of participation. The high level of education allows the community to optimize the source of their livelihoods while maintaining forest sustainability. Communities with a high level of knowledge tend to have a high level of concern for protected forests. This shows that community knowledge affects one's view; if one's knowledge is high, the better one judges something, and the better one understands. This means that a person's level of knowledge can affect his mindset or level of consciousness, especially in maintaining good forest conditions in line with [39]; the higher the level of public knowledge, the better and higher the level of public awareness. The excellent health condition of mangrove forests can also provide the development of various kinds of products from mangrove forests. According to [40], the state of a healthy mangrove forest can provide potential that can be developed, one of which is the use of environmental services. Communities can have an understanding that mangrove forests provide many benefits for the environment and their lives. Mangroves also play an essential role in protecting the coast from storms, sea animal spawning grounds, abrasion barriers, and so on [41].

Community knowledge is intended as how much knowledge and insight are possessed by people who are active in managing and utilizing mangrove forests. According to [39], the higher the level of public knowledge, the better and higher the level of public awareness. The community around the forest who actively participates in extension activities and discussions will have an open insight into the management decisions made in managing forest land. People who have high knowledge can preserve the forest so that the health condition of the forest becomes healthy [42]. The experience and habits of managing forests from generation to generation by the community are also a factor in the wise use of mangrove forests [43].

Housing and environmental indicators have a strong positive relationship with forest health, with a correlation value of 0.822 at a significance level of 0.05 (Table 9). This relationship shows that the higher the housing and environmental indicators, the better the health condition of the forest, and vice versa, the higher the health condition of the forest will affect the housing and environmental conditions of the community around the mangrove forest. According to [44] that the Utilization of forest resources is generally for their own needs or interests and public buildings in the village and community craft materials. Most of the housing and community levels are in good condition. The distribution of housing and community levels around the mangrove forest is presented in Figure 5.

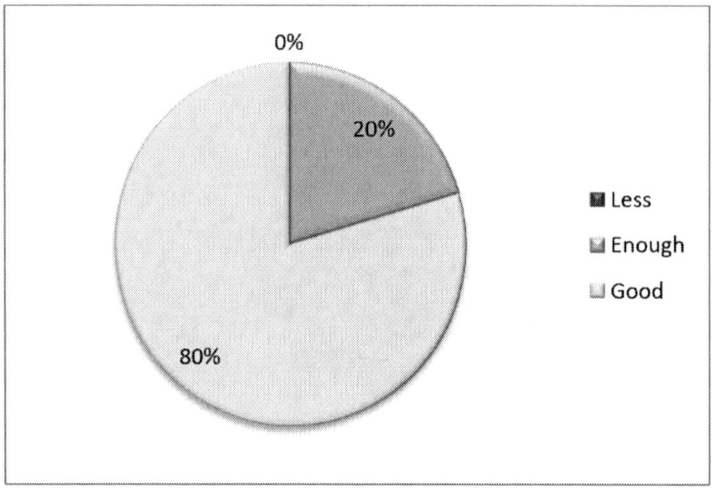

Figure 5. Distribution of housing and neighborhood levels

The community views the forest as a business land and provider of various daily needs. It maintains a stable environmental condition so that the condition of the forest must be well preserved. Good housing conditions and the environment around the mangrove forest also affect the health condition of the mangrove forest. This is in line with [45] the community's interaction with the forest has been going on for quite a long time because the existence of the forest has provided many benefits for people's lives. The surrounding community depends on the resources in the mangrove forest and can provide added value to their lives. These natural products can come from the natural environment and come from the environment cultivated by humans [46]. The community views the mangrove forest in Margasari Village as business land and a provider of various daily needs.

4. Conclusion

The conclusions of this research are, indicators of the level of community welfare that positively affect the health of mangrove forests are indicators of health and nutrition, education, and indicators of housing and environment. This positive relationship occurs because the existence of the community around the mangrove forest is inseparable from the health condition of the mangrove forest, as well as the existence of mangrove forest, which cannot be separated from the condition of the surrounding community that uses its forest products, both animal and vegetable. This shows that people's dependence on forests needs to pay attention to their forests' health condition to be maintained.

Authorships

All authors in this paper are main authors.

References

[1] Carugati, L., Gatto, B., Rastelli, E., Lo Martire, M., Coral, C., Greco, S., & Danovaro, R. 2018 Impact of mangrove forests degradation on biodiversity and ecosystem functioning *Scientific Reports* **8** 1 pp 1–11

[2] Ariftia, R.I., Qurniati, R., & Hermawati, S. 2014 Nilai ekonomi total hutan mangrove Desa Margasari Kecamatan Labuhan Maringgai Kabupaten Lampung Timur *Jurnal Sylva Lestari* **2** 3 pp 19-28

[3] Putra, A.K., Bakri, S., & Kurniawan, B. 2015 Peranan ekosistem hutan mangrove pada imunitas terhadap malaria: studi di Kecamatan Labuhan Maringgai Kabupaten Lampung Timur *Jurnal Sylva Lestari* **3** 2 pp 67-78

[4] Kustanti, A., Nugroho, B.M., Nurrochmat, D.R., & Okimoto, Y. 2014 Evolusi hak kepemilikan dalam pengelolaan ekosistem hutan mangrove di Lampung Mangrove Center *Jurnal Risalah Kebijakan Pertanian dan Lingkungan* **1** 3 pp 143-158

[5] Giri, C., Pengra, B., Zhu, Z., Singh, A., & Tieszen, L.L. 2007 Monitoring mangrove forest dynamics of the Sundarbansin Bangladesh and India using multi-temporal satellite data from 1973 to 2000 *Estuar. Coast. Shelf Sci* **73** pp 91–100.

[6] Kustanti, A. 2011 *Manajemen Hutan Mangrove*. Buku. Institut Pertanian Bogor Press. Bogor. 248 hlm.

[7] Sumardi dan Widyastuti, S. M. 2007 *Dasar-Dasar Perlindungan Hutan*. Buku. Gadjah Mada University Press.Yogyakarta. 228 hlm.

[8] Ardiansyah, F. and Safe'i, R. 2021 Analysis of Changes in Health of Coastal Mangrove Forest on the East Coast of Lampung. *Annual Conference on Health and Food Science Technology 2020.* pp 1-7.

[9] Rikto. 2010 Tipe Kerusakan Pohon Hutan Kota (Studi Kasus: Hutan Kota Bentuk Jalur Hijau, Kota Bogor-Jawa Barat). Skripsi. Departemen Konservasi Sumberdaya Hutan dan Ekowisata, Fakultas Kehutanan. IPB. Bogor. 128 hlm.

[10] Wullf S., Cornelia, R, Anna, H. R, Soren, H., Goran, S. 2013 On the possibility to monitor and assess forest damage with in largescale monitoring programmes –Asimulation study. *Journal of Silva Fennica.* **47** 3 pp 1–18.

[11] Ritohardoyo, S., & Ardi, B.G. 2014. Arahan kebijakan pengelolaan hutan mangrove: kasus Pesisir Kecamatan Teluk Pakedai, Kabupaten Kuburaya, Provinsi Kalimantan Barat. *Jurnal Geografi* **11** 1 pp 43-57

[12] Mangold, R. 1997 Forest Health Monitoring: Field Method Guide. Book. USDA Forest Service.New York. 197 hlm.

[13] Woodall, C.W., Michael, C. A., William, A. B., John, W. C., Sarah J., Charles,H. P., Kadonna, C. R., Beth,K. S., Gretchen, C. S., Borys,T., & Susan,W. 2011Status and future of the forest health indicators program of the USA *Journal Environ Monit Assess* **177** 1 pp 419–436

[14] Safe'i, R., Hardjanto, Supriyanto and Sundawati, L. 2014 Value of Vitality Status in Monoculture and Agroforestry Planting Systems of the Community Forests *International Journal of Sciences:Basic and Applied Research* **18** 1 pp 340–353

[15] Safe'i, R., Hardjanto., Supriyanto., Sundawati, L. 2015 Pengembangan metode penilaian kesehatan hutan rakyat sengon (*Falcataria moluccana* (miq.) barneby & j.w. grimes) *Jurnal Penelitian Hutan Tanaman* **12** 3 pp 175-187

[16] Badan Pusat Statistik. 2014 *Lampung Tengah dalam Angka*. Book. Badan Pusat Statistik Provinsi Lampung. Bandar Lampung.

[17] Sugiyono. 2007. *Statistik Non Parametrik untuk Penelitian*. Book. Cetakan kelima. CV Alfabeta.Bandung. 403 hlm

[18] Hartati, F. 2020. *Nilai Manfaat Langsung dan Eksistensi Ekosistem Mangrove di Provinsi Lampung*. Skripsi. Fakultas Pertanian Universitas Lampung. Lampung. 76 hlm.

[19] Sarwono, J. 2006. Analisis Data Penelitian Menggunakan SPSS. Buku. Andi. Yogyakarta. 248 hlm.

[20] Prasetyo, D., Darmawan, A., and Dewi, B.S. 2019 Persepsi wisatawan dan individu kunci tentang pengelolaan ekowisata di *Lampung Mangrove Center. Jurnal Sylva Lestari* **7** 1 pp 22-29

[21] Supriyanto., Indiyanto., Bintoro, A. 2014 Inventarisasi jenis tumbuhan obat di hutan mangrove Desa Margasari Kecamatan Labuhan Maringgai Kabupaten Lampung Timur *Jurnal Sylva Lestari* **2** 1 pp 67-76

[22] Cesario, A.E., Yuwono, S.B., and Qurniati, R. 2015 Partisipasi kelompok masyarakat dalam pelestarian hutan mangrove di Desa Margasari Kecamatan Labuhan Maringgai Kabupaten Lampung Timur *Jurnal Sylva Lestari* **3** 2 pp 21-30

[23] Desmania, D., Harianto, S.P., and Herwanti, S. 2018 Partisipasi kelompok wanita cinta bahari dalam upaya konservasi mangrove. *Jurnal Sylva Lestari* **6** 3 pp 28-35

[24] Alfrida, A., and Noor, T. I. 2017 Analisis pendapatan dan tingkat kesejahteraan rumah tangga petani padi sawah berdasarkan luas lahan *Jurnal Ilmiah Mahasiswa Agroinfo Galuh* **4** 3 pp 426-433

[25] Pratama, A.B., Yuwono, S.B., & Hilmanto, R. 2015 Pengelolaan hutan rakyat oleh kelompok pemilik hutan rakyat di Desa Bandar Dalam Kecamatan Sidomulyo Kabupaten Lampung Selatan *Journal Sylva Lestari* **3** 2 pp 99-112

[26] Rochmah, S.F., Safe'i. R., Bintoro, A., Kaskoyo, H. and Rahmat, A. 2021 The effect of forest health on social conditions of the community *The 1st Universitas Lampung International Conference on Science, Technology and Environment 2020* pp 1-7

[27] Marliah, A., Hidayat, T. & Husna, N. 2012 Pengaruh Varietas dan Jarak Tanam Terhadap Pertumbuhan Kedelai (Glycine max (L.) Merrill) *J. Agrista* **16** pp 22-28

[28] Dung, B. X., & Thanh, D. T. K. 2021 Runoff and Soil Erosion Response to Clear Cutting Period of Acacia Plantation in A Headwater Mountain of Vietnam *Applied Research in Science and Technology* **1**1 pp 12–25

[29] Safe'i, R., Latumahina, F. S., Suroso, E., & Warsono. 2020 Identification of durian tree health (Durio zibethinus) in the prospective nusantara garden wan Abdul Rachman Lampung Indonesia *Plant Cell Biotechnology and Molecular Biology* **21** 42 pp 103–110

[30] Solbergh S., Naesset, E., Lange, H., & Bollandsas, O. 2015 Remote sensing of forest health *Journal Remote Sensing and Spatial Information* **36** 4 pp 161-166

[31] Apriliyani, Y., Safe'i, R., Kaskoyo, H. Wulandari, C., Febryano, G.I. 2020 Analisis penilaian kesehatan hutan mangrove di Kabupaten Lampung Timur *Jurnal Hutan Tropis* **8** 2 pp 123-130

[32] Nursin, A., Wardah., Yusran. 2014 Sifat kimia tanah pada berbagai zonasi hutan mangrove di Desa Tumpapa Kecamatan Balinggi Kabupaten Parigi Moutong *Jurnal Warta Rimba* **2** 1 pp 17-23.

[33] Safe'i, R., Latumahina, F.S., Dewi, B.S., & Ardiansyah, F. 2021. Short communication: Assesing the state and change of forest health of the proposed arboretum in Wan Abdul Rachman Grand Forest Park, Lampung, Indonesia. *Journal of Biodiversitas,* 22(4), 2072-2077.

[34] Arwanda, E.R., Safe'i, R., Kaskoyo, H., & Herwanti, S. 2021. Identifikasi Kerusakan Pohon pada Hutan Tanaman Rakyat PIL, Kabupaten Bangka, Provinsi Kepulauan Bangka Belitung, Indonesia. *Agro Bali: Agriculutral Journal.* 4(3): 351-361.

[35] Mulia, R. M. 2005. *Kesehatan Lingkungan*. Book. Penerbit Graha Ilmu. Yogyakarta. 132 hlm.

[36] Widyanto, A.F., Yuniarno, S., Kuswanto. 2015. Polusi air tanah akibat limbah industry dan limbah rumah tangga. *Jurnal Kesehatan Masyarakat.* 10(2): 246-254.

[37] Egwu, O.C., Jennifer, U.O., Goretti, A.C.M., Uchechukwu, O., & Marksydney E.U. 2021. Toxic Elements and microbial Loads in African Giant Land Snail (Archachatina margenata) Reared with Waste Contaminated Soil. *Applied Research in Science and Technology*, 1(1), 26–35.

[38] Suraimah, Thamrin, E., & Iskandar, A.M. 2019 Persepsi masyarakat terhadap keberadaan hutan mangrove di Dusun Setingga Asin Desa Sebubus Kecamatan Paloh Kabupaten Sambas. *Jurnal Hutan Lestari.* **7** 1 pp 482-491

[39] Saputra, D.K., Zainal, S., Riyono, J.K. 2015 Studi tingkat kepedulian masyarakat sekitar hutan terhadap hutan lindung gunung pemancing - gunung ambawang Kabupaten Kubu Raya. *Jurnal Hutan Lestari.* **3** 4 pp 569-578

[40] Norsidi., Suherdiyanto., & Manu, H, Y. 2021 Persepsi masyarakat mengenal pengembangan ekowisata hutan mangrove di Desa Pasir Kecamatan Mempawah Hilir. *Jurnal Kajian Ilmu dan Pendidikan Geografi.* **5** 1 pp 94-102

[41] Siahaya, M.E., Salampessy, M.L., Febryano, I.G., Rositah, E., Silamon, R.F. dan Ichsan, A.C. 2016 Partisipasi masyarakat lokal dalam konservasi hutan mangrove di Wilayah Tarakan, Kalimantan Utara. *Jurnal Nusa Sylva.* **16** 1 pp 12-17

[42] Harimurti C.S. 2016 *Analisis Pengetahuan Kognitif Petani Hutan dalam Pelaksanakan Program Pengelolaan Hutan Bersama Masyarakat (PHBM) di Desa Jomblang, Kecamatan Jepon, Kabupaten Blora*. Tesis. Universitas Brawijaya.
[43] Safira, G. C., Wulandari, C., Kaskoyo, H. 2017 Kajian pengetahuan ekologi lokal dalam konservasi tanah dan air di sekitar Taman Hutan Raya Wan Abdul Rachman (Studi kasus di Desa Bogorejo Kecamatan Gedong Tataan). *Jurnal Sylva Lestari*. **5** 2 pp 23-29
[44] Baharuddin. 2006 *Kajian Interaksi Masyarakat Desa sekitar Taman Nasional Gunung Rinjani Provinsi Nusa Tenggara Barat (Studi Kasus di Desa Panggalangan, Desa Loloan, Desa Sembalun Lawang).*Thesis. Sekolah Pasca Sarjana Institut Pertanian Bogor. Bogor.
[45] Girsang, R.E. 2006 *Pemanfaatan Sumberdaya Hutan oleh Masyarakat Sekitar Hutan Jati di BPKH Bancar, KPH Jatirogo, Peum Perhutani Unit II Jawa Timur*. Skripsi. Fakultas Kehutanan Institut Pertanian Bogor.
[46] Ningrum, N.H.S. 2014 *Kajian Interaksi Masyarakat Sekitar Dengan Sumberdaya Hutan di Hutan Lindung Gunung Slamet KPH Banyumas Timur*. Skripsi. Fakultas Kehutanan Institut Pertanian Bogor.

IOP Conf. Series: Earth and Environmental Science 1027 (2022) 012034 doi:10.1088/1755-1315/1027/1/012034

Farmers perceptions of paddy rice cultivation technology in Jakarta city

Chery Soraya Ammatillah[1], Susi Sutardi[1], Syarifah Aminah[2] and Dan Nurmalinda[1]

[1] Jakarta Assessment Institute for Agricultural Technology, Indonesia
[2] Indonesian Center for Food Crops Research and Development, Indonesia

* Corresponding Author: chery.soraya@yahoo.com(Chery Soraya Ammatillah)

Abstract. Rice cultivation technology from the IAARD has been introduced to several rice farmers in DKI Jakarta. The introduction carried out to optimize rice productivity amid all the limited urban agricultural resources. Farmers in urban areas have different characteristics. Easy access to information and access to farming locations should influence farmers' perceptions of the technology introduced. Therefore, this research aims to determine the characteristics of rice farmers in Jakarta and to find out the farmers' perceptions of introducing rice cultivation technology. The study was conducted in DKI Jakarta involving 25 farmers. Qualitative descriptive analysis was used to determine the characteristics of rice farmers in Jakarta. Quantitative analysis was used to determine perception of paddy rice cultivation technology. Perception analysis with a Likert scale approach was used to determine farmers' perceptions. The results show that although most of the rice farmers in Jakarta are not owners of cultivated rice fields and have limited land in various aspects, they have good accessibility. The analysis results also show that most respondent farmers have good/positive perceptions of introducing rice cultivation technology.

1. Introduction

Agricultural activities in urban areas will continue despite all the limitations that exist because humans always need food. DKI Jakarta is one of the big cities, in Indonesia whose people are still working on the agricultural business. Agricultural cultivation business activities are still ongoing in DKI Jakarta, including rice cultivation activities. The background of rice cultivation activities in Jakarta, among others, is due to the high demand for rice, the limited education and expertise possessed by farming actors, and the urgent need for households.

Rice cultivation activities in Jakarta are still ongoing today, especially in West Jakarta, North Jakarta, and East Jakarta. Limited land and all available resources are not an obstacle to the activities of rice farmers in Jakarta. The average rice productivity in Jakarta in 2019 was 5.396 ton of GKG / ha [1], compared to the average national rice productivity of 5.114 ton of GKK / ha [2], the number is still relatively good. The productivity obtained shows that Jakarta still has the potential to produce rice optimally.

Good rice productivity achieved is thought to be due to access to technology, access to counselling and easy access to inputs and infrastructure in Jakarta. Farmers who live near urban areas have a higher level of technology and relatively improved technical efficiency than farmers who live far from the city core [3] .

IAARD (Indonesian Agency for Agricultural Research and Development) rice cultivation technology innovation has been introduced to several rice farmers in DKI Jakarta. IAARD rice cultivation technology innovation has been introduced to several rice farmers in DKI Jakarta. The introduction of this technology aims to increase Jakarta farmers' rice production amid limited land and agricultural human resources. The location of rice fields is close to various vital accesses to farming. The information available and accessible to get is hoped to influence farmers' perceptions and make it easier for farmers to learn and accept the technology packages. The introduction of technology has been given, but it is unknown how rice farmers' perception of technology. Therefore, this study aims to determine rice farmers' characteristics in Jakarta and determine farmers' perceptions of the introduced rice cultivation technology.

2. Methods

The research was carried out in October-December 2019, involving 25 rice farmers, in six farmer groups spread across three cities in the DKI Jakarta province, namely North Jakarta (Rorotan Village, Cilincing District), West Jakarta (Semanan Village and Pegadungan Village, Kali Deres District) and East Jakarta (Ujung Menteng Village, Cakung District). The samples in each area were determined purposively.

The data used are primary and secondary. Primary data obtained through surveys, direct interviews with farmers. Meanwhile, secondary data were obtained from the Central Bureau of Statistics. The types of primary data collected consist of data on farmer characteristics and farmers' perceptions of technology. Meanwhile, secondary data consists of rice production in Jakarta and Indonesia, obtained from the Central Bureau of Statistics.

Farmers' characteristics data collected include age (years), length of formal education (years), number of household members (souls), farming experience (years), arable land area (ha), land ownership, livelihoods and distance to farming locations. These data were analyzed using qualitative descriptive analysis.

Farmers' perception data collected is data on farmers' perceptions of rice cultivation technology: Farmers' perceptions of new high yielding varieties (productivity and price). Farmers' perceptions of organic matter application technology in the field. Farmers' perceptions of technology for planting young seedlings. Farmers' perceptions of the technology for setting the number of plant populations. Farmers' perception of balanced fertilization technology. Farmers' perceptions of the technology for controlling plant pest organisms (OPT) with an integrated pest control (IPM) approach. Farmers' perception of timely harvest technology. The data obtained were then analyzed using a five-tiered Likert scale. Score 1 stated disagree, score two stated disagreed, score 3 expressed doubt, score four stated agree, and score five stated strongly agree. The formula used to measure perception is explained as follows [4]:

$$K = \frac{n}{N} \times 100\ \%$$

Information:
K = Perception constant
n = Number of respondents who stated (person)
N = Number of Respondents (person)

3. Results

3.1. Characteristics of Respondent Farmers

Rice farmers in Jakarta have spread across three administrative city areas: the North Jakarta area, concentrated in the Cilincing District, West Jakarta, concentrated in the Kalideres District, and East Jakarta, concentrated in Cakung District. The internal characteristics of rice farmers can be seen in Table1.

The characteristics of farmers in Jakarta are undoubtedly different from farmers in rural areas in general. Household farmers in urban areas have a greater variety of occupations and higher labor expenditure than rural areas. The average income earned is also greater than rural farmers [5]. From interviews with respondent farmers, most farmers have more than one type of work, namely 67% of farmers (Table 1).

Farmers who are in and close to urban areas have better economic opportunities in market access and access to production information than those living in rural areas [3]. The ease of accessibility of farmers in Jakarta can be seen in table 1. The average distance of farming to strategic accesses is quite close. Accessibility is one of the factors that influence farmers in applying a technology [6].

Rice farming in Jakarta is classified as commercial farming because it is oriented towards market demand, no longer oriented towards meeting household food needs. Most of the products produced are sold, not for their own consumption. Market-oriented (commercial) rice farming is more efficient than farming which is only oriented towards meeting household (subsistence) food needs [3].

The average respondent's rice farmer is 45 years old. This age is still classified as the productive age. In the vulnerable period of productive age, it is hoped that farmers can quickly grasp the technology information provided, have a good perception of technology, adopt technology, and apply this technology to their farms. Age affects farmers' perceptions of technology [7], [8] and also affects technology adoption [9], [10].

Table 1. Internal Characteristics of Rice Farmers.

	Characteristic	Score		
		Max	*Min*	*Average*
1.	Age (years)	65	29	45
2.	Education (years)	12	0	6
3.	Number of household member (soul)	5	1	3
4.	Farming experience (years)	40	1	17
5.	Arable land area (ha)	4,50	0,25	1,38
6.	Land ownership			
	- Owned (%)			4,00
	- Non owned (%)			96,00
7.	Livelihood			
	- Have more than one livelihood (%)			68,00
	- Have one livelihood (%)			32,00
8.	Farming Location Distance:			
	- To residence (km)	1,00	0,00	0,38
	- To the nearest highway (km)	2,00	0,00	0,39
	- To the input store (km)	5,00	1,00	2,18
	- To the sales market (km)	4,00	0,00	0,98
	- To sources of capital (km)	20,00	0,00	1,65

Source: Primary data processed, 2019

The level of education affects farmers' perceptions of technology [8]. Education is also one of the factors affecting the adoption of technology by farmers [11]. Saridewi and Siregar [12] stated that in general, highly educated farmers will have a better way of thinking, thus enabling them to act more rationally and easily accept changes in managing their farming. The average formal education of the respondent farmers is six years or equivalent to elementary school education. Low education is the background of farmers who still choose agriculture as their main source of livelihood. The level of education of respondent farmers is still low, indicating the need for intensive training, counseling, and mentoring both formally and informally to increase farmers' knowledge and optimize the farming management.

The low level of education of farmers requires a strategy in disseminating new technology. Things that can do include, among others, the delivery of information on new technological innovations with methods that are easy to understand, routine, and intensive technology guidance and assistance. Optimizing the role of extension workers and community leaders as motivators. Conduct technology demonstration plots, technology degrees, field meetings, or group meetings [13].

Education also affects farmers' cultivation techniques. A good level of farmer education is thought to observe the complexity of a new technological innovation [14]. Even though they are farming in the city, rice farming cultivated by the respondent's farmers still uses simple cultivation techniques. Modern agricultural machine tools are not widely used. Apart from the educational factor, this is also because most farmers are not landowners. The rice fields that they work on can be evicted at any time so that if they have to invest in agricultural equipment at a high cost, considered very risky.

The average number of household members is three people. Their family members assist some rice farmers in Jakarta in farming. The family member who usually helps is the wife. Children as family members rarely help in agriculture. The small number of farmer's children who participate in farming is the child's wish and the willingness of their parents because the farmers want their children to no longer make agriculture their profession and primary source of livelihood.

The average rice farming experience of the respondent farmers is quite long, namely, for 17 years. Most farmers have carried out rice farming from generation to generation. The cultivation knowledge they obtained was obtained from their parents. The difference is, in the past, their parents worked on privately owned land. Now they are working on cultivated land, which developer control. Most respondent farmers are residents of Jakarta. Some farmers, especially in the Cilincing area, are native Jakarta residents with the Betawi ethnicity.

The average rice field area cultivated by the respondent farmers was 1.38 ha (Table 1.). However, 96.0% of the farmers who became respondents were not landowners (Table 1.). Most farmers rent developer land and private land. Land lease systems that are most applicable include a profit-sharing system, which is paid every harvest and a fixed rental system, which is paid annually. Most farmers said that the land rental payments were quite profitable because the rental fees charged were not too high.

3.2. Farmers' Perceptions of Technology

Perception has a significant influence on agricultural management decision-making [15]. One of the factors that influence farmer adoption is the perception of the farmers themselves. The results show that perceptions of the usefulness and convenience of technology have a positive and significant effect on farmers' attitudes and adoption intentions [16]. The perceptions of respondent farmers on introduced technology can be seen from several variables. The overall results of the analysis of farmers' perceptions of the introduced rice cultivation technology innovation of IAARD show that most farmers in Jakarta have a good/positive perception of this technology, with an average value of 55,00% who agree and 5,55% agree strongly (Table 2.).

Farmers' perceptions of new high-yielding varieties show good/positive perceptions. Most farmers strongly agreed (8.00%) and approved (80.00%) that the new superior productivity was higher than the existing varieties planted by farmers. Farmers who expressed disagreement argued that the current varieties they planted, such as Cibatu and Kemuning, had higher productivity than the introduced varieties. However, farmers said that the grain selling price of existing varieties was lower than high-yielding varieties, the quality of rice produced was also lower than high-yielding varieties.

High-yielding varieties are one of the technology components of Integrated Crop Management for lowland rice, which has a high adoption rate [17]. Of the several components of recommended technology, high-yielding varieties are technologies that are quickly adopted by farmers [18]. The availability of high-yielding varieties seeds needs attention because this is one factor that influences the adoption of high-yielding varieties [19].

DKI Jakarta has the characteristics of paddy fields that are slightly different from other regions. The average rice productivity in Jakarta in 2019 was 5.396 ton of GKG / ha [1]. The condition of Jakarta, which is routinely inundated by floods, impacts rice fields, which are also frequently inundated by flooding, making it vulnerable to the fall of rice plants. Therefore, the characteristics of several new high-yielding varieties, which are somewhat resistant to falling, are very popular with farmers. Another characteristic of rice fields in Jakarta is the location of the rice fields that are scattered. These characteristics impact the types of pests that often attack. Birds are the primary pests of rice plants in Jakarta. Several new high-yielding varieties have been introduced have erect flag leaves and panicles located far below the leaves. This condition makes rice plants relatively safe from bird disturbances, thereby minimizing yield loss before harvest. Some of the advantages of the high-yielding varieties mentioned earlier impact the excellent perception of respondent farmers of the introduced high-yielding varieties.

Farmers' perceptions of the price of seeds, most farmers disagree (68.00%) with the statement that the prices of quality seeds are relatively cheap. This means that most farmers have the perception that quality seeds are expensive. Farmers who have limited capital will think about using quality seeds even though they know quality seeds give high productivity.

Table 2. Farmers' perceptions of lowland rice cultivation technology innovation

No.	Statement Variable	Perception of farmers (%)					Total (%)
		Strongly agree	Agree	Doubtful	Disagree less	Disagree	
1	The productivity of high-yielding varicties is relatively higher	8,00	80,00	0,00	0,00	12,00	100
2	Prices of quality seeds are relatively cheap	8,00	4,00	20,00	68,00	0,00	100
3	The use of organic matter increases rice productivity	4,00	60,00	12,00	12,00	12,00	100
4	Plant young seedlings 15-18 days after sowing to increase the number of productive rice tillers	4,00	60,00	16,00	12,00	8,00	100
5	The number of plant populations in the legowo row is relatively high	4,00	56,00	28,00	8,00	4,00	100
6	Fertilization based on soil nutrient requirements saves production costs	0,00	64,00	36,00	0,00	0,00	100
7	Pest control using the IPM approach increases crop production	0,00	44,00	56,00	0,00	0,00	100
8	Harvest on time and the grain is immediately threshed, are the key factors in overcoming yield loss	16,00	72,00	12,00	0,00	0,00	100
	Average	5,50	55,00	22,50	12,50	4,50	100

Most respondents strongly agreed (4.00%) and approved (60.00%) that organic matter increased rice productivity. Thus, the perception of rice farmers towards organic matter technology in rice cultivation is good/positive. However, although farmers' perceptions are good about organic matter technology, in practice, farmers still find it challenging to apply organic matter intensively. This is due to the lack of availability of organic fertilizers in Jakarta. Rice farmers in Jakarta need intensive assistance in organic fertilizer processing technology that is easy, inexpensive, and time saving. Training and extension intensity are the factors that most influence the application of a technology [20].

The technology of planting young seedlings 15-18 days after sowing has good perception. 4.00% of farmers strongly agree, and 60.00% agree that planting young seedlings will increase productive rice tillers. However, with the characteristics of paddy fields in Jakarta that are endemic to snails, young seeds are difficult for farmers to apply. Most farmers use relatively old seedlings at the time of planting to avoid snail pests. The existing snail pests are usually controlled using snail-killing drugs or controlled mechanically.

The farmers' perceptions of the legowo row planting technique were quite good/positive. Most farmers strongly agree (4.00%) and agree (56.00%) that the number of plant populations in the legowo row is relatively high. The constraints felt by the respondent farmers when applying legowo row planting were the higher planting costs and the longer planting time, especially in the legowo row 2: 1 application. Muh et al. [21] in Their research results also state that the obstacles felt by farmers in applying the legowo row planting technique are the higher planting costs incurred and the longer planting time required.

Fertilization based on the need for soil nutrients gives a good perception (64,00%) because farmers think it can save production costs. However, this technology is also challenging to apply due to the unavailability of soil nutrient measuring devices (PUTS) and Leaf Color Charts (BWD). The unavailability of measuring tools for nutrient levels makes farmers fertilize according to habit.

As many as 56.00% of farmers were still doubtful that pest control using the IPM (Integrated Pest Management) approach could increase crop production. Respondent farmers still firmly believe that the intensive use of chemical pesticides can control rice pests. Therefore, there is still a need for further outreach and counseling on techniques for controlling plant-disturbing organisms using an integrated pest management approach to farmers.

Perceptions of post-harvest technology, harvest technology on time, and the grain are immediately threshed are pretty good/ positive, with numbers strongly agree (16.00%) and agree (72.00%). Most respondent farmers have applied this technology. When the harvest age is sufficient, harvest soon immediately, and after the rice plants are harvested, the grain is immediately threshed using a rice thresher machine. Gathering on time at the optimum age is very important to reduce yield losses [22]. Post-harvest handling promptly threshing the grain must be done by farmers because delaying the rebellion of grain will result in a decrease in rice quality [23].

4. Conclusion

Rice farming in Jakarta is classified as commercial farming, which is market and profit oriented. The results of the study show that although most rice farmers in Jakarta are not owners of cultivated rice fields and have limited in various aspects, they have several advantages, including access to information, good location, and infrastructure. Good market access also impacts the ease with which farmers can sell their produce and get a reasonable selling price.

Farmers' perceptions of rice cultivation technology show good or positive perceptions. The overall results of the analysis of farmers' perceptions of the introduced rice cultivation technology innovation of IAARD show that most farmers in Jakarta have a good/positive perception of this technology, with an average value of 55,00% who agree and 5,55% agree strongly. However, although some technologies have good perceptions, there are some obstacles in their application. Therefore, intensive technical assistance is still very much needed, primarily to provide solutions to the obstacles faced.

Rice farming in the city of Jakarta still has a significant enough potential to be developed. Support from the government and related stakeholders is needed, especially in providing perpetual land, increasing the added value of products, and strengthening market networks. Rice cultivation in Jakarta should not only be production-oriented because urban rice fields are not only a means of production but also a means of recreation, education, and commercialization.

Reference

[1] Central Bureau of Statistics 2020 *Provinsi DKI Jakarta Dalam Angka*. DKI Jakarta: BPS-

Statistics of DKI Jakarta Province
[2] Central Bureau of Statistics 2020 "Harvested area of rice production and productivity by province," DKI Jakarta
[3] Piya S, Action P, Kiminami A, Yagi H 2012 "Comparing the technical efficiency of rice farms in urban and rural areas : A Case Study from Nepal," *Trends Agric. Econ.*. DOI; 10.3923/tae.2012 pp 1–14
[4] Hendayana R 2016 *Persepsi dan Adopsi Teknologi Jakarta*. Jakarta: IARRD Pres
[5] Ratnaningtyas S and Purwana W 2011 "Comparison of income and labor productivity of rural and urban farmer households: A case in wetland agro ecosystem based on rice fields in Tarogong Kaler district, Garut regency," *Sosiohumaniora* **13** pp 251–262
[6] Fachrista IA, Hendayana R, Risfaheri 2013 "Socio-economic factors that determine the adoption of integrated crop management (PTT) for paddy rice in Bangka Belitung," *Inform. Pertan.* **22** pp 113–120
[7] Fakkhong S, Suwanmaneepong S, Mankeb P 2016 "Farmer's perceptions towards economic sustainability of rice farming in peri-urban area, Bangkok, Thailand," *Int. J. Agric. Technol.* **12** pp 1759–1772
[8] Pinthukas N 2015 "Farmers perception and adaptation in organic vegetable production for sustainable livelihood in Chiang Mai Province," *Agric. Agric. Sci. Procedia* **5** pp 46–51
[9] Malá Z and Malý M 2016 "The determinants of adopting organic farming practices : a case study in the Czech Republic," *Agric. Econ.* **59** pp 19–28
[10] Farid A, Romadi U, Witono D 2018 "Factors affecting farmer adoption in the implementation of the jajar legowo planting system in Sukosari Village, Kasembon District, Malang Regency, East Java Province," *J. Penyul.* **14**
[11] Ghimire R, Wen-chi H, Shrestha RB 2015 "Factors affecting adoption of improved rice varieties among rural farm households in Central Nepal," *Rice Sci.* **22** pp 35–43
[12] Saridewi TR and Siregar AN 2010 "The relationship between the role of extension workers and the adoption of technology by farmers on increasing rice production in Tasikmalaya district," *J. Penyul. Pertan.* **5** pp 55–61
[13] Indraningsih KS 2017 "Agricultural innovation dissemination strategy to support agricultural development," *Forum Penelit. Agro Ekon.* **35** pp 107–123
[14] Fachrista IA and Mamik S 2014 "Farmers perceptions and adoption rates of innovations in integrated crop management technology for paddy rice," *Agriekonomika* **3** pp 1–10
[15] Abdul-Gafar A, Xu S, Yu W 2016 "Perceptions of rice farmers towards production constraints: case study of Niger State of Nigeria and Hainan of China," *J. Agric. Chem. Environ.* **05** pp 20–30
[16] Ashari, Sharifuddin J, Mohammed Z, Terano R 2018 "Paddy farmer's perception and factors influencing attitude and intention on adoption of organic rice farming," *Int. Food Res. J.* **25** pp S135–S145
[17] Fachrista IA and Hendayana R 2013 "Socio economic factor as determinant for adoption of integrated crops management (icm) of paddy at Bangka Belitung," *Inform. Pertan.* **22** pp 113–120
[18] Sumarno, Kartasasmita UG, Zaini Z, Hakim L 2009 "The gap of technology adoption and gap of rice yield," *Iptek Tanam. Pangan* **4** pp 116–130
[19] Kadar L, Siregar H, Kumala I 2016 "Factors influencing the adoption of superior white corn varieties in Grobogan Regency, Central Java," *Inform. Pertan.* **25** pp 215–220
[20] Abubakar H, Garba Y, Gana A, Jacob I 2019 "Factors influencing adoption of rice improved production practices by farmers in adopted villages ," *Adv. Plants Agric. Res.* **9** pp 183–189
[21] Muh A, S SB, Warda, Zainal A 2017 "Analysis of farmers' perceptions of the application of jajar legowo planting for lowland rice in Southeast Sulawesi," *J. Pengkaj. dan Pengemb. Teknol. Pertan* **20** pp 197–208
[22] Iswari K 2012 "Readiness of rice harvest and postharvest technology in reducing yield loss and

improving rice quality," *J. Litbang Pertan.* **31** pp 58-67
[23] Lesmayati S, Sutrisno, Rokhani H 2013 "The effect of delay time and threshing methods on yield and quality of local rice grain of Karang Dukuh Varieties in South Kalimantan," *J. Pengkaj. dan Pengemb. Teknol. Pertan.* **16** pp 159–169

Spatial Structure and Community Perception of Pejaten Village on Policy Determining Rural Industry Area in Tabanan, Bali

NU Vipriyanti[1], SPK Surata[1], IP Sujana[1], E Handayani[1] and DA Puspawati[1]

[1] Universitas Mahasaraswati Denpasar

Corresponding Author: mangtiutari@yahoo.com

Abstract. Tabanan Regency is well known as Bali's rice granary. However, one of the villages located in an agricultural area, Pejaten Village, is famous as its tile industry. The Tabanan Regency Government has designated Pejaten Village as a rural industrial area. The successful implementation of these policies needs to be supported by preparing human and natural resources that have adequate competence and creative power. This study aims to analyze (1) community knowledge and perception about rural industria area; (2) the spatial structure of industrial village based on the Hindu's concept of balance called *Tri Hita Karana*. The method of determining respondents was done by stratified random sampling. The strata are carried out based on the scale of the tile industry. Methods of data analysis using: (1) likert scala and (2) GIS. The results showed that (1) the community had quite positive attitudes and moderate perception categories regarding the policy of determining rural industrial area of Pejaten Village and (2) the spatial structure of common space had adapted from a rural space structure to an urban spatial structure. It is necessary to conduct a more intensive approach to increase moderate perception of the community.

1. Introduction

Pejaten is a unique village located in 18 km from Denpasar city, Bali. Administratively, this village located at Kediri District, Tabanan Regency, Bali Province and has 8 *banjar or small community resident*, namely *Banjar* Dukuh, Pejaten, Simpangan, Baleran, Dalem, Pamesan, Badung and Pangkung. The landscape is flat at 300 asl. It is a village with a mixed area, does not have a watershed although it is located between two rivers.

Pejaten is very well known not only in Bali but also abroad as the center for the pottery industry. The total village area is 217 Hectares (Ha) with the largest land use as rice fields and plantations. However, the agriculture sector does not contribute significantly to rural community and household income. Most of rural community income comes from the pottery industry sector [1]. Recently, Tabanan Regency has designated this village as rural industrial area. Although the uniqueness design and quality have recognized worldwide but pride in recognition alone is not sufficient enough because there is a responsibility inherent in that recognition. There was a negative impact arise from its industry. Effort to reduce this require community involvement in future rural planning

Many studies have been carried out in Pejaten Village, but the research was conducted partially and lack of community involvement so that the risk of negative environmental impact is relatively high and finally take effect on sustainability of this rural business. The involvement plays an important role and depend on their knowledge, behavior and attitude toward the policy. There are two problem formulations that require careful such as: in what level was the community's knowledge and attitude toward the rural industrial estate policy? and how was the spatial structure of rural industrial area based on the *Hindu's* concept of balance called *Tri Hita Karana*? Based on the problem, the aims of this research are as follow:

1. Assessing community perceptions and attitude of the designation of Pejaten Village as a rural Industrial Area Policy (KIP).
2. Analysing the spatial structure of rural industrial area based on the Hindu's concept of balance called *Tri Hita Karana.*

2. Theoretical Framework

2.1 *Rural Industrial Estates: Utilization of Space. As stated in* Government Regulation Number 24 of 2009 [2] that the aim of industrial estate development are to accelerate industrial growth in the regions, provide convenience for industrial activities, encourage industrial activities to be located in industrial areas, and increase industrial development efforts that are environmentally sound. In addition, it is also stated that an Industrial Estate is an area where industrial activities are concentrated which is equipped with supporting infrastructure and facilities which are developed and managed by an Industrial Estate Company that already has an Industrial Estate Business License. Companies in industrial areas are required to provide / control land, prepare land site plans, prepare technical area plans, Amdal, order industrial areas, develop land, and build infrastructure including installation of necessary installations / equipment.

The determination of rural industrial areas does provide an increase in community income but is also accompanied by direct influence on the surrounding environment (water and air pollution), or indirectly (flooding). Environmental pollution consists of air, water and soil pollution. Not every industry causes these three contaminants. This depends on the type of industry, for example the textile industry has a major contribution to water pollution. In addition, there are also industries that cause pollution in the form of smoke, odors, wastewater, chemical reactions and noise that reduce environmental quality. It has been stipulated that:

a. Industrial companies are required to carry out efforts to balance and conserve natural resources and prevent damage and pollution to the environment as a result of their industrial activities.
b. The government shall arrange and provide guidance in the form of guidance and counseling regarding the implementation of preventing damage and overcoming pollution of the environment due to industrial activities.

In assessing the feasibility of developing industrial estates, the variables of consideration can be grouped into 2 factors, namely internal and external factors.

2.1.1. Internal Factors. Internal factors are defined as factors that take into account the feasibility of industrial development from the point of view of industrial activities alone, as follows: (a) The amount of land demand; (b) The trend of growing industry types, (c) Various environmental problems that arise as a result of existing industrial growth; (d) Availability of infrastructure; (e) Availability of Human Resources (HR)

2.1.2. External Factors. Several external factors that are considered in the assessment of the feasibility of developing industrial estates are as follows:

a. Hinterland conditions
b. Competition with Other Regions
c. Strategic Location of the Macroeconomic System

2.2 Perceptions and Attitudes of the Community. According to [3], attitude is a person's closed response to certain stimuli or objects that already involve the opinion and emotional factors concerned (happy-displeased, agree-disagree, good-bad, and so on). Meanwhile, according to [4] opinion, attitude is a tendency that comes from within an individual to behave in certain patterns, towards an object due to his stance and feelings towards that object. Based on the opinion of [5] attitude is an organization of opinions, one's beliefs about objects or situations are relatively steady. which is accompanied by a certain feeling and provides a basis for the person to make a response or behave in a certain way he chooses.

2.2.1. Components of Attitude. The attitude that is shown by an individual towards the object has a structure consisting of several components. The components in the attitude structure, namely:

(1) The cognitive component, which is an individual's belief and understanding of an object through the process of seeing, hearing and feeling. The belief and understanding that is formed provides information and knowledge about these objects.
(2) The affective component, which is the component related to the subjective emotional problems of the individual towards something.
(3) Behavioral or conative components, namely the tendency to behave in an individual towards the object it faces.

2.2.2. Nature of Attitude. Attitudes can be positive and negative:

(1) A positive attitude, there is a tendency for action to approach, like, expect certain objects.
(2) Negative attitude, there is a tendency to stay away, avoid, hate, dislike certain objects.

Attitude measurement can be done directly or indirectly. It can be directly stated how the respondent's opinion or statement on an object. Indirectly, it can be done by using hypothetical questions and then asking the respondent's opinion through a questionnaire.

3. Research Method

This research used the mixed method in analyzing data related to spatial structure and community perceptions of the rural industrial area policy. Respondents were determined used stratified random sampling starting from determination of the location of the *banjar* and followed by the determination of respondents by proportional random sampling. Total amount of respondents are 250 people and they were asked to complete a questionnaire by choosing the most appropriate answer that best fits their opinion. Each option has a specific value. The value of the expected choice is 3 otherwise the answer that is not expected is 1. Perception analysis was carried out using a *likert* scale as follow: low, moderate, high level of perception and negative, neutral and positive attitude.

4. Result and Discussion

Pejaten Village tends to have the characteristics of an urban area compared to a village area indicated by the existence of various public facilities such as educational, economic and health facilities as an indicator component of the Human Development Index. The total population of Pejaten Village is 4,098 people, consisting of 2,099 men and 2,059 women including 1,190 families. The biggest source of livelihood for people is industry (entrepreneurship). Although most of the land use is for agriculture, there are only 58 farmers in Pejaten Village. This strengthens the designation of Pejaten Village as a Rural Industrial Area surrounded by agricultural villages.

Figure 1. Pejaten Village Map

Most of the population of productive age, but not all work because there are 1,460 people who are still in school or are looking for work. The number of poor families is 157 of the existing 1150 families. This condition certainly must be a concern so that the number of poor families can be reduced continuously. The structure of the population according to livelihoods shows that most of the population depends on the craftsmen sector (27.2%), the private sector at 12.1%, followed by farm laborers at 0.8% for productive age at 61.6%, while non earning 38.4%.

Economic sectors that are developing in addition to the industrial sector are the trade sector, the manufacturing sector, the tourism sector and the service sector such as carpenters, tailors, blacksmiths, odd shops and suppliers of building materials. In the service sector, what stands out is the growth of microfinance institutions / institutions in the form of cooperatives, Village Credit Institutions (LPD), and BUM-Desa as supporters of the village economy. Village markets have an important role in supporting increased community income.

The population structure according to the level of education shows the quality of human resources owned by Pejaten village. As much as 66 percent of the population is at the age of 7 to 18 years of primary education (primary and secondary school education, 9.2 percent have tertiary education, 16.2 percent are not yet in school.

Educational infrastructure and facilities in Pejaten Village are available in a fairly good category. Pejaten Village has 1 early childhood education institution (*PAUD*), 2 Kindergarten Education institutions. In addition, it also has 3 elementary schools and has 1 yunior high school. The social life of the Pejaten village community is based an agrarian society lifestyle. Familiarity among the members of the community is still visible in everyday life. For example, some residents still adhere to the principle of mutual cooperation, such as holding community service two days before the full moon, they will help each other in cleaning the entire area of the temple and the area of each banjar. Please help the people of Pejaten village not only when someone is happy but also if there are some among the residents who experience grief or *Ngaben*.

4.1. Community Perception and Attitude. The results showed that the knowledge of the people of Pejaten Village on the policy of rural industrial areas (KIP) is in the moderate category. Likewise, the attitude of the people belonging to the positive category. Table 1. Showed that 89.2 percent respondent have a moderate level of knowledge. According to demography characteristic, most of respondent are 84 percent responden are under 45 years old that make a good access to KIP information suitable with [6] that stated community understand about the importance's of environment sustainability for their livelihood.

Table 1. Respondent distribution based on knowledge of KIP Policy, 2021

No	Catagory	Freqwency	Percentage (%)
1	high	22	8.8
2	moderate	223	89.2
3	low	5	2.0
	Total	250	100

Most of respondent have a positive attitude toward KIP policy. The attitude assessment emphasize on the respondent act toward KIP program. The result indicated that collective action plays an important role in policy implementation at grassroot level [7],[8].

Table 2. Respondent Distribution based on Level of Attitude toward KIP Policy

No	Catagory	People Attitude	Percentage (%)
1	Positive	243	97.2
2	Neutral	7	2.8
3	Negative	0	0
	Jumlah	250	100

The younger generation is less interested in developing the industrial sector, they are more interested in the world of tourism. Table 3 showed that the average value of people's attitudes at the age of 41-55 years is the highest compared to younger ages. Likewise, men have a higher average score than women, indicating that there is a tendency for men to have a more positive attitude. Most of tile factory workers (85%) are migrants from Java and Lombok Island. The business scale was categorized as a small business scale or a household business that lack of modern management and technology.

Table 3. Demography Characteristic of Respondent, 2021

Demography Variable		Total Respondent	Attitude	
			Mean	Standar Deviasi
Age	17-29 year	136	57,36	2,16
	30-42 year	74	65,25	3,15
	43-55 year	40	70,50	0,71
Sex	Male	164	64,70	5,76
	Female	86	63,08	5,85

4.2. The spatial structure of rural industrial estate based on the Hindu's concept of balance called Tri Hita Karana.

There are 2 concepts in its management which are referred to as the DUWE concept. Natural resources in the village are divided into 3, namely private, communal and government resources. Each group has its own characteristics and is different for each type of soil and water resources. The Pejaten Village community manages the *duwe* for the greatest amount of community welfare with *perarem* (unwritten agreement) and *awig awig* (written agreement).

IOP Conf. Series: Earth and Environmental Science 1027 (2022) 012035 doi:10.1088/1755-1315/1027/1/012035

Figure 2. Hulu-Teben (Head and Tail) Concept

The concept of Ulu Teben (Head-Tail), as a form of division of village space/region which can be distinguished as follows [9]:

1 Upstream Zone (southwest of the settlement), the village spiritual activity facilities.
2 Intermediate Zone (middle part), is a residential area with residential buildings and public service facilities such as community halls, *Dadya*/family temples, schools, and fields.
3 *Teben* (tail) Zone, is the lowest part which is used as a cemetery (*setra* area, *Prajapat*i temple and plantation).

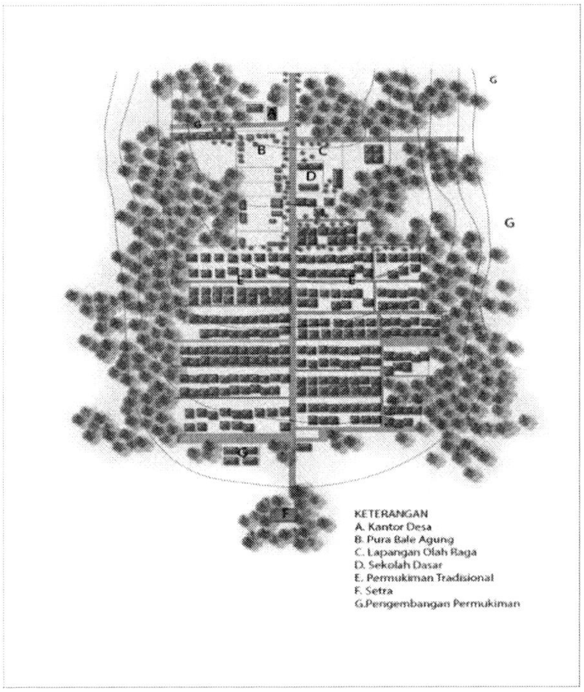

Figure 3. Spatial Map Village, 2021

The main village road that extends from north to south is a confirmation of the linear pattern of the village which functions as the main circulation route. The main village road is the core of the Spatial Pattern which functions as a center for public space orientation during the implementation of traditional ceremonies (ritual ceremonies) [10]. Village streets and alleys are the orientation direction of each yard which is perpendicular and intersects symmetrically. Road tracks are formed from environmental patterns that are adapted to village conditions and the height and low of the site. The yard functions more as a place to live to hold ceremonies and relate to the family, while in an effort to fulfill their daily needs they cultivate gardens outside the village (basic residence) [11], [12], [13]. Limited land and the desire to be close to the main road led to housing development towards *luanan/ulu*, head of spatial structure, but still no building in the *luanan* temple area, before head area. The residential area is in the village core zone which is in the area around the village's main road which stretches obliquely from the southwest to the northeast. The outermost area of the village is filled with residents' plantation.

5. Conclusion. There are two conclusions of this research as follow:

1 The community had quite positive attitudes and moderate perception categories regarding the policy of determining rural industrial area of Pejaten Village.
2 The spatial structure of common space had adapted from a rural space structure to an urban spatial structure.

References

[1] Statistik K. Monography Pejaten 2020 2019;2019(37):1–16. Available from: http://sippa.ciptakarya.pu.go.id/sippa_online/ws_file/dokumen/rpi2jm/DOCRPIJM_1536548764Bab_2-Profil_KabKota_(tbn)_2018.pdf
[2] Government Regulation No 24, year 2009 1390; pp 99–117
[3] Notoatmodjo S. Promosi Kesehatan & Ilmu Perilaku. Jakarta: Rineka Cipta. 2012.
[4] Koentjaraningrat 2013 *Pengantar Ilmu Antropologi*. pp 43
[5] Walgito B 2013 Pengantar Psikologi Umum. Rajawali Perss. pp 252
[6] Nanlohy H, Bambang AN, Ambaryanto A, and Hutabarat S 2014 Analisis Persepsi Masyarakat terhadap Pengelolaan Kawasan Mangrove Teluk Kotania. *Jurnal Wilayah dan Lingkungan*. **2** 1:89
[7] Vipriyanti, N.U, Emy Handayani, Yulianti Sapanca, and Dewa Ayu Puspawatu. Model Penguatan Subak Di Era New Normal: Analisis Peran Collective Action Pada Penerapan Sistem Terintegrasi Itik Padi Di Tabanan-Bali: http://ojs.uho.ac.id/index.php/JIMDP/article/view/14166
[8] Surata SPK, and Vipriyanti NU 2018 The subak cultural landscape as environmental education: Knowledge, attitudes, and experiences of Balinese teachers, student teachers, and students. *Journal Environment Education*. **49** 1 pp 59–70
[9] Suyasa, I. W. A, I Komang Gede Santhyasa, and Wahyudi Arimbawa 2019 Tata Ruang Pemukiman Sentra Industri Genteng di Desa pejaten Tabanan. *Jurnal SPACE*. **1** 2 pp 28-34. Fakultas Teknik UNHI.
[10] Parashita, N.LL, and I Made Adi Kampana 2017 Perkembangan Desa Wisata Pangsan sebagai Destinasi Pariwisata di Kecamatan Petang Badung. *Jurnal Destinasi Pariwisata* **5** 1 pp 34-39. p-ISSN: 2338-8811, e-ISSN: 2548-8937. Universitas Udayana.
[11] Kasuma, I Putu Agus Wira 2009 Persepsi Masyakarat Adat sebagai Dasar Perumusan Konsep Tata Ruang Desa Adat Penglipuran, Bali. Tugas Akhir Program Studi Perencanaan Wilayah dan Kota ITS. Surabaya.
[12] Suyasa, I Nyoman 2006 Strategi Pelestarian Pusat Kota Bangli Berdasarkan Prinsip-Prinsip Ruang Tradisional Bali. Tugas Akhir Program Studi Perencanaan Wilayah dan Kota ITS, Surabaya.

[13] Kasuma, I Putu Agus Wira, and Iwan Suprijanto 2012 Karakteristik Ruang Tradisional pada Desa Adta Penglipuran, Bali. *Jurnal Permukiman* **7** 1 (2012) pp 40-50. p-ISSN: 1907-4352

The impact of transplants machine on acceleration of innovations adoption in rice cultivation in Lampung Province

F Y Adriyani [1] **and K K Rangga** [2]

[1] The Assessment Institute of Agricultural Technology, Indonesia
[2] University of Lampung, Indonesia

Corresponding Author: fyadriyani@gmail.com (Fauziah Yulia Adriyani)

Abstract. The use of machinery in agriculture is mainly to solve the problem with the scarcity of labor. In fact, this innovation is not automatically accepted by farmers, especially in areas with large labor force. This study aims to determine the impact of transplants machines on rice cultivation. This paper was written as literature study and based on the author's experience since 2014-2020 as extension worker in rice cultivation assessment in Lampung province. The result shows that the use of rice transplants machine has a particularly positive impact on increasing the application of innovation such as improved nursery techniques, use of young seedlings, and use of legowo planting systems. In addition, the use of transplants machines also causes changes in society and has a negative impact such as the emergence of competition between farm workers and transplants machine service providers.

Keywords: Technology; Transplanter; Paddy; Legowo; Adoption.

1. Introduction

The agricultural sector in general and in particular the conventional management of plant cultivation requires a large workforce, starting from land processing to harvesting and post-harvesting. Even there is increasing in population, seems that it will not cause problems in employment. Agricultural sector always faces problems with the need for labor. The amount of labor in land preparation and planting activity in rice cultivation is higher (370.7 hours) than plant protection (194.8 hours) and harvesting (109.2 hours) activity [1]. This fact could impact to the time for planting in rice cultivation that is mainly depend on planting workers.

The problem of labor scarcity in the agricultural sector is likely to increase with the development of science and changes in mindset. It is undeniable that the interest of the younger generation to work as farmers is decreasing and there is a tendency to choose to work as factory workers or private companies.

To overcome this, the use of machines is an alternative solution to the problem. Machine could replace human labor. However, even farmers know that the use of machine could decrease amount of labor but farmers doesn't directly move to machine in order to replace human labor, for example only a few farmers use transplanting machine in rice cultivation [1]. Furthermore, technology does not necessarily solve all problems in the community. In fact, the existence of technology might be the cause of some problems for example is the use of rice planting machines (transplanters). The use of transplanter has a positive impact, especially in increasing the application of innovation and productivity which ends in increasing welfare. In addition, the use of transplanter can also bring about changes in society that can have a negative impact if they are not wise to respond.

The following is a discussion of how the application of an innovation has an impact on the pattern of people's lives.

2. Materials and methods

This paper was written as literature study and based on the author's experience since 2014-2020 as extension worker in rice cultivation assessment in Central Lampung Regency (Seputih raman, Seputih Banyak and Seputih Agung District), East Lampung Regency (Raman Utara and Purbolinggo district), South Lampung Regency (Candipuro District), Tanggamus Regency (Gisting District) and Pringsewu Regency (Gading Rejo and Ambarawa District) Lampung province.

3. The impact of rice transplanting machine (Transplanter)

3.1. Increase the adoption of other innovation.

Innovation has some characteristics that influence someone to adopt it and one of characteristics is complexity [2]. Reduction in production cost is encouraging the adoption of innovation in rice cultivation meanwhile complexity and compatibility is obstructed factor in adoption [3]. Complexity is related to difficulty. Farmers usually avoid innovation that is difficult to use and understand. However, there are some innovation that could make user easier to use the other innovation.

3.1.1. Nursery. Nursery is main activity for most farmers in rice cultivation. However, many farmers don't pay attention for how doing good nursery. Many research showed that density and age of seed in nursery are influenced the growth and yield. Low density nursery could give a good place for seed with low competition for nutrient and light so it will be vigorous [4]. Productive tillers are increasing when we use vigorous seed and width nursery area (low density seedling nursery) through lower competition among seed [5]. However, farmers are reluctant to carry out nursery with low density. Low density in nursery means more area for nursery. Based on the results of interviews with farmers, farmers' hesitancy is as a result of their observations and experiences that the rice that is planting in ex field nursery will be grow poorly. So, the wider the nursery, the more land that is not good for planting. In addition, another cause is the cost of making wider nursery will be increase (increasing production costs). Through the use of a transplanter, farmers could solve their problems in nursery width. The use of a transplanter requires a special nursery model that is using a tray called the dapog system. In this nursery, the nursery can be done in the yard and doesn't decrease quality of rice field.

Figure 1. Nursery system in implementation of rice transplanting machine

3.1.2. Young seedlings. The age of rice seedlings at transplanting has a direct effect to plant growth and yield [5]. More productive tillers and higher yield unit-1 area will be obtained by planting healthy and vigorous seedlings, because mortality rate after transplanting will be decreased [6]. Furthermore, based on several studies that have been carried out, the advantages of planting young seedlings are that plants can grow better with the number of tillers tend to be more. Other study showed interaction between seedling density in nursery, transplanting age of seed, and variety. High seedling density should not transplant at older age because the yield will be reduced drastically and the higher yield could be reached if farmers transplanted young seedling from low seedling density nursery [7]. The effect of seedlings age will be different by varieties. In planting Dendang varieties, 14 DAS seedlings tended to show better in shoot dry weight [8]. Other result showed that in planting local varieties, higher productivity will be obtained when use 21 DAS seedlings [9]. However, the advantages of using these innovations are not necessarily able to increase the application of young seedlings by farmers. Problems encountered in the use of young seedlings include (1) limited availability of planting labor, (2) most planting labor reject young seedling because its size is smaller than old seedling, (3) some area is not suitable for young seedling because the presence of snail pests and rainfed with uncertainty rainfall [7]. The use of a transplanter can anticipate obstacles in the application of young seedlings because the transplanter can substitute and reduce the amount of labor. Farmers won't be depending on planting labor and could use young seedlings. Additionally, the use of transplanter need some requirement for seed such as seed height so that will be force farmers to use young seedling.

3.1.3. Implementation of Jajar Legowo Planting System. The jajar legowo planting system is one of the technological engineering that has become a government program to be developed in order to increase rice productivity to support the P2BN Program. The jajar legowo 2:1 and 4:1 rice planting systems have been introduced in Central Lampung Regency since 2000 and have been shown to increase rice productivity and income by 15-20% compared to the conventional planting system. However, until now the development of the adoption of the row legowo planting system every year is still less than 10% of the total existing planting area. Some research showed that farmers are doubtful in applying jajar legowo planting system [10], the knowledge about jajar legowo doesn't associated with implementation of this innovation [11]. Based on the interview with farmers, the obstructing factor in adoption of Jajar legowo planting system is the jajar legowo planting system still used human labor. The population in jajar legowo planting system increase about 30% and as a consequence of that, labor need extra time and energy to planting with this system. Moreover, even though the planting costs are increased, the planting labor tends to still choose the conventional planting system. The adoption rate for jajar legowo planting system is slow, farmers has difficulty to implement this innovation because of a few labor want and able to use this planting system and planting legowo jajar could spend higher costs [10]. In addition, the jajar legowo planting system is influenced by the complexity [12]. The scarcity of planting labor is main influence the implementation of jejer legowo planting system. With these considerations in mind, the Agency for Agricultural Research and Development has introduced a jajar legowo rice planting machine, named Indo Jarwo Transplanter 2:1. The operation of the transplanter is relatively easy and simple and could be operated by 2 - 3 people with the division of tasks for one person as an operator, one provider and transporter of seeds and one embroiderer for empty clumps moreover it can be completed within 5-6 hours ha^{-1} with an operational cost of Rp.500,000 [13].

Figure 2. The using of rice transplanting machine and "jajar legowo" implementation

3.2. Farmers Income

The presence of the rice planting system with the jajar legowo planting system is a solution from the labor aspect, reducing work costs, and accelerating the work process. This planting machine is very prospective to be socialized in order to increase the planting area of jajar legowo rice which is an innovation that could increase rice productivity.

In areas with limited manpower, especially planting workers, the presence of rice planting machines (transplanters) is welcomed by farmers and other technology users, with the hope of solving the problem of scarcity of planting labor. Transplanters in Jarwo Super cultivation only require 2 - 3 workers with a planting time of 5 - 6 hours /ha (Kiswanto, 2014) it means a transplanter could reach 1 ha planting area per day this is much more efficient in terms the amount of labor needed in conventional planting. Although the planting cost by using a transplanter is the same as conventional (human labor) that is around Rp. 600,000-800,000, but for planting with human labor there is an additional labor cost to make lines and divide the seeds which is around Rp. 200,000 and an additional cost for removing the seeds from nursery bed which is around Rp. 400,000. These additional costs showed that using a transplanter has lower cost production for planting that is about Rp, 600,000. The study showed that the application of Jarwo Transplanter as a rice-planting machine could increase 23.64% of productivity and 39.40% of farmers' income comparing with the conventional method of planting. Moreover, the application of Jarwo Transplanter as rice planting machine more efficient with the R/C ratio of 2.71 and BEP Rp.1.550/kg, meanwhile the conventional planting has the R/C Ratio 2.27 and BEP 1,850/kg. The Marginal B/C Ratio 14.90 indicated the rice planting by using Jarwo Transplanter machine had opportunities to be developed in a wider area [14].

4. Strengths and weaknesses in using transplanting machines

Another study showed that the efficiency of the use of labor with the jarwo transplanter machine in tidal land was 73.63% by reducing planting costs by Rp. 1,472,500 due to manual labor savings of 37 HOK/ha [15]. Rice productivity with the use of a transplanter is higher than the manual method, which is 6.80 tons/ha compared to 6.41 tons/ha because the distance and depth of planting using a transplanter machine can be more uniform so that plant growth can be more optimal [16].

Figure 3. Comparison labor in conventional rice planting and using rice transplanting machine

In addition to advantages, there are weaknesses in the use of transplanters that is related to its adoption. The adoption of rice transplanter in Lampung Province is still low (17,07%) [17]. The reason they are reluctant to use transplanter planting tools is because it could change the existing cultural system (shifting the planting force). The use of transplanters in areas with an adequate planting labor can create new problems. The problem is mainly because of the competition between planting workers and the use of transplanters which in turn has an impact on the income of the farming family from planting workers. Actually, the income of the farming family is not decrease. This is based on the fact that the planting labor come from farming families. The reduction in family income from planting labor can be substituted by an increase in farm income margins on account of the reduction in planting costs. In addition, planting workers who lose their jobs can be transferred to workers for making nurseries for the dapog system. In this case, the planting laborers, who are mostly women, still have other income with jobs that are physically easier.

Another problem face to adoption of transplanter is the price of transplanter is relatively expensive so it is not affordable for individual farmers furthermore an analysis showed that benefit reduction or increase in cost of transplanters can be a reason for entrepreneur to be away from adopting transplanters because its profit is not as their expectation [18]. Rice transplanter machine technology also need some suitable requirement both a technical side or land area as well as socio-cultural or customs [19]. The innovation which has high compatibility will be adopt faster by farmers [20].

The rice planting machine (transplanter) is an innovative technology, in its application it must pay attention to the social, economic and cultural conditions of local farmers, to avoid the eviction of farmer workers. Therefore, the development of the Indo Jarwo Transplanter machine is currently concentrated in areas where the workforce is limited.

5. Conclusion

In conclusion, the use of rice transplants machine has a particularly positive impact on increasing the application of innovation such as improved nursery techniques, use of young seedlings, and use of legowo planting systems. The advantages of the transplanter include (1) it can speed up planting time, reduce planting costs, (2) overcome the scarcity of planting labor, (3) use only 2 workers, (4) reduce production costs and be more efficient, (5) easy to operate, (6) uniform spacing and depth of planting so that plant growth can be optimal and uniform. In addition, the use of transplants machines also causes changes in society and has a negative impact such as the emergence of competition between farm workers and transplants machine service providers.

References

[1] Nikkhah A, Kougir-Chegini Z, Kosari-Moghadam A and Nikkhah F 2017 Musculoskeletal Disorders, Energy Use, and Costs of Human Labor for Paddy Cultivation in Northern Iran *International Journal of Agricultural Management and Development (IJAMAD)* **7**(4) 439-46

[2] Rogers E M 1962 Diffusion of Innovations *The Free Press of Glencoe New York* p 367

[3] Zaman N B K, Ali J and Othman Z 2019 Faktor Penerimaan Teknik Penanaman Padi Secara Sistem Intensifikasi Padi *Jurnal Dunia Perniagaan* **1**(2) 11-20

[4] Paul S K, Akter M, Sarkar S K and Sarkar M A R 2018 Effect of nursery seeding density, age of seedling and number of seedlings hill on the performance of short duration transplanted Aus rice (cv. Parija) *Bangladesh Agril Univ* **16**(2) 215–220

[5] Adhikari B B, B Mehera, and Haefele S 2013 Impact of Rice Nursery Nutrient Management, Seeding Density and Seedling Age on Yield and Yield Attributes *American Journal of Plant Sciences,* **4**(12) 146-55

[6] Sarwa N, Maqsood M, Wajid S A, and Anwar-ul-Haq M 2011 Impact of Nursery Seeding Density, Nitrogen, and Seedling Age on Yield and Yield Attributes of Fine Rice *Chilean Journal of Agricultural Research* **71**(3) 343-49

[7] Subedi R 2013 Nursery Management Influences Yield and Yield Attributes of Rainfed Lowland Rice *Journal of Sustainable Society* **2**(4) pp 86-91

[8] Afrinda R and Kurniasih B 2021 Effects of seedling age on the growth and yield of two rice (Oryza sativa L.) varieties transplanted in Saline Coastal Area of Baros, Yogyakarta. *Ilmu Pertanian (Agricultural Science)* **6**(1) pp 38–46

[9] Kurniasari I and Proyoga A 2018 Pengaruh Umur Transplantasi Benih Terhadap Pertumbuhan dan Produktivitas Varietas Lokal Jenis Padi Merah (Oryza Sativa L) *Agrotech Research Journal* **2**(1) 11-5

[10] Vitriawati, Sapja A, and Eny L 2019 The Relationship of Farmers Socio-Economic Conditions, The Role of Chairman of Groups and Authors With The Innovation Level of Super Legowo Rice Cultivation *Russian Journal of Agricultural and Socio-Economic Sciences* **1**(85) 121-27

[11] Sunandar B, Hapsari H and Sulistyowati L 2020 Tingkat Adopsi Tanam Jajar Legowo 2:1 pada Petani Padi di Kabupaten Purwakarta *Jurnal Pemikiran Masyarakat Ilmiah Berwawasan Agribisnis.* **6**(2) 500-18.

[12] Zuriani and Martina 2016 Analisis Adopsi Inovasi Penyuluhan Pertanian di Kabupaten Aceh Utara dalam Mendukung Kedaulatan Pangan *AGRISEP* **15**(2) 143-50

[13] Kiswanto, Adriyani, F Y and Wijayanto B 2014 Pemasyarakatan Inovasi Indo Jarwo Transplanter dalam Mendukung Program Peningkatan Produksi Beras Nasional (P2BN) di Kabupaten Lampung Tengah *Laporan akhir Balai Pengkajian Teknlogi Pertanian Lampung*

[14] Kiswanto 2015 Pemasyarakatan dan Peluang Adopsi Inovasi Jarwo Transplanter Mendukung Swasembada Beras di Lampung *Laporan akhir Balai Pengkajian Teknlogi Pertanian Lampung*

[15] Umar S and Pangaribuan S 2017 Evaluasi Penggunaan Mesin Tanam Bibit Padi (Rice Transplanter) Sistem Jajar Legowo di Lahan Pasang Surut *Jurnal Teknik Pertanian Lampung* **6**(2) 105–14.

[16] Sahara D, Kushartanti E and Suhendrata T 2013 Kinerja Usahatani Padi dengan Mesin Transplanter dalam Rangka Efisiensi Tenaga Kerja SEPA **10**(1) pp 55-62

[17] Adriyani F Y and Kiswanto 2018 Aksi Kolektif Sebagai Kunci Percepatan Adopsi Inovasi Jajar Legowo Super: Kasus Di Provinsi Lampung *Bunga rampai: Diseminasi Inovasi dan Adopsi Teknologi dalam Perspektif Pengembangan Pertanian Perdesaan* 147-59

[18] Acharya P, Regmi P P, Gauchan D, Bahadur KC D and Gopal Bahadur KC 2020 Benefit Cost Analysis of Adoption of Small Farm Machineries for Rice Cultivation in Nepal *International Journal of Applied Science and Biotechnology* **8**(4) 448-53

[19] Fujiarta P I, Sarjana I D G R, Putra I G S A 2019 Faktor yang Berkaitan dengan Tahapan Adopsi Petani terhadap Teknologi Mesin Rice Transplanter (Kasus pada Enam Subak di Kabupaten

IOP Conf. Series: Earth and Environmental Science 1027 (2022) 012036 doi:10.1088/1755-1315/1027/1/012036

Tabanan) *E-Jurnal Agribisnis dan Agrowisata* **8**(1) pp 29-38
[20] Wijayanto B, Kiswanto, Adriyani F Y and Santri N 2018 Diseminasi Inovasi Teknologi Jarwo Super: Kasus di Lampung *Bunga rampai: Diseminasi Inovasi dan Adopsi Teknologi dalam Perspektif Pengembangan Pertanian Perdesaan* pp 75-94

Effect of Consumer Skepticism on Consumer Intention in Purchasing Green Product

Rinaldi Bursan[1], DriyaWiryawan[1], Habibbullah Jimad[1], Indah Listiana[1*], Maya Riantini[1], Helvi Yanfika[1], RAD Widyastuti[1], Abdul Mutolib[2] and Dina Arini Adipathy[3]

[1]University of Lampung, Indonesia
[2]University of Siliwangi, Indonesia
[3]Magister Management University of Lampung, Indonesia

* Corresponding Author: indahlistiana1@gmail.com

Abstract. The image of environmentally friendly products is often seen as bad, consumer doubt that environmentally friendly products are really made with environmentally friendly raw materials. This attitude is known as consumer skepticism (doubt) towards products with an environmentally friendly label. Consumer skepticism occurs because product advertisements contain confusing information as well as consumer insecurity about the raw material for products that the company claims are environmentally friendly. The aims of this study aims to determine the effect of consumer skepticism on green purchase intention. This study uses a theoretical approach to consumer behavior control and attitude, perceived value and environmental consciousness associated with consumer intentions to buy green products in Bandar Lampung. The study was conducted on April until June 2020. The sample in this research were consumers who had been buying green product. The research sample was calculated using the Lemeshow method, with a margin of error of 5%, the minimum number of samples was 236 samples. The analysis tool used is regression model. The results of this study are the reduction of consumer skepticism about the intention to buy due to the use of social media to promote green products. Meanwhile, the influence of consumer value perceptions and environmental awareness will increase on the purchase intention of green products. The implication of this research is that companies must pay attention to communication about green products that are produced so as to reduce consumer doubt. Another impact if consumers consume green products will lead to environmentally friendly perceptions and attitudes.

1. Introduction

Environmentally friendly products are classified as new products made to overcome the existing problem of global warming, as well as to anticipate further environmental damage [1]. These products usually have the 3R (reused, reduce and recycle) mark on the product packaging and can be clearly identified by consumers. As a form of innovation, this environmentally friendly product takes a long time to socialize its benefits and be adopted by the wider community [2] [3]. The target of this eco-friendly product covers all consumer groups. Various studies on environmentally friendly products are associated with consumer attitudes resulting in consumer skepticism, because they doubt the content of the raw materials so they doubt consumers to make purchases [1] [4] [5] [6].

Consumers will associate environmentally friendly products with the company's awareness of environmental sustainability. Consumers give positive responses to companies that produce environmentally friendly products so that they have the intention to make purchases [7] [8] [9] [10]. Companies that maintain environmental sustainability will lead to positive perceptions of consumers who have a view of consuming environmentally friendly products. Consumers' perceptions of the value of environmentally friendly products will cause consumers to make purchase intentions for the products produced by the company [2] [6] [11] [12] [13].

Companies that produce environmentally friendly products will be appreciated by consumers who have the awareness to consume environmentally friendly products. This product has more value in terms of self-esteem, because someone who uses the product is considered to adhere to environmentally conscious behavior that is better than others. Thus, consumers who use environmentally friendly products will feel an increase in self-confidence [14] [15] [16].

The effectiveness of this environmentally friendly product begins with consumer awareness of the existence and function of the product. Consumer awareness of a product is usually used as an indicator of the success of the product's performance [6]. Because, after awareness is owned by consumers, then consumers will try the product until they finally decide to become regular consumers or not [17].

2. Theoretical Background

Skepticism is the attitude of a person's doubts in understanding a certain situation or event. This attitude of doubt arises due to the lack of information that a person gets about an object [18]. Meanwhile, according to [19], it is said that skeptics believe that all claims of any truth seem vague or doubtful, those who skeptics advocate hanging decisions. This is reinforced by the opinion of that consumer doubt arises because information about a product is incomplete so that the communication made by the company is not well received by the company [20].

Research conducted by Awamate and Deb (2021) results in consumer skepticism in a product that will reduce consumer buying intentions on organic food products. The attitude of consumer skepticism arises because of doubts about raw materials, incomplete information on packaging and communications made by the company confuse consumers. Another cause of the decline in consumer purchase intentions for environmentally friendly products due to skepticism is that the price of environmentally friendly products is perceived to be more expensive than other products and the benefits obtained both to health and to the environment have not been communicated properly by the company [1] [6]. Based on the opinion of Awamate and Deb [4], and Chen and Chang [6], the first hypothesis of this study is formulated as follows:

H1: consumer skepticism has a negative effect on consumer purchase intentions

Consumers who buy environmentally friendly products are based on consumer awareness of the benefits of the product not only on individual consumers but also on environmental preservation. Consumer awareness of the environment causes consumers to consume these products, because they can be reused and can reduce waste and can decompose quickly [10] An environmentally conscious attitude will provide knowledge to consumers, and form attitudes and environmental awareness so that it causes purchase intentions to make purchasing behavior for environmentally friendly products, although sometimes consumers will pay more. Another thing that consumers face is that the variety of environmentally friendly products is still limited [7].

Research conducted by Bhaduri and Copeland [8] describes that consumers who have environmental awareness will buy products that are also environmentally friendly, although at a higher price. Silva [16] resulted in a higher consumer awareness attitude towards the environment, the higher the purchase intention for the product. So that the second hypothesis is formulated as follows:

H2: environmental awareness has a positive effect on consumer purchase intentions

An environmentally conscious attitude will shape consumers' perceptions of the value of environmentally friendly products. Consumers will form congruence between perceived value and environmentally friendly products. The stronger the consumer's congruence with the product, the higher the purchase intention [6] [12] The consumer's perception of the value of environmentally friendly products is obtained after consumers evaluate the product or feel the benefits.

This perception will foster consumer purchase intentions for environmentally friendly products. Mimouni and Parguel [21] resulted in the perception of consumer value for environmentally friendly products causing consumers to seek information about the product, thereby generating purchase intentions. This is reinforced by research conducted by Sindhuri [22] which states that the purchase intention of environmentally friendly products is carried out by consumers who have a positive perception of the benefits of these products. Thus, the third hypothesis is formulated as follows:

H3: perceived value of environmentally friendly products has a positive effect on consumer purchase intentions

3. Methodology

Type of research is a survey and is carried out April to June 2020. This research was conducted in Bandar Lampung City by distributing questionnaires to consumers who made purchases of environmentally friendly products. The questionnaire given is given the initial question whether the respondent has ever made a purchase of environmentally friendly products. This question is intended so that consumers are able to provide answers based on their knowledge of environmentally friendly products. The distribution of the questionnaire was carried out online based on the consideration of the Covid-19 pandemic which was still ongoing at the time this research was conducted.

This research is descriptive of causality in order to know the relationship between the independent variable and the dependent variable. The population of this study is unknown, so the number of samples used in this study refers to the opinion of the sample using the Lemeshow method [23], with a margin of error of 5%, obtained a minimum sample size of 236 samples. Data analysis used descriptive analysis to determine the characteristics of research respondents and to find out the demographic background of the selected sample. Multiple linear regression was used to determine the effect of the independent variable on the dependent variable

This study has 3 independent variables, namely skepticism, environmental, consciousness and green perceived value and 1 dependent variable, namely purchase intention. The indicators used for both independent and bound variables refer to Chen and Chang [6]. All research indicators use a 5-point Likert scale. All indicators have been tested for validity using factor analysis and reliability of the indicator is calculated by the Cronbach alpha method [23].

4. Result and Discussion

4. 1. Profile of Responden

The majority of consumers who consume environmentally friendly products are the young age group between 17 – 26 years as much as 42% and the age group 27 – 36% as much as 37%. If you pay attention to the age group, it seems that most of the millennial groups are characterized by easy access to information via smartphones and care for the environment. Profession that consume environmentally friendly products are entrepreneurs as much as 31%, private employees as much as 25% and civil servants as much as 16%. If you pay attention to the respondent's work, it can be seen that he has a good profession so that he has the ability to make purchases of environmentally friendly products. S1 education dominates as much as 36%, then S2 as much as 25% and S3 as much as 11%. Consumers have a good level making it easier to find information and evaluate environmentally friendly products.

4. 2. Hypotehesisi Testing

All hypotheses were tested by comparing the p value with the value of $\alpha = 5\%$. The results of hypothesis testing are in Table 1 below:

Tabel 1. Hypothesis Testing

	β	t-value	p-value	R	R^2	F-value	F-sig
Consumer Skepticism	-0.209	-2.714	0.011	0.670	0.450	26.129	0.000
Environmental Consciousness	0.526	4.216	0.005				
Green Perceived Value	0.157	2.661	0.019				

Note : significant at 0.05 level

The results of the regression calculation show that the dominance of the environmentally conscious attitude variable on consumers' purchase intentions for environmentally friendly products is significantly 0.526. If the results of the regression coefficients are compared with the tabulated data, there is consistency in the answers. Consumers qualitatively state that consumers consume environmentally friendly products because of the company's strong commitment to preserving the environment and consumers realize that environmentally friendly products will preserve the environment.

The results of the regression calculations show that the dominance of the skepticism/doubtful attitude variable towards consumers' purchase intentions is only -0.197. This negative value implies that companies producing environmentally friendly products are able to suppress consumer skepticism or doubt, so that consumers' purchase intentions will increase. The study findings provide empirical evidence that its effect on customer intention to buy environmentally friendly products is strongly influenced by scepticism. Consumers who have a high level of skepticism towards environmentally friendly products, they tend to have lower attention and lower knowledge about environmental issues. This is due to the reluctance of consumers to seek information and knowledge of the benefits of environmentally friendly products.

The variable that has the smallest effect is the perception variable on the value of environmental awareness, which is only 0.157. The consumer's perception of the value of environmentally friendly products is formed from the environmental campaigns carried out by the company and its benefits to society. This means that the campaign carried out by the company has not been fully effective in shaping consumers' perceptions of the value of environmentally friendly products.

The results of this study indicate that consumer skepticism, environmental awareness and consumer value perceptions have a significant influence on purchase intention. This study also shows the same results as the research of Chen and Chang [6] and Awate and Deb [4], the difference is only in the value of the variable effect coefficient. This is presumably due to differences in respondent characteristics, differences in the number of respondents and differences in analytical tools.

5. Conclusion

The attitude of consumer scepticism towards environmentally friendly products is due to the lack of information about the benefits of these products on the environment. This causes consumers to have doubts about whether the product really uses environmentally friendly materials so that it can reduce the negative impact on the environment (reuse, reduce, recycle). Meanwhile, an environmentally conscious attitude will form congruence with environmentally friendly products. The more congruent causes an increase in consumer purchase intentions for environmentally friendly products such as food and beverages products. Meanwhile, consumers' perceptions of value will be formed if doubts about environmentally friendly products can be eliminated, so that consumers will change their purchase intentions into purchasing decisions.

Acknowledgment
This research was conducted by a lecturer at the Faculty of Economics and Business, Department of Management and the Faculty of Agriculture, University of Lampung. We appreciate those who have facilitated this research. In addition to that, we also thank the reviewers for their corrections and suggestions.

References
[1] Bae, M. 2018 Overcoming Skepticism toward Cause-Related Marketing Claims: The Role of Consumers' Attributions and a Temporary State of Skepticism *Journal of Consumer Marketing* **35** 2 pp 194–207 doi:10.1108/JCM-06-2016-1827
[2] Bursan, Rinaldi and Listiana, Indah and Ardeno, R, Bangsawan, Satria and Jimad, Habibullah and Mutolib, Abdul 2020 Consumer attitude toward using eco-friendly plastic bags: A green marketing approach. In: ULICoSTE 2020, 18 – 19 November 2020, Bandar Lampung.
[3] Listiana, Indah and Sumardjo, Sumardjo and Dwi, Sadono and Prabowo, Tjitropronoto 2018 Affecting Factors the Capacity of Freelance Extension Agents and Its Impacts on Farmers *International Journal of Business and Social Science International Journal of Business and Social Science* **9** 1 pp 1-10 ISSN 2219-1993
[4] Amawate, V and M. Deb 2021 Antecedents and consequences of consumer skepticism toward cause-related marketing: Gender as moderator and attitude as mediator *Journal of Marketing.*
[5] Anuar, M. M., K. Omar, and O. Mohamad. 2013 Does Skepticism Influence Consumers Intention to Purchase Cause-Related Products? *International Journal of Business and Social Science* **4** 5 pp 94-98
[6] Chen, Y.-S. and Chang, C.-H. 2012 Enhance green purchase intentions. The roles of green perceived value, green perceived risk, and green trust *Management Decision* **50** 3 pp 502-520
[7] Barbarossa, C. and Pastore, A. 2015 Why environmentally conscious consumers do not purchase green products: A cognitive mapping approach *Qualitative Market Research* **18** 2 pp 188-209 https://doi.org/10.1108/QMR-06-2012-0030
[8] Bhaduri, G. and Copeland, L. 2021 Going green? How skepticism and information transparency influence consumers' brand evaluations for familiar and unfamiliar brands *Journal of Fashion Marketing and Management* **25** 1 pp 80-98. https://doi.org/10.1108/JFMM-08-2019-0175
[9] Listiana, Indah and Hudoyo, Agus and Prayitno, Rio Tedy and Mutolib, Abdul and Yanfika, Helvi and Rahmat, Ali 2020 Adoption Level of Environmentally Friendly Paddy Cultivated Innovation in Pringsewu District, Lampung Province, Indonesia. *Journal of Physics: Conference Series* 2020 **1467** pp 1-8 ISSN Print: 1742-6588 Online: 1467-012025
[10] Copeland, L. and Bhaduri, G. 2020 Consumer relationship with pro-environmental apparel brands: effect of knowledge, skepticism and brand familiarity *Journal of Product & Brand Management* **29** 1 pp 1-14 https://doi.org/10.1108/JPBM-03-2018-1794
[11] Albayrak, T., Aksoy, Ş. and Caber, M. 2013 The effect of environmental concern and scepticism on green purchase behavior *Marketing Intelligence & Planning* **31** 1 pp 27-39 https://doi.org/10.1108/02634501311292902
[12] Shin, S. and Ki, E.-J. 2019 The effects of congruency of environmental issue and product category and green reputation on consumer responses toward green advertising *Management Decision* **57** 3 pp 606-620 https://doi.org/10.1108/MD-01-2017-0043
[13] Ulusoy, E. and Barretta, P.G. 2016 How green are you, really? Consumers' skepticism toward brands with green claims, *Journal of Global Responsibility* **7** 1 pp 72-83. https://doi.org/10.1108/JGR-11-2015-0021

[14] Golob, U., Kos Koklic, M., Podnar, K. and Zabkar, V. 2018 The role of environmentally conscious purchase behaviour and green scepticism in organic food consumption *British Food Journal* **120** 10 pp 2411-2424 https://doi.org/10.1108/BFJ-08-2017-0457

[15] Priporas, C.-V., Kamenidou, I. (E)., Nguyen, N. and Shams, R. 2020 The impact of the macro-environment on consumer scepticism towards cause-related marketing: Insights from an economic crisis setting *International Marketing Review* **37** 5 pp 841-861 https://doi.org/10.1108/IMR-04-2019-0124

[16] Silva, M.E., Sousa-Filho, J.M.d., Yamim, A.P. and Diógenes, A.P. 2020 Exploring nuances of green skepticism in different economies *Marketing Intelligence & Planning* **38** 4 pp 449-463 https://doi.org/10.1108/MIP-10-2018-0435

[17] Zhang, L. and Hanks, L. 2017 Consumer skepticism towards CSR messages: The joint effects of processing fluency, individuals' need for cognition and mood, *International Journal of Contemporary Hospitality Management* **29** 8 pp 2070-2084 https://doi.org/10.1108/IJCHM-11-2015-0666

[18] Ju, I. 2017 The effects of advertising skepticism in consumer prescription drug advertising: The moderating role of risk disclosure location *International Journal of Pharmaceutical and Healthcare Marketing* **11** 4 pp 395-411. https://doi.org/10.1108/IJPHM-10-2016-0054

[19] Newman, K.P. and Trump, R.K. 2019 Reducing skepticism about corporate social responsibility: roles of gender and agentic-communal orientations, *Journal of Consumer Marketing* **36** 1 pp 189-196. https://doi.org/10.1108/JCM-02-2018-2577

[20] Pomering, A. and Johnson, L.W. 2009 Advertising corporate social responsibility initiatives to communicate corporate image: Inhibiting scepticism to enhance persuasion, *Corporate Communications: An International Journal* **14** 4 pp 420-439. https://doi.org/10.1108/13563280910998

[21] Mimouni Chaabane, A. and Parguel, B. 2016 The double-edge effect of retailers' cause-related marketing: When scepticism cools the warm-glow effect, *International Journal of Retail & Distribution Management* **44** 6 pp 607-626 https://doi.org/10.1108/IJRDM-08-2015-0126

[22] Sindhuri P, Julianna P, Timo O, Friederike V, Werner W 2017 The Effect of Consumer Scepticism on the Perceived Value of a Sustainable Hotel Booking *J. Tourism Hospit* **6** 312 doi:10.4172/2167-0269.1000312

[23] Lemeshow, S., Hosmer, D. W., Klar, J., & Lwanga, S. K. 1990 Adequacy of sample size in health studies. Chichester: World Health Organization: John Wiley & Sons

AUTHOR INDEX